Mechanical engineering science

Third edition

John Hannah
BSc(Eng), CEng, FIMechE, FIEE

M. J. Hillier
MSc(Eng), DIC, BSc

LONGMAN

Addison Wesley Longman Limited
Edinburgh Gate
Harlow
Essex CM20 2JE
England
and Associated Companies throughout the world

Visit Addison Wesley Longman on the world wide web at:
http://www.awl-he.com

First published 1962
Second edition published 1991
Third edition published 1999

ISBN 0 582 32675 3

British Library Cataloguing-in-Publication Data
A catalogue record for this book is available from the British Library

Library of Congress Cataloging-in-Publication Data
Hannah, John, 1920–
    Mechanical engineering science / John Hannah, M.J. Hillier. — 3rd
ed.
        p.   cm.
    ISBN 0–582–32675–3
    1. Mechanical engineering.   I. Hillier, M.J. (Marcus Jeans)
II. Title.
TJ146.H257   1999
621—dc21   98–55735
                                                          CIP

Typeset in 9/12pt Stone Serif by 35
Produced by Addison Wesley Longman Singapore (Pte) Ltd.
Printed in Singapore

Mechanical engineering science

 **LEARNING SUPPORT SERVICES**

Please return
on or before
the last date
stamped below

 City College NORWICH

A FINE WILL BE CHARGED ON OVERDUE ITEMS

Contents

··

Preface to third edition

The first metric edition of *Mechanical Engineering Science* was introduced in 1971 when SI units were adopted for the primary system of weights and measures. The text has been updated over the years to take account of ever-changing technologies and the redesigning of the engineering courses followed by students at this level. The continuing success of the book justifies maintaining its original character and retaining the same approach to the teaching of basic mechanics and science for the young engineer.

In this third edition, in addition to a new format and the redrawing of all the diagrams, further examples and problems have been added and amendments made throughout. The work on gravitation and satellites has been extended and now merits a separate chapter. The chapter on heat exchangers and steam plant has been updated and extended to include the modern combined-cycle gas turbine plant and renewable energy sources; in addition, the important topic of the sources, effects and control of emissions of pollutants is now dealt with in some detail.

Although not designed to cover any specific unit of study, the book has been found useful in a range of courses such as Foundation, Access, work-related GNVQ and, in particular, the early years of the National Certificate and Diploma courses of the Business and Technician Council (BTEC) in Engineering, Motor Vehicle Work and Science, as well as on related City and Guilds of London (CGLI) courses, both at home and overseas. The chapters on statics, dynamics and fluids are intended to lead in to the authors' *Applied Mechanics*, and the chapters on heat engines serve as a broad introduction to a subsequent course in applied thermodynamics. The two books taken together will be found to cover much of the requirements of a number of BTEC units up to Level III. Sixth form students and those embarking on an engineering course from studies mainly involving physics, should benefit from working through the texts in regard to engineering applications and approach.

I have been responsible for all the changes and revisions since prior to the first metric edition when my co-author Mr M. J. Hillier went abroad and no longer collaborated on the book.

My thanks are due to the users of the book and to those in the power generation, space and engineering industries and professional institutions who provided information and advice.

I am indebted to two colleagues of long standing, Mr R. C. Stephens for his overall invaluable contribution and ever-ready assistance on this edition as on previous editions and Mr S. Dunn on this occasion for his contributions and suggestions on the science topics.

Objectives

- To provide a fully up-to-date, practical and thorough text in SI units covering the principles of the basic mechanics and science required by the young engineer.
- To emphasize the engineering aspects of the subject, the importance of the assumptions that have to be made in engineering analysis and the limitations these assumptions impose.
- Through the use of particular topics (i) to broaden the general science awareness of the engineering student and (ii) to illustrate that mechanical engineering expertise is relevant and often crucial to the work of broadly-based teams of engineers and scientists engaged in a wide range of industries.
- To keep formulae to a minimum, and to provide carefully chosen worked examples, a diversity of practice problems and clear well-designed diagrams, enabling the student to analyze and solve problems at this level.
- To equip the student for entry to a more advanced engineering course.

Note on SI units

....

SI is the abbreviation, in all languages, for the full title 'Système International d'Unités', which is the rationalized form of the metric system of units agreed internationally. Of the seven fundamental or **base** units, five will be met with in this book, i.e. the *metre* (length), *second* (time), *kilogram* (mass), *kelvin* (temperature) and *mole* (amount of substance). The sole **derived** unit for measuring work or energy is the *joule* and that for force is the *newton*. These names, 'joule' and 'newton', are examples of certain derived units being given special names and symbols. The SI is a coherent system of units since the product of any two unit quantities in the system is the unit of the resultant quantity. For example, unit velocity (metre per second) results when unit length (metre) is divided by unit time (second). Normally calculations in the text are carried out by converting all given quantities to these base units, but on occasion it has been found convenient to work in multiple or sub-multiple units. The kilojoule and kilonewton are particularly convenient. A few non-SI units whose use is accepted have been used where appropriate, for example, the *bar* (and its multiples) as a unit of pressure and the *knot*, a unit of speed, in aerial and marine navigation work.

The SI system has a number of prefixes to the base unit indicating multiples and sub-multiples. Those that generally meet the needs of this text are: *milli* (m) $= 10^{-3}$; *centi* (c) $= 10^{-2}$; *hecto* (h) $= 10^{2}$; *kilo* (k) $= 10^{3}$; *mega* (M) $= 10^{6}$; and *giga* (G) $= 10^{9}$. Exceptionally, in some areas of work where very large quantities are involved, engineers use *tera* (T) $= 10^{12}$. Also, in today's micro- and nano-technologies, a mechanical engineer may be dealing with the manufacture and design of miniaturized components, actuators, etc. and at this level, the common prefixes in use are: *micro* (μ) $= 10^{-6}$; *nano* (n) $= 10^{-9}$; and *pico* (p) $= 10^{-12}$.

For full information on SI units reference should be made to *SI International System of Units*, R. J. Bell and D. T. Goldman (National Physical Laboratory), published by H.M. Stationery Office (1986), and to British Standards No. 5555 and No. 350 Part I.

Chapter 1

Statics

The study of mechanics may be conveniently divided into two parts: (i) **statics**, the study of forces on bodies at rest or in steady motion and (ii) **dynamics**, the study of forces on bodies whose motion is changing. We begin in the first chapter with some principles of statics and as a first step, it is necessary to define the basic quantities of length, time, mass, force and weight and their units. A vector, a quantity specified by a magnitude and a direction, is also defined, followed by resolution of forces and the concepts of the principle of concurrency and conditions of equilibrium of a system of forces. The application of these concepts, with the aid of free-body diagrams, is demonstrated in a wide range of worked examples.

1.1 Length

Length is measured as a multiple of a fixed standard unit of measurement. The SI unit of length is the *metre* (m), which has been defined precisely in terms of time and the speed of light of known frequency.

In mechanics we use also the following multiple and sub-multiple units:

$$1 \text{ millimetre (mm)} = \frac{1}{1000} \quad \text{or} \quad 10^{-3}\,\text{m}$$

$$1 \text{ centimetre (cm)} = \frac{1}{100} \quad \text{or} \quad 10^{-2}\,\text{m}$$

$$1 \text{ decimetre (dm)} = \frac{1}{10} \quad \text{or} \quad 10^{-1}\,\text{m}$$

$$1 \text{ kilometre (km)} = 1000 \quad \text{or} \quad 10^{3}\,\text{m}$$

1.2 Time

The SI unit of time is the *second* (s), and in mechanics we also use the *minute* (min), and the *hour* (h). The definition of the second is now independent of any motion of the earth and is based on the frequency of the radiation of the caesium atom in undergoing a particular change.

1.3 Mass

The mass of a body is a measure of the quantity of matter it contains. The base SI unit of mass is the *kilogram* (kg). The following units of mass are also in use:

$$1 \text{ milligram} = \frac{1}{1000} \quad \text{or} \quad 10^{-3}\,\text{g} = 10^{-6}\,\text{kg}$$

$$1 \text{ gram (g)} = \frac{1}{1000} \quad \text{or} \quad 10^{-3}\,\text{kg}$$

$$1 \text{ megagram (Mg)} = 1\,000\,000 \quad \text{or} \quad 10^{6}\,\text{g} = 10^{3}\,\text{kg} = 1\,\text{tonne (t)}$$

The base unit for mass is the only SI base unit that already contains a prefix in its name. The decimal multiples and sub-multiples of the unit of mass are formed by attaching prefixes to the word 'gram'. Thus the milligram is 10^{-3} gram and therefore $10^{-6}\,kg$. Similarly, the prefix M = 10^6 represents in megagrams a multiple of a gram and not a multiple of the base unit; the megagram is therefore 1000 times the base unit, i.e. $1000\,kg$, also called the *tonne* (t). For mass *see also* page 176.

1.4 Force

A *force* is defined as that which causes or tends to cause a *change in motion* of a body. If a change of motion is prevented, the force will cause a deformation or change in shape of the body to which it is applied. For example, a spring will extend or compress under load. In other cases the body may be so rigid that the deformation may not be apparent to the naked eye. Nevertheless the change in shape under load always occurs.

1.5 Units of force

The derived SI unit of force is the *newton* (N) which is defined in terms of the change in motion such a force produces in a standard mass of 1 kg (*see* page 176). The following multiple forms of the newton are used:

$$1 \text{ kilonewton (kN)} = 1000 \quad \text{or} \quad 10^{3}\,\text{N}$$

$$1 \text{ meganewton (MN)} = 1\,000\,000 \quad \text{or} \quad 10^{6}\,\text{N}$$

$$1 \text{ giganewton (GN)} = 1\,000\,000\,000 \quad \text{or} \quad 10^{9}\,\text{N}$$

1.6 Magnitude of a force

Our experience of force is similar to our experience of *weight* – the pull of the Earth on a body. We say that force and weight are of the same nature. It is natural therefore to relate the magnitude of a force to the equivalent weight of a given body. For example, a force of 1 N is roughly equivalent to the weight of a medium-sized apple.

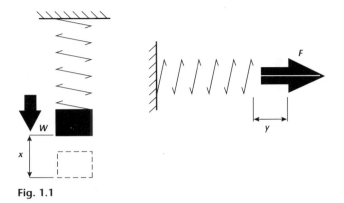

Fig. 1.1

If a standard mass is supported from a given spring and causes it to extend through a distance of 10 mm, and if a given force of magnitude denoted by F will extend the *same* spring through a distance of 20 mm, we say that the force F has twice the magnitude of the weight of the standard mass. Similarly, if a mass m kilograms of weight W newtons extends a spring a distance x, Fig. 1.1, then a force of F newtons will extend the *same* spring a distance y according to the ratio

$$\frac{F}{W} = \frac{y}{x}$$

This method of measuring force is based on the fact that a spring extends a distance proportional to the magnitude of the weight of the mass suspended from it, provided that the spring is not 'overstretched'. Alternatively, a force may be measured by the change of motion it produces in a given body (Ch. 8).

1.7 Mass and weight

It is necessary now to distinguish clearly between the mass and weight of a body. In Chapter 8 a force of 1 N is defined in terms of the change in motion (acceleration) it produces in a body of mass 1 kg and the relationship between force, mass and acceleration is derived. It is shown that if a body of mass 1 kg falls freely it moves downwards with an acceleration due to gravity equal to 9·8 metres per second per second, or 9·8 m/s², approximately. The relationship between mass m, weight W and acceleration due to gravity g, of a body is given by

$$W = mg$$

i.e. W (N) $= m$ (kg) $\times$ 9·8 (m/s²)

Thus, *to find the weight of a body in newtons multiply its mass in kilograms by 9·8.* For example,

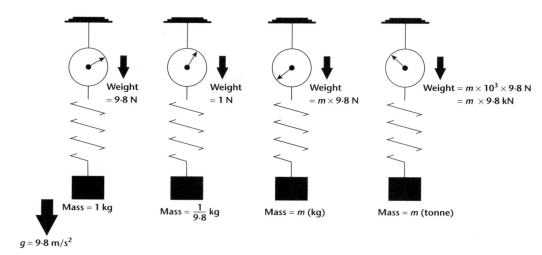

$g = 9.8 \text{ m/s}^2$

Fig. 1.2   Relation between mass and weight

The weight of a body of mass 56 kg $= 56 \times 9{\cdot}8\,\text{N}$

The weight of a body of mass 56 g $= 56 \times 10^{-3} \times 9{\cdot}8\,\text{N}$

The weight of a body of mass 56 Mg
   or   56 tonnes $= 56 \times 10^{3} \times 9{\cdot}8\,\text{N}$

$= 56 \times 9{\cdot}8\,\text{kN}$

Note that the weight of a body of mass m tonnes is $m \times 9{\cdot}8\,\text{kN}$.

A spring balance may be calibrated to measure either the mass (kg) or the weight (N) of a body, Fig. 1.2 (*see also* page 177).

1.8 Dead loads

In statics the vertically downwards force due to the weight of 'dead loads' must always be taken into account, and this force of gravity acts through the centre of gravity of the load. A load may be given in force units, N, kN, MN or GN. A 'load' may also be specified in mass units, kg or Mg (tonne), and in this case the corresponding *weight* of the mass must be found before carrying out calculations involving forces.

1.9 The rigid body

In statics it is often convenient to consider the effects of forces that act on a *rigid body*, i.e. a body whose change in shape is negligibly small. The spring used to measure force in section 1.6 is, of course, not a rigid body.

The perfectly rigid body, which suffers no deformation under the action of any force, does not exist. Nevertheless it is a very useful idea by which we simplify many practical problems. In this chapter all bodies will be assumed rigid unless otherwise stated.

1.10 Line of action; sense

The *line of action* of a force is a line drawn through the point of application of the force and along the direction in which the force acts. The *sense* in which a force acts is important, e.g. the line of action of a force may be in the vertical direction, but its sense may be vertically *upwards* or vertically *downwards*.

A force applied to a rigid body has the same effect at all points on the line of action. Hence for many purposes it does not matter where the point of application may be, so long as the line of action is known. If the body is not rigid this rule does not hold. For example, the extension of a spring due to a load suspended from the free end is greater than if the load were hung from a point halfway along its length. This does not mean that the magnitude of the force is altered but that its effect on a *non-rigid* body depends on the point of application as well as on the line of action.

The effect of a force at a point which is *not* on its line of action depends on the position of that point relative to the line of action. There is then a 'turning effect' which will be studied later in Chapter 2.

1.11 Specification of a force; vector quantity

To describe a force completely, we need to know three things about it:

1. The magnitude.
2. The point of application.
3. The direction (line of action and sense).

Such a quantity, having magnitude, direction and sense, is known as a *vector quantity*. It may be represented by a line **ab**, drawn to scale, whose length is proportional to the magnitude of the force and whose direction is parallel to the line of action of the force. The sense of the force vector is indicated by an arrow-head, Fig. 1.3.

Fig. 1.3

1.12 Contact forces; smooth surfaces; rollers; knife-edges

A *perfectly smooth* surface is a surface that offers no resistance to sliding parallel to the surface. *The force (or reaction) R exerted by such a surface must be at right angles (normal) to the surface.* If it were not normal to the surface, the reaction would tend to resist or assist sliding. In practice it is not possible for the surfaces to be so smooth that there are no frictional forces resisting sliding, but in many cases it may be a fair approximation to reality. The assumption of smooth surfaces simplifies the solution to many practical problems. Friction is dealt with in Chapter 4.

Similarly, when a smooth rigid roller lies on a smooth flat or curved surface, Fig. 1.4(a), the force exerted by the surface on the roller must be along the normal to the surface of the roller at the point of contact. The direction of the reaction of a surface to a knife-edge is also normal to the surface, Fig. 1.4(b).

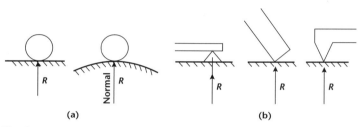

Fig. 1.4

1.13 Smooth pin-joints

In Fig. 1.5 a link A is attached by a pin or hinged joints to links B and C. If the pin is perfectly smooth, there is no resistance to rotation at the hinge. One link is free to rotate relative to the other. Each link can then transmit a force *only along its length*. Pin-joint in this text refers to a *smooth* joint.

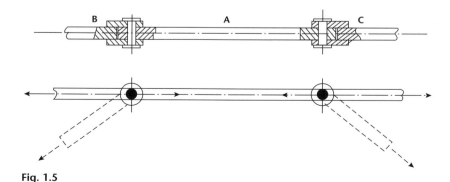

Fig. 1.5

1.14 Equilibrium

A body is said to be *in equilibrium* under the action of a system of forces if all the forces on it are in *balance*, i.e. their total net effect is to leave the position of the body, or its motion, unchanged.

The conditions to be satisfied in order that a body shall be in equilibrium under the forces acting on it depend on whether there are two, three or more forces to be considered. It is convenient to consider each case in turn.

A body cannot remain in equilibrium under the action of a single force.

Strictly speaking a single force cannot exist alone without changing the motion of the body to which it is applied. In statics all forces come in pairs – an *action* and a *reaction*. Nevertheless, when considering the equilibrium of a system of forces, it is usually convenient to consider each force as existing separately. As will be seen, it is then possible to describe each force graphically by a suitable vector to some scale.

Now, a body is in equilibrium under the action of *two* forces provided that:

- the forces are of *equal magnitude*;
- the forces have the *same* line of action but *opposite* sense.

In general a force may be thought of as a *pull* or a *push*. A pull on a spring anchored at one end will extend the spring. The pull, together with the *reaction* of the fixed end *on* the spring, forms a pair of equal and opposite forces. The spring is said to be in *tension*. Similarly, a push on a spring will compress it, if it is fixed or anchored at the other end. The push, together with the equal and opposite reaction of the fixed end on the spring, are compressive forces and the spring is said to be in *compression*.

Tension is produced by two equal and opposite pulls.
Compression is produced by two equal and opposite pushes.

1.15 Triangle of forces: principle of concurrence

In order that a body shall be in equilibrium under the action of *three* forces *in the same plane* the conditions are:

- the lines of action of the three forces must pass through the same point;
- the forces may be represented by the three sides of a triangle, *taken in order*, each side of which is parallel to the line of action of a force.

The first condition is called the *principle of concurrence*. The second condition gives the *triangle of forces*. We shall not attempt to prove these conditions since they are easily verified by a simple experiment. We shall illustrate them, however, by an example.

In Fig. 1.6, the link OA is pin-jointed at O and is subject to a force P in the sense A to O. The link OB is also pinned at O and is subject to a force Q in the sense O to B.

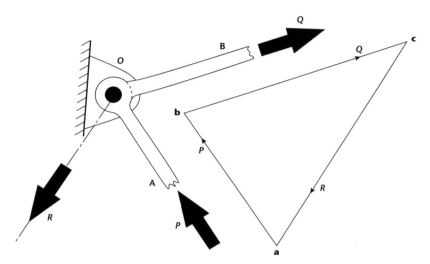

Fig. 1.6

The forces P and Q acting on the pin at O are not in the same straight line, hence a third force is required for equilibrium. Since P and Q pass through the point O the third force must also pass through O (principle of concurrence). This third force is the reaction R which the bearing exerts *on* the pin.

The triangle of forces is drawn as follows:

- From any point **a** draw a line **ab** parallel to AO to represent the force P.
- From point **b** draw a line **bc** parallel to OB to represent the force Q.
- Join **ca**.

The line **ca** represents, in magnitude and direction, the third force R required for equilibrium. The sense of R is from **c** to **a**.

Then **abc** is the triangle of forces for the pin O. The sense of each force is shown by the arrows drawn on the figure. Note that the arrows point the 'same way round' the figure, i.e. they point, in succession, from **a** to **b**, **b** to **c**, and **c** to **a**.

Each side of the triangle represents a force, and is called a *vector*. The successive drawing of the vectors, **ab**, **bc** and **ca** is called the 'addition' of vectors.

Note that, correctly drawn, each vector joins the succeeding vector 'head to tail', i.e. the tail of vector **bc** is joined to the head of the vector **ab** at point **b**. This is the meaning of the words 'taken in order' in the second condition for equilibrium. However, when drawing the triangle of forces, it is not necessary to pick out the forces P, Q, R in any special order. For example, it may be more convenient to draw the triangle in the order Q, P, R.

The triangle of forces shown in Fig. 1.6 was drawn close to the sketch of the forces acting in order to show clearly the relation between each force and the vector representing it. In practice the force triangle should be drawn well clear and *to as large a scale as possible*.

Note on equilibrium of bodies in steady motion. A body 'at rest' or in 'steady motion' can be taken to mean not being accelerated, i.e. it is either stationary, moving in a straight line with constant speed, or rotating at a constant rate.

Velocity and acceleration will be dealt with later where it will be shown that if there is no acceleration of a body there is no out-of-balance force acting and the body is said to be in equilibrium. A stationary body is clearly not accelerating and is said to be in *static* equilibrium. When a body has steady motion there is no out-of-balance force acting, e.g. an aeroplane or ship 'cruising' at a constant or steady speed, or a block sliding with constant speed between guides. Such bodies are also in equilibrium and this is sometimes called *dynamic* equilibrium. Thus the triangle of forces can be applied to a body subject to three forces maintaining it in steady motion. The general conditions of equilibrium for a stationary body to be met with in the following paragraphs also apply to the equilibrium of a body in steady motion.

Example

A light ladder stands on a rough horizontal floor and leans against a 'perfectly smooth' vertical wall. The ladder makes 30° with the vertical. A man of mass 95 kg stands halfway up the ladder. Assuming the ladder does not slip on the floor, find (a) the reaction between ladder and wall and (b) the direction and magnitude of the reaction of the floor on the ladder.

Solution (Fig. 1.7)
The reaction at the rough floor is unknown both in magnitude and direction. The reaction of the wall to the top of the ladder is perpendicular to the wall since the wall is perfectly smooth.

The forces acting *on* the ladder are:

- the load due to the weight of the man *vertically downwards*, i.e. $95 \times 9.8 = 931$ N;
- the *horizontal* reaction N of the wall;
- the reaction R of the floor.

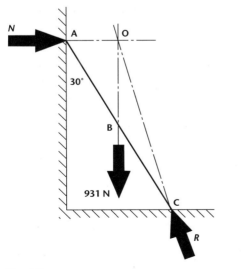
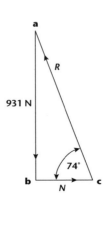

Fig. 1.7

These three forces are in equilibrium, hence their lines of action pass through a common point O (to be found).

To determine O, draw the diagram to scale. Draw AO normal to the wall at A (along the line of action of N), and draw BO vertical through B (along the line of action of the 931 N load). The point of intersection O is the point of concurrency and determines the direction of the reaction R, which must act along CO.

Construct the *triangle of forces* as follows:

- Draw **ab** to scale, parallel to OB and of length proportional to 931 N.
- Draw **bc** parallel to AO and **ca** parallel to CO.
- The point **c** completes the triangle.

By construction,

$N = $ **bc** $= 268$ N

$R = $ **ca** $= 969$ N

The direction of R is 74° to the horizontal and its sense is from **c** to **a**, i.e. from C to O as shown.

Free-body diagram. In Fig. 1.7 the wall and ground may be removed to show the ladder with the forces N, R and weight of the man acting on it. This is called a *free-body diagram.* Such a diagram is a useful concept in the solution of problems and shows the body isolated completely from its surroundings with supports and holding devices removed. The magnitude and direction of all the forces acting at a particular instant are shown, assumptions being made where necessary. Every force must be shown including individual external loads, forces due to gravity and reactions at supports and fixings. In Fig. 1.7 the weight of the man was known in magnitude, direction and sense, N in direction only, and R was assumed to act upwards to the left. The solution to the problem confirms or otherwise the assumptions made. It is essential in constructing a free-body diagram to know the kind of force exerted between bodies in contact and through the various connections and types of support (*see* section 1.12 and the authors' book *Applied Mechanics*).

Example

The snap-action toggle clamp shown in Fig. 1.8(a) exerts a vertical force of 300 N on the workpiece when the line AB makes 30° with the horizontal. AC is vertical and DA horizontal.

Determine the magnitude and direction of the reaction at the pivot, C. DA = 125 mm, AC = 150 mm.

Solution

The forces acting *on* the link DAC are, Fig. 1.8(b):

- the upwards vertical load of 300 N at D;
- the reaction Q at pin A exerted along BA; this force is inclined at 30° to the horizontal;
- the reaction R at the pivot C (unknown in both magnitude and direction).

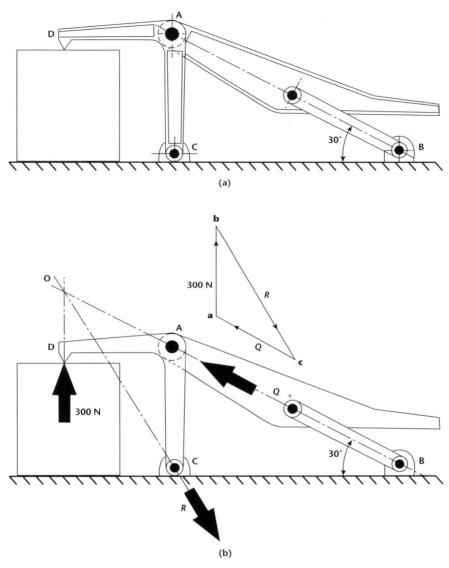

Fig. 1.8

These three forces are in balance, hence their lines of action pass through a common point O. To find point O, draw the mechanism to scale. Draw DO vertically from D and AO along the line BAO. The intersection of AO and DO determines point O. The reaction *R* must pass through O, hence *R* acts along the line OC.

By construction, OC is at **61°** to the horizontal.

Construct the *triangle of forces* as follows:

- Draw **ab** to represent 300 N acting upwards at D.
- Draw through point **b** a line **bc** parallel to OC.
- Through point **a** draw a line **ac** parallel to OB.

The intersection of **ac** and **bc** determines the point **c** and completes the triangle.

Then **bc**, in the sense **b** to **c**, represents the force R at C.

By construction,

$R = $ **bc** $ = 510$ N

and its direction is from O to C.

Note that this is the force exerted by the pivot bearing *on* the link. The force exerted on the pivot at C *by* the link is equal and opposite to this.

Similarly, magnitude of $Q = $ **ca** $ = 289$ N in the sense A to O.

Example

A light rigid link AB, length 600 mm, is part of a mechanism, Fig. 1.9. It is freely pinned at A and is maintained in equilibrium by a force of 120 N acting at D as shown where AD = 350 mm, a force at the pin-joint A, and by a force at B acting along ZZ. Find the forces on the link at A and B.

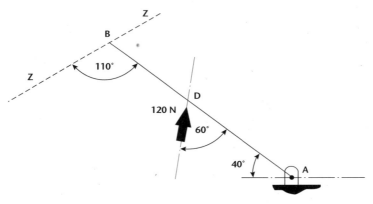

Fig. 1.9

Solution

Link AB is in equilibrium under three external forces, thus

- a force at B acting along ZZ, magnitude and sense unknown;
- a force of 120 N at D, known in sense and direction;
- a force at A, due to the support, completely unknown.

These three forces must be concurrent. They meet in point X where the line of action of the 120 N force cuts ZZ, Fig. 1.10(a). Join AX. Then the line of action of

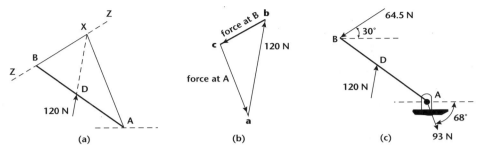

Fig. 1.10   (a) Point of concurrency X; (b) triangle of forces; (c) forces on AB

the force at A is in direction AX. The triangle of forces **abc** is drawn as follows, Fig. 1.10(b):

- draw **ab** to represent the 120 N force;
- through **a** draw a line parallel to AX;
- through **b** draw a line parallel to ZZ;
- the intersection point **c** then completes the triangle.

The arrowhead is placed on **ab** from **a** to **b** in the direction of the 120 N force, then following on with the arrowheads the force at B is from **b** to **c**, and the force at A is from **c** to **a**. By measurement:

ca = reaction at A = 93 N at 68° to the horizontal, downwards to the right;
bc = force at B = 64.5 N at 30° to the horizontal, downwards to the left.

Figure 1.10(c) shows the forces acting on the link AB.

Problems

1.1   A light ladder leans against a perfectly smooth vertical wall at an angle of 30° to the horizontal. A load of 800 N is placed three-quarters of the way up the ladder. If the ladder rests on a rough horizontal surface which prevents slipping, find the magnitude and direction of the reaction between the ladder and the ground.   (1·31 kN at 38° to the horizontal)

1.2   A cylindrical roller of mass 40 kg is drawn up a smooth plane inclined at 30° to the horizontal by a force *F*. Find the magnitude of *F* and the reaction between the roller and the plane (a) if *F* is horizontal and (b) if *F* is acting along a line parallel to the plane.   ((a) 226 N, 453 N; (b) 196 N, 340 N)

1.3   ABC, Fig. 1.11, is a toggle mechanism. Draw the triangle of forces for the pin-joint B under the loading shown and hence find the force in link AB. The 60 N force at B acts at right angles to AC. AB = 30 cm, BC = 20 cm.

Similarly, find the vertical force *F* required at A to maintain equilibrium. The surfaces at A are perfectly smooth. What is the reaction *N* at A?

(138; 136; 24 N)

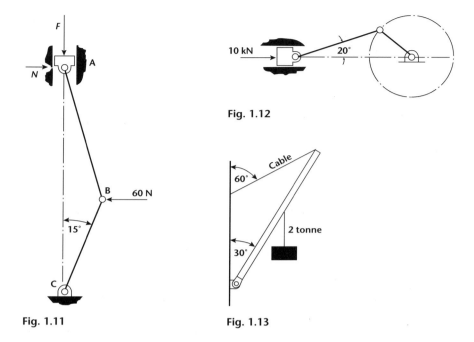

Fig. 1.12

Fig. 1.11   Fig. 1.13

1.4 Figure 1.12 shows an engine mechanism. The force on the piston is 10 kN. Neglecting friction find, for the position shown, the force in the connecting rod and the side thrust between piston and cylinder.   (10·7; 3·64 kN)

1.5 Figure 1.13 shows a jib crane pin-jointed to a wall at its lower end and supported in the position shown by a cable attached to the upper end. The jib carries a loads of 2 tonnes at its midpoint. Find the tension in the cable and the magnitude and direction of the reaction at the bottom pin.

(9·8 kN; 26 kN at 19° to vertical)

1.6 A light link AC in a mechanism, Fig. 1.14, is subject to a force of 350 N acting at its midpoint B in the direction shown. The link is maintained in equilibrium by a force at C parallel to ZZ and by an unknown force at end A. Find the magnitude and direction of the force at A and the magnitude and sense of the force at C.   (364 N at 14° to horizontal; 101 N downward)

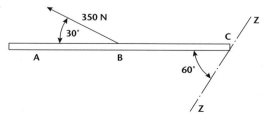

Fig. 1.14

1.7 A platform, of weight 70 N, is suspended by a cable from a fixed point and a horizontal pull of 50 N is applied to the platform, displacing it horizontally. Find the angle made by the cable with the vertical.   (35·5°)

1.8 Figure 1.15 shows a light link AB of a mechanism which is subject to a force of 500 N applied at its midpoint C and to an unknown force at the end B. The other end A is pinned freely to a light block which moves in smooth fixed guides. Find the force required at B to maintain the link in equilibrium.
(377 N, at 20° to the horizontal, downwards to the right)

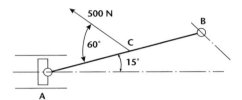

Fig. 1.15

1.9 The jib crane shown in Fig. 1.16 carries a load M kg at the end C. If the maximum permissible force in the jib is 60 kN, and the maximum permissible force in the tie is 35 kN, find the maximum safe value of the load M.
(If force in jib = 60 kN, M = 2450 kg; if force in tie = 35 kN,
M = 1930 kg: hence maximum safe load = 1930 kg)

1.10 In the wall crane shown in Fig. 1.17, find the maximum mass that can be carried safely at the point B if the permissible loads in tie and jib are 16 kN and 24 kN respectively. What is the reaction at A when the crane carries the maximum load? Neglect weight of jib.
(1050 kg; reaction at A = force in jib = 12·3 kN)

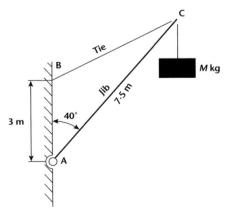

Fig. 1.16

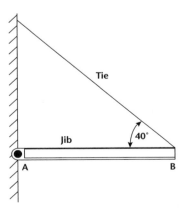

Fig. 1.17

1.11 For the jib crane shown in Fig. 1.17, find the reaction at A when a load of 2000 kg is carried at the midpoint of the jib. Neglect weight of jib.

> (15·2 kN upwards at 40° to the horizontal; note that the reaction at A is not equal and opposite to the thrust in the jib along AB, for the load is not applied at a pin-joint nor at an end of a member)

1.12 Figure 1.18 shows an aeroplane descending in a steady glide at an angle θ to the horizontal. The engine cuts out and equilibrium is maintained by three forces L, D and W as shown, assumed to act through the same point. If $W = 120$ kN, and $\theta = 20°$, find the forces L and D. If $D = 70$ kN what would then be the value of θ and the magnitude of L whose direction remains at right angles to that of D?   (113 kN; 41 kN; 36°; 97·5 kN)

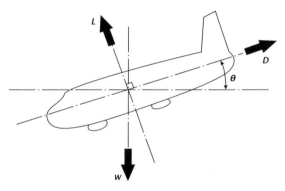

Fig. 1.18

1.13 A uniform link of length 4 m and weight 80 kN is freely pinned at each end to light blocks sliding in smooth guides, Fig. 1.19. For the position shown a force F maintains equilibrium. Find the force F and the total force exerted on the lower block.

> (45·4 kN; 92 kN at 60·4° to horizontal, upwards to the left)

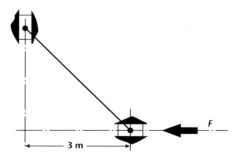

3 m

Fig. 1.19

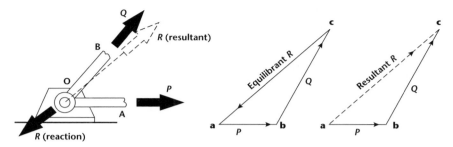

Fig. 1.20

(1.16) Resultant and equilibrant

Consider again the two forces P and Q, Fig. 1.20, acting along the links OA, OB respectively, together with the reaction R at the bearing O.

- Draw **ab** to represent P.
- Draw **bc** to represent Q.
- Join **c** to **a** to complete the triangle of forces **abc**.

Then **ca**, in the sense **c** to **a**, represents the reaction R exerted by the bearing *on* the pin at O. This force R is also known as the *equilibrant* of the two forces P and Q since it is the *force required to balance them*.

The joint O is in equilibrium under the action of the three forces, P, Q and R.

Now, the forces P and Q may be replaced by a *single force* which has the same effect on the pin at O as the two forces acting together. If it has the same effect it must be in equilibrium with the force R, represented by **ca**, i.e. it must produce the same reaction at the pin. The single force required must therefore be equal and opposite to the equilibrant **ca** and is represented by **ac** in the diagram.

That is, the single force R which may replace P and Q acting together is the force **ac**, in the sense **a** to **c**, and is called the *resultant* of P and Q. The resultant is the single force which has the same 'result' as P and Q together.

(1.17) Parallelogram of forces

The magnitude and direction of the resultant of two forces P and Q, Fig. 1.20, may also be obtained by the following construction.

Complete the parallelogram **abcd**, where **ad** is parallel to **bc** and **dc** is parallel to **ab**, Fig. 1.21.

Then, since **ad** = **bc** and **ad** is parallel to **bc**, the forces P and Q may also be represented by the sides **ab**, **ad** of the parallelogram **abcd**. As before, **ac** is the resultant R and is the diagonal drawn from the common point **a** of the vectors **ab**, **ad**. The parallelogram **abcd** is called the *parallelogram of forces*.

This construction is equivalent to drawing the triangle of forces **abc**. When drawing the parallelogram it should be remembered that vectors **ab** and **ad**

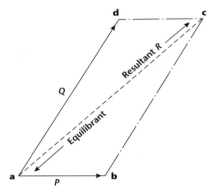

Fig. 1.21

representing forces P and Q are drawn *from one point*. When drawing the triangle of forces **abc**, P and Q are represented by vectors **ab** and **bc** drawn *in order, head to tail*. The triangle of forces is of course one-half of the parallelogram of forces.

(**Example**)

The resultant of two forces acting on a gusset plate is 1·45 kN acting at 30° to the horizontal, Fig. 1.22(a). One of the forces is known to be 300 N acting horizontally as shown. Find the magnitude and direction of the second force P.

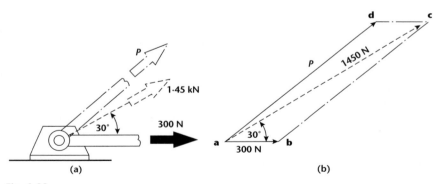

Fig. 1.22

Solution
The 300 N force and the unknown force P form a parallelogram of forces in which the resultant force of 1450 N is the diagonal.

● Draw **ab** to represent the 300 N force to scale, Fig. 1.22(b).
● Draw **ac** to represent the 1450 N resultant (drawn *from* **a**).
● Join **bc** and complete the parallelogram **abcd**.

Then **ad** represents the force P in magnitude and direction. By construction, length **ad** = 1200 N

angle **cad** = 7°11′

Hence P = **1·2 kN** and acts at **37·2°** to the horizontal, as shown.

Problems

1.14 A force of 930 N acts on a pillar in a direction due N. A force of 300 N on the same pillar acts 30° W of N. Determine the magnitude and direction of the resultant force. (1·2 kN; 7·2° W of N)

1.15 A force P acts on a body at an angle θ to a horizontal line OX. P, together with a pull of 1·5 kN at 30° to, and below, OX, has the same effect as a force of 5 kN at 45° to, and above, OX.

Find the magnitude of P and the angle θ. (4·84 kN; 62·5°)

1.16 A load is to be moved using a small tractor and also a cable from a winch. Lack of space means that tractor and cable must pull in different directions. The tractor exerts a pull of 3 kN due N. The winch cable pulls with a force of 10 kN. If it is required to move the load towards the NW, what should be the direction of pull of the cable? (57·3° W of N)

1.17 A capstan holds a rope which laps it by an angle of 150° as shown in Fig. 1.23. The tensions in the two sides of the rope are T_1 and T_2. What is the magnitude of the resultant force on the capstan when T_1 = 4 kN and T_2 = 1·75 kN? (5·6 kN)

1.18 A belt laps a pulley as shown in Fig. 1.24, the angle of contact of the belt on the pulley being 240°. The belt tensions are 2·5 kN and 1 kN. Find the resultant force R on the pulley. If the weight of the pulley is 500 N, find the resultant of R and the dead weight. (3·13 kN; 3 kN)

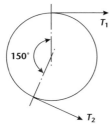

Fig. 1.23

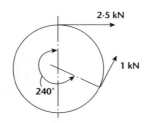

Fig. 1.24

1.19 A vehicle rounding a curve at speed is subject to two forces acting at the point G, Fig. 1.25. One force is 10 kN vertically downwards, the other 4 kN horizontally. Find the resultant force on the vehicle. Where does the line of action of the resultant cut AB?

(10·8 kN; 300 mm from midpoint towards B)

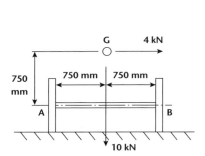

Fig. 1.25

Fig. 1.26

1.20 Figure 1.26 shows the cross-section of a retaining wall. The resultant force on the vertical face acts horizontally and is 100 kN per metre length of wall. The weight of wall per metre length is 170 kN. These two forces act at the point G as shown. Find the resultant force on the wall per metre length and state where its line of action cuts the base AB.

(198 kN; 1·93 m from A, to the right)

1.21 Figure 1.27 shows an aeroplane banking at an angle as it turns steadily in a horizontal circle. The forces on the plane are the lifting force L due the surrounding air and the weight W acting as shown. Find the resultant of these two forces when $W = 45$ kN and $L = 70$ kN (*see* page 242).

(53·6 kN, acting radially inwards towards the centre of the turning circle.)

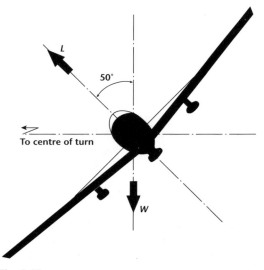

Fig. 1.27

1.18 Resolution of forces

In the previous section we have shown how to combine two given forces into a single force – the resultant – which may replace the two original forces completely. In Fig. 1.21 the force R, represented by **ac**, was shown to be equivalent to the two forces represented by the vectors **ab** and **bc**.

We require now to carry out the reverse process, that is, given the *single* force **ac**, to find the equivalent *two* forces **ab** and **bc**. Evidently we may draw any number of vectors **ab** and **bc** which would give the same resultant vector **ac**. In order to fix these two vectors we must *specify the directions* along which both **ab** and **bc** act.

Hence, given the force **ac** and the directions along which the two required forces are to act, we find the forces by the following construction, Fig. 1.28(a):

- Draw **ab′** from **a** in the required direction.
- Draw **cb″** from **c** in the second given direction.
- The intersection of **ab′** and **cb″** fixes the point **b**.
- Then **ab** and **bc** are the vectors representing the two forces required.

The forces **ab** and **bc** are known as the *components* of the force R in the given directions. (Alternatively, given one component in direction *and* magnitude the other component can be found.)

1.19 Special case: two directions at right angles

A special case of great importance arises when the two given directions are at right angles. In Fig. 1.28(b), forces X and Y are the components of the force R in two directions at right angles, in this case, horizontally and vertically. X, Y and R are represented by vectors **ab**, **bc** and **ac** respectively. Since the angle **abc** is a right angle we can obtain the components X and Y by using the trigonometrical ratios. If θ is the angle between R and the direction of X, then

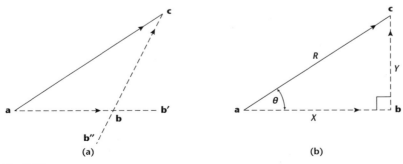

(a)   (b)

Fig. 1.28

$X = \mathbf{ab}$ and $Y = \mathbf{bc}$

$\quad = \mathbf{ac} \cos \theta$ $\quad = \mathbf{ac} \sin \theta$

$\quad = R \cos \theta$ $\quad = R \sin \theta$

These two formulae are important and should be remembered. It should be noted also that, when drawing the vectors **ab** and **bc**, broken lines have been used. This is to emphasize the fact that **abc** does *not* now represent a triangle of forces. We are not dealing with three independent forces but with the components of a *single force*.

(1.20) Resolution and equilibrium of forces

Consider three forces P, Q and R acting as shown in Fig. 1.29(a). If the forces are in equilibrium they may be represented by a closed triangle **abc**, Fig. 1.29(b), where $\mathbf{ab} = P$, $\mathbf{bc} = Q$ and $\mathbf{ca} = R$.

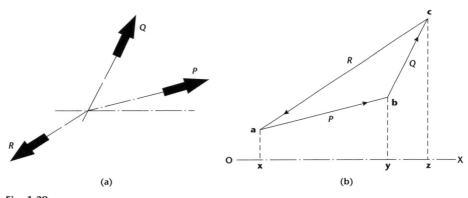

(a)   (b)

Fig. 1.29

Now consider the effects (components) of these three forces along any one direction such as the horizontal OX. The component of P along OX is given by the horizontal projection of **ab** on line OX, Fig. 1.29(b), that is, by **xy**. Similarly the horizontal projection of **bc** is **yz** and the horizontal projection of **ca** is **zx**.

Since the triangle of forces closes, then, when we move around the triangle **abc** *in order*, we also move along the straight line OX in order, that is, in the order **xy**–**yz**–**zx**. Now **xy**, **yz** and **zx** are vectors and the sum of these vectors must be zero, since we have returned to the same point **z**. Note that **xy** and **yz** act from left to right and are positive whereas the force **zx** acts from right to left and is negative. Thus the *algebraic* sum of the projections of the vectors **ab**, **bc**, **ca** is zero, or

$\mathbf{xy} + \mathbf{yz} - \mathbf{zx} = 0$

Hence, *if three forces are in equilibrium, the algebraic sum of their components along any given direction is zero.*

The truth of this is evident when we consider what happens if the algebraic sum of the components in one direction is *not* zero. In this case there will be a resultant out-of-balance force having a component along OX. If such a force exists, the body on which the force acts will tend to move in the direction of the force. Hence, in this case, neither the body nor the system of forces can be in equilibrium.

For equilibrium, therefore, there must be no resultant force along any given direction.

In practice, we must consider the algebraic sum of the component forces along any *two* directions. For, even if the components of the forces along OX are in equilibrium, we cannot know that there is a balance of forces along any other direction. However, if there is no component force in either the OX or OY directions at right angles to one another then there is no *resultant* force on the body, and it is therefore in equilibrium.

1.21 Conditions of equilibrium

Thus, for forces whose lines of action all lie in the same plane and all pass through the same point, the *conditions of equilibrium* are:

- the algebraic sum of the resolved components of all the forces along any direction OX must be zero;
- the algebraic sum of the resolved components along any other direction must be zero. (For convenience, the second direction chosen is usually at right-angles to OX.)

These rules are true also for any number of forces acting at a point and are not limited to three forces. The application of the rules to cases where more than three forces act on a body is dealt with in Chapter 3.

The rules are true only for forces that all act in *one plane* (the plane of the paper), i.e. the forces must be *coplanar*.

When the forces do not act at a point the rules are true, but are not sufficient for complete equilibrium. When forces do not act at the same point, turning effects are introduced. These turning effects are considered later, but it is sufficient to note now that, when several forces all act at a point, there is no turning effect about that point.

Example

The roller shown in Fig. 1.30 has a mass $m = 3$ tonne and is held on a smooth plane inclined at $\theta°$ to the horizontal by a pull P parallel to the plane.

If $\theta = 30°$, calculate (a) the pull P and (b) the normal reaction N of the inclined plane on the roller.

Solution

The weight W of the roller acts vertically downwards. This force can be resolved into two components, one parallel to the incline, the other at right angles to the incline.

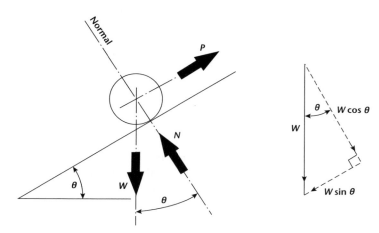

Fig. 1.30

The components of W are

$W \sin \theta$ down the incline
$W \cos \theta$ normal to the incline

and $W = mg = 3 \times 10^3 \times 9 \cdot 8 = 29\,400$ N.

The pull P up the incline must balance the component of the weight down the slope, i.e.

$P = W \sin \theta$

$= 29\,400 \times \sin 30° = 14\,700\,\text{N} = \mathbf{14 \cdot 7\,kN}$

The reaction N of the inclined plane is equal and opposite to the component of W normal to the slope, i.e.

$N = W \cos \theta$

$= 29\,400 \times \cos 30° = 25\,500\,\text{N} = \mathbf{25 \cdot 5\,kN}$

(If the pull is applied in a horizontal direction, what is its magnitude? *Ans*: 1·7 kN.)

Example

In the part of a mechanism shown in Fig. 1.31 the force P on the piston is 6 kN for the position shown.

Find, for this position (a) the force Q in the rod and (b) the reaction N between the piston and the guides. Neglect friction.

Solution

The reaction N of the guides on the piston is vertical, since friction is neglected.

Resolving the force Q exerted by the rod on the piston, we have

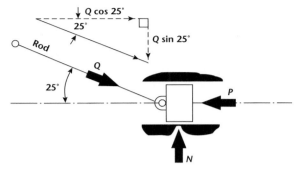

Fig. 1.31

Horizontal component = $Q \cos 25°$ (left to right)

Vertical component = $Q \sin 25°$ (downward)

For equilibrium of horizontal forces *on* the piston,

$P = Q \cos 25°$

Therefore

$$Q = \frac{6}{\cos 25°} = 6 \cdot 62 \, \text{kN}$$

For equilibrium of vertical forces on the piston,

$N = Q \sin 25° = 6 \cdot 62 \sin 25° = 2 \cdot 8 \, \text{kN}$

The student should confirm this result by drawing the triangle of forces.

Note: It is not correct to write (as many students do) that $Q = P \cos 25°$. This statement neglects the fact that Q must be in equilibrium with *both P and N*.

1.22 Flexible cables: smooth pulleys

A perfectly flexible cable or cord offers no resistance to bending or compression. *Therefore a flexible cable can only carry a tensile load.*

A flexible cable wrapped around a pulley carried on frictionless bearings has the same tension in it on both sides of the pulley.

Example

Figure 1.32 shows two flexible cables attached at a fixed point A. One passes over a pulley B and supports a load of 20 kg. The other passes over a pulley C and supports a load of 10 kg. Find the resultant force at A. Neglect friction.

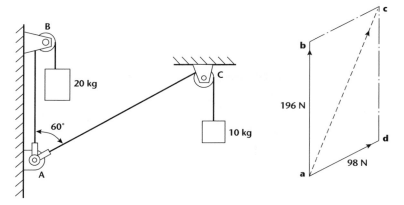

Fig. 1.32

Solution

Consider the cable at B. It supports a load of 20 kg. Therefore the tension in the cable is $20 \times 9.8 = 196$ N, to balance the dead weight. The cable is flexible so that the force in the cable is 196 N along its length. Therefore the cable pulls upwards on the fixing at A with a force of 196 N.

Similarly, the cable at C must provide an upward force of 98 N to support the dead load of 10 kg. Hence this cable pulls at A with a force of 98 N.

Combining the two forces at A by the parallelogram of forces, Fig. 1.32, we have

resultant force at A = **ac**

$$= 259 \,\text{N (by construction)}$$

Problems

1.22 A wagon is drawn steadily up a slope of 20° to the horizontal by a cable inclined at 25° to, and above, the plane. If the pull in the cable is 20 kN, calculate (a) the components of the pull parallel and perpendicular to the plane, (b) the horizontal and vertical components of the pull.
   (18·1, 8·45 kN; 14·1, 14·1 kN)

1.23 The light roller A, Fig. 1.33, is drawn up the smooth plane by the horizontal force *F*. The roller is attached to a light spring anchored at B. The stiffness of the spring is such that a force of 1 kN extends it 100 mm. Calculate the force *F* when the spring has extended 15 mm. What is then the normal reaction of the plane on the roller?   (212 N; 150 N)

1.24 Two light rods AB, BC are pivoted together at B, Fig. 1.34. Both rods lie in a vertical plane. The end C is pinned to a fixed bearing and end A is pinned to a roller which is free to move along a smooth horizontal surface. If a load of 5 t is hung from B find, for the position shown, (a) the horizontal force *F*

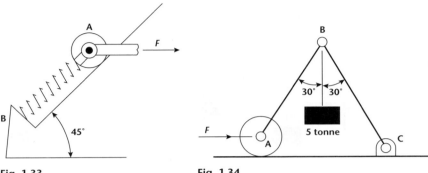

Fig. 1.33   Fig. 1.34

required at A to prevent the roller moving outwards, (b) the force in link AB and (c) the vertical reaction on the roller.   (14·2 kN; 28·3 kN; 24·5 kN)

1.25 In the crank and connecting rod mechanism shown in Fig. 1.35, the force P on the piston is 1000 N and the friction force F opposing motion of the piston is 70 N. The connecting rod is at 15° to the line of stroke XX of the piston. Calculate (a) the force Q in the rod, (b) the normal reaction N of the cylinder on the piston and (c) the resultant reaction of the cylinder on the piston. Is the connecting rod in tension or compression?

(963 N; 249 N; 259 N; compression)

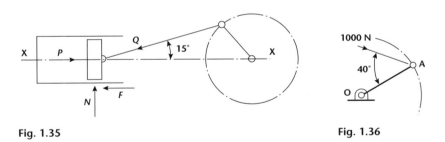

Fig. 1.35   Fig. 1.36

1.26 Figure 1.36 shows the resultant force acting on the crankpin of a crank OA. Find the radial and tangential forces on the crank by resolving the resultant force into its components.   (766 N; 643 N)

1.27 Figure 1.37 shows a rigid beam fixed at one end and supporting a vertical load of 100 kg at the free end A. At A is attached a flexible cable passing over a smooth pulley B which supports a load of 250 kg. Calculate the resultant force on the beam at A.   (1890 N, 23·5° to horizontal, upwards to the right)

1.28 A load of 10 t hangs from a pulley A, Fig. 1.38. A flexible wire is threaded through the pulley and attached at one end to a roof girder at B. The other end of the wire passes over the pulley C. The load is supported by the pull

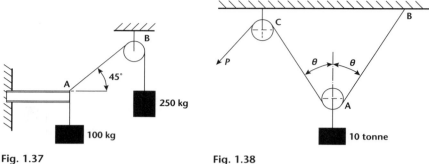

Fig. 1.37   Fig. 1.38

P on the free end of the wire. The angle between the two parts of the wire at A is 2θ. Calculate the pull *P* (a) when $\theta = 0°$ and (b) when $\theta = 60°$. Neglect pulley friction and the weight of the hanging pulley.   (49 kN; 98 kN)

1.29 Figure 1.39 shows a power shovel digging rock. The boom FB is at 50° to the horizontal. The dipper handle DE is at right angles to the boom and pivots freely in the block D. The cable ABC is at 20° to the boom and passes over a pulley at the top of the boom, the portion BC of the cable being vertical. The dipper teeth exert a thrust of 120 kN on the rock in a direction at right angles to the handle DE.

Find (a) the pull in the cable; (b) the thrust in the dipper handle; and (c) the thrust on the boom along BD. Neglect the weight of handle and dipper.
(157, 101, 267 kN)

1.30 The cutting tool shown in Fig. 1.40, when cutting mild steel on a planing machine, cuts square to the horizontal surface of the workpiece. The rake angle is 15° as shown. The forces on the tool are 1·5 kN horizontally and 500 N vertically upwards. Calculate (a) the resultant force on the tool and (b) the components of the resultant force normal and parallel to the cutting face AB.   (1580 N; 1320, 871 N)

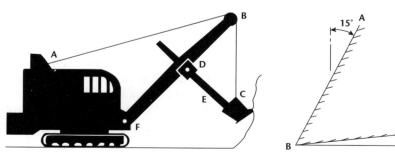

Fig. 1.39   Fig. 1.40

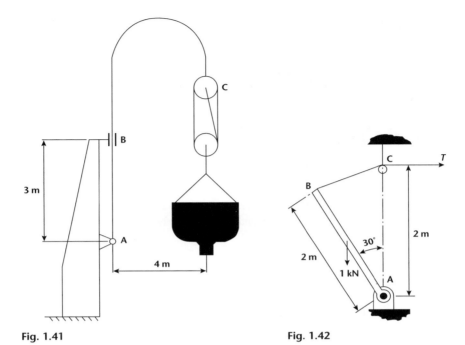

Fig. 1.41   **Fig. 1.42**

1.31 ABC, Fig. 1.41, represents one of a pair of lifeboat davits. A is a pin-joint and B a bearing. If the mass of the boat is 1 tonne, calculate the resultant force at A and B respectively.

(8·16 kN at 53·1° to the vertical; 6·54 kN horizontal)

1.32 The trapdoor AB, Fig. 1.42, is hinged at A and is raised by the rope BC passing over the small pulley C. For the position shown, find the tension T in the rope. Neglect friction and assume the trapdoor to be uniform.

(259 N)

Chapter 2

Moment, couple and torque

··

We must now take account of the fact that a force may produce a turning moment or a couple when combined with another parallel force. These turning effects are considered and the principle of moments is introduced and applied to forces acting on bodies at rest or in steady motion.

2.1 Effects of a force

As we have already seen, the effect of a force along its own line of action is simply a push or a pull.

We now consider the effect of a force at a point on a rigid body *not* on the line of action of the force. The effects of the force on the body are

- to tend to move the body, and
- to tend to rotate the body.

Thus, in particular, the effect at any point O of a force *F* acting along AB, Fig. 2.1, is to tend to move the point O parallel to the line of action AB and also to produce a turning effect about O.

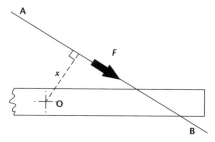

Fig. 2.1

2.2 Turning effect; moment

We measure the turning effect at point O, or the tendency of the force to cause rotation about O, by the product

$F \times x$

where F = magnitude of the force

x = *perpendicular distance* of the line of action AB from O.

This turning effect is called a *moment* and the distance x is called the *moment arm*. A moment always refers to the turning effect *about a particular point*. Thus, the moment of F about O is Fx.

In Fig. 2.1, the line of action AB is to the right of point O, the force F acts in the sense A to B, and hence the turning effect is clockwise.

When the force is in *newtons* and the moment arm in *metres*, the moment is measured in *newton metres* (N m). This form is the derived SI unit. Other forms met with are kN m and MN m.

2.3 Resultant moment

When more than one force acts on a body the total turning effect about *any given point* of all the forces acting is the sum of all the separate turning effects. But, since some forces exert clockwise and others anticlockwise moments about a point, we must take account of *sign*. Hence, *the total or resultant moment about any given point is the algebraic sum of the moments of all the forces about the same point.*

2.4 Balance of moments

If a body is at rest under the action of a number of coplanar forces there must be a balance of moments, otherwise the unbalanced resultant moment would cause rotation of the body. Thus the sum of the clockwise moments about any point must equal the sum of the anticlockwise moments about the same point or, for equilibrium,

Clockwise moments = anticlockwise moments

This is sometimes called the *principle of moments*.

The same principle applies if the body is rotating at a *steady* speed under the action of a number of coplanar forces. An unbalanced resultant moment would cause a *change* in the rotation of the body.

2.5 Couple

Consider two parallel forces P, Q acting in the opposite sense, Fig. 2.2. The effect of both P and Q at any point O in the same plane is: (a) to produce a turning effect and (b) to tend to move O parallel to their respective lines of action AB and CD.

Due to P, moment = Pa clockwise
Due to Q, moment = Qb clockwise

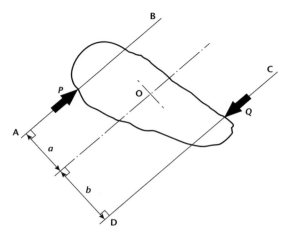

Fig. 2.2

The resultant moment about O is therefore clockwise and of magnitude $(Pa + Qb)$.

Suppose P to be greater than Q; then the body will tend to move in the direction of the force P and at the same time tend to turn clockwise.

Now let $Q = P$, Fig. 2.3, so that we have two *equal*, parallel forces P acting in opposite directions. Such a pair of forces is said to produce a *couple*.

The resultant moment about O is now

$$Pa + Pb = P(a + b)$$

$$= Px$$

where x is the perpendicular distance between the lines of action of the forces.

This resultant moment is independent of the position of the point O. Therefore *the magnitude or moment of a couple is the same about any point in the plane.*

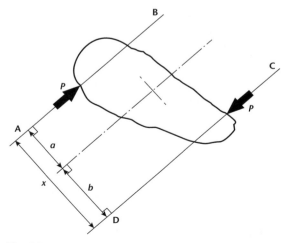

Fig. 2.3

Further, since the forces are *equal and opposite*, there is no tendency for the body to move in the direction AB. *The effect of a couple is therefore to tend to cause rotation only.*

Thus we have the definitions:

A couple is formed by two equal parallel forces which act in opposite directions.

The moment of a couple is the product of one of the forces and the perpendicular distance between the lines of action of the forces.

To specify a couple completely we must know

- its moment,
- its direction of rotation,
- the plane in which it acts.

An important point to note is that, if we wish to replace or balance a couple acting on a body, we can *only* do so by another couple, i.e. by another set of equal parallel, oppositely directed forces. *We cannot replace or balance a couple by a single force.*

The magnitude of a couple is sometimes called a *torque*. The term *torque* is usually used when dealing with the turning moment exerted on a shaft.

Typical examples of couples are as follows:

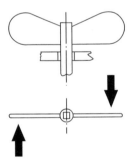

Fig. 2.4

- rotating the winding key of a clockwork mechanism, Fig. 2.4;
- rotating a 'butterfly' nut by hand;
- turning a tap-wrench, provided equal forces are exerted at both arms of the wrench;
- the torque exerted by an electric motor or transmitted by a shaft.

Note that, in each case, the couple exerts no sideways force on the spindle shaft or bearing.

Now we consider a case where such a sideways force exists.

2.6 Couple and force combination

A *single* force F applied at right angles to a spanner and at a distance x from the nut, provides a turning moment Fx about the nut, Fig. 2.5(a). In addition, for

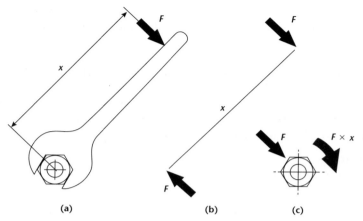

(a)   (b)   (c)

Fig. 2.5

balance of forces on the spanner the *reaction* of the nut is a force F opposite to the applied force, Fig. 2.5(b).

The two forces F on the spanner form a couple of moment Fx.

The force *on* the nut is equal and opposite to the reaction of the nut to the spanner. It has, therefore, the same magnitude and direction as the applied force F, but its line of action passes through the nut.

The effect of the force at the end of the spanner is to produce at the nut a force F together with a couple Fx, Fig. 2.5(c). Note that we now have a sideways force at the nut.

Figure 2.6 shows a further example of a force-and-couple effect. A shaft carries a heavy wheel of weight W overhanging a bearing by a distance x. The effect of W at the section of the shaft at the bearing is to produce a direct load W at the bearing, together with a couple of moment Wx tending to bend the shaft.

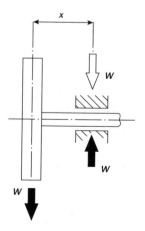

Fig. 2.6

The second force providing the couple in this case is the upward reaction of the bearing on the shaft.

Example

A deep rigid beam AD is fastened by a stud at B to the ledge shown in Fig. 2.7. It carries a vertical load of 1·5 kN as shown. Calculate the force F in the stud (a) if the load acts downwards and (b) if the load acts upwards. If the safe load in the stud is 2·5 kN, what is the maximum distance from C that the 1·5 kN load may act in case (a)?

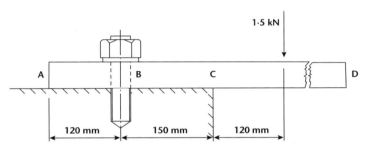

Fig. 2.7

Solution

(a) The vertical downward load tends to tilt the beam about the edge C. Hence the anticlockwise moment of the load F in the stud about C must balance the clockwise moment of the 1·5 kN load about the same point. Thus, taking moments about C,

$$F \times 150 = 1{\cdot}5 \times 120 \quad \therefore F = \textbf{1·2 kN}$$

(b) If the load is acting upwards the beam will tend to pivot about end A. Equating clockwise and anticlockwise moments about A,

$$F \times 120 = 1{\cdot}5 \times 390 \quad \therefore F = \textbf{4·88 kN}$$

Let x millimetres be the distance from C of the 1·5 kN load when the force in the stud is 2·5 kN; then, taking moments about C,

$$2{\cdot}5 \times 150 = 1{\cdot}5 \times x \quad \therefore x = \textbf{250 mm}$$

If this distance of 250 mm is exceeded, the force in the stud will exceed 2·5 kN.

In practice, since the beam is not perfectly rigid, it may not tend to pivot about an edge as assumed, but the method of calculation used here is sufficient for most practical purposes.

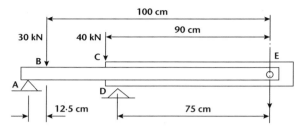

Fig. 2.8

The compound lever shown in Fig. 2.8 is made up of two levers AE and CE, pinned at a common point E. AE carries a 30 kN load at B and rests on a knife edge at A. CE carries a 40 kN load at C and rests on a knife edge at D. Calculate the vertical force required at the pin E to hold the two levers horizontal.

Solution
The vertical force required at E is made up of two parts:

1. The force F_1 required to hold the lever AE.
2. The force F_2 required to hold the lever CE.

For AE, taking moments about A,

$$F_1 \times 112 \cdot 5 = 30 \times 12 \cdot 5$$

Therefore

$$F_1 = 3 \cdot 33 \, \text{kN, } upward$$

For CE, taking moments about D,

$$F_2 \times 75 = 40 \times 15$$

Therefore

$$F_2 = 8 \, \text{kN, } downward$$

Thus the net *downward* force required at E is

$$F_2 - F_1 = 8 - 3 \cdot 33 = \mathbf{4 \cdot 67 \, kN}$$

Figure 2.9(a) shows the four main forces L, W, T and D, maintaining a Cessna aircraft in equilibrium when cruising at a steady speed in straight level flight. There is also a small vertical force P at the tailplane, sense unknown. If W = 60 kN and T = 10 kN, both acting through G and the lines of action of L and W are a distance x apart, find the force P required for equilibrium when x = 100 mm. The forces T and D are opposite and parallel. What value of x would result in zero tailplane load?

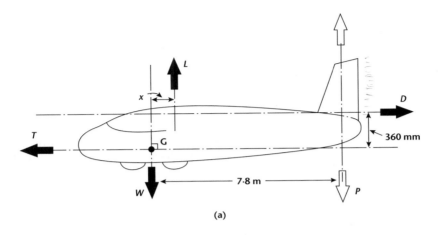

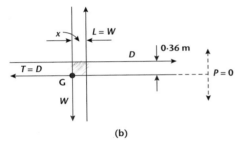

(b)

Fig. 2.9

Solution

For balance of horizontal and vertical forces (assuming P to act downwards), we have

$$T = D = 10 \times 10^3 \, \text{N}$$

and

$$L = W + P$$

Taking moments about G, thereby eliminating T and W, then

$$L \times x = D \times 0.36 + P \times 7.8$$

i.e. $(W + P) \times 0.1 = 0.36 \, D + 7.8 \, P$

i.e. $(60 \times 10^3 + P) \times 0.1 = 0.36 \times 10 \times 10^3 + 7.8 \, P$

therefore

$$P = 312 \, \text{N}$$

P is positive, therefore its sense is downwards as assumed. If P is to be zero, then $L = W$ and the two opposite and parallel forces form a couple. Also, the equal and opposite parallel forces T and D form a couple causing a nose-up pitching moment on the plane. For balance therefore the forces L and W must produce a *nose-down moment* and hence the line of action of L must lie behind that of W as in Fig. 2.9(b). Thus, equating the pitching moments,

$$L \times x = D \times 0\cdot36$$

i.e. $60 \times 10^3 \times x = 10 \times 10^3 \times 0\cdot36$

therefore $x = 0\cdot06\,\text{m}$ or   **60 mm**

The magnitude and position of the four main forces alter during flight and the correcting load P must be continually altered to maintain equilibrium (*see* page 189).

(*see* page 189).

Example

The rectangle shown in Fig. 2.10(a) has AB = 2 m, BC = 1·5 m, and diagonal AC = 2·5 m. For the system of forces represented, find the magnitude, sense and direction of the resultant force and state its position.

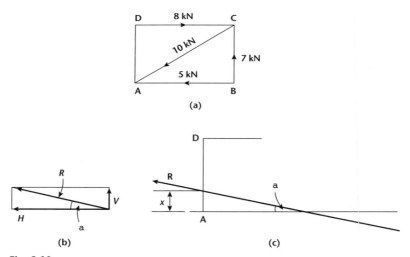

Fig. 2.10

Solution

Let V and H be the vertical and horizontal components of the resultant force R. Then, resolving forces vertically (upwards positive), and horizontally (positive to the right), we have

$$V = 7 - 10 \sin \text{B}\hat{\text{A}}\text{C}$$

$$= 7 - \left(10 \times \frac{1\cdot5}{2\cdot5}\right)$$

$$= 1\,\text{kN (upwards)}$$

and $H = 8 - 5 - 10 \cos \hat{BAC}$

$$= 3 - \left(10 \times \frac{2}{2 \cdot 5}\right)$$

$$= -5 \, \text{kN (to the left)}$$

The resultant R and its components are shown in Fig. 2.10(b).

$$R = \sqrt{(H^2 + V^2)}$$

$$= \sqrt{(5^2 + 1^2)}$$

$$= 5 \cdot 1 \, \text{kN}$$

$$\tan \alpha = \frac{V}{H} = \frac{1}{5 \cdot 1} = 0 \cdot 196$$

$$\therefore \quad \alpha = 11 \cdot 1°$$

R is now known in magnitude, sense and direction. Assume that its line of action is in the position shown in Fig. 2.10(c), cutting AD at a distance x from A. The moment of the resultant R about A must be the same as that due to the given forces, i.e.

$$7 \times \text{AB} - 8 \times \text{AD} = R \times x \cos \alpha$$

i.e. $7 \times 2 - 8 \times 1 \cdot 5 = 5 \cdot 1 \times x \cos 11 \cdot 1°$

i.e. $x = \mathbf{0 \cdot 4 \, m}$

x is positive, therefore the assumed position is correct. Note that the 10 kN and 5 kN forces pass through the point A, therefore their moment about A is zero. The resultant is **5·1 kN**, with its line of action at **11·1°** to the horizontal, cutting AD at a point **0·4 m** from A, above A, and acting upwards. The equilibrant, i.e. the balancing force, is equal and opposite to R.

Problems

2.1   For the bracket shown in Fig. 2.11 calculate the force in the bolt B due to a load F of 300 N on the bolt A (a) when F is in the direction shown and (b) when F is in the reverse direction to that shown.   (450 N; 900 N)

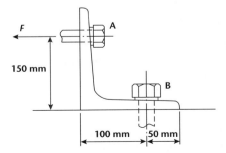

Fig. 2.11

2.2   The arm shown in Fig. 2.12 is keyed to a 15 cm diameter shaft; otherwise it is free to rotate on the shaft. Calculate the force on the key A if the load *W* is 200 N.   (2·4 kN)

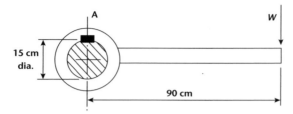

Fig. 2.12

2.3   The beam of Fig. 2.13 is built-in to the wall and carries two loads as shown. Calculate the resultant moment of these forces about a point O in the beam at the wall.   (1·26 kN m clockwise)

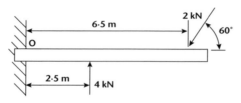

Fig. 2.13

2.4   Calculate the vertical reactions at the supports of the framework shown in Fig. 2.14. The frame is made up of equilateral triangles.   (4 kN; 5 kN)

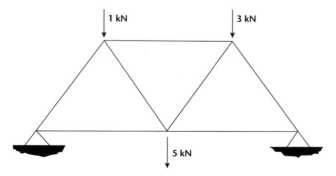

Fig. 2.14

2.5   Calculate the horizontal force *F* which will just pull the roller over the ledge shown in Fig. 2.15. The roller is 60 cm in diameter and of mass 1 tonne.
                                                                       (11 kN)

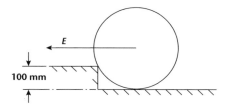

Fig. 2.15

2.6 Figure 2.16 is a diagrammatic sketch of the weighing system of a compound-lever testing machine. Rigid links AD, EG, HJ and HK are horizontal. Links DE and FH are vertical. AD rests on a knife edge at C; EG rests on a knife edge at G; HJ and HK rest on knife edges at J and K respectively. Joints at D, E, F and H may be assumed freely pinned.

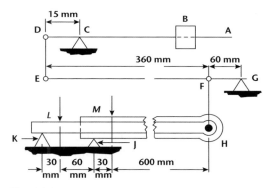

Fig. 2.16

The load on the machine is applied at L and M and may be assumed to be divided equally between these two points. The load is weighed by a movable poise at B of mass 30 kg.

Calculate the value of the load on the machine when B is 240 mm from C.

(75 620 kg)

2.7 The light lever AB shown in Fig. 2.17 is hinged at A and supported by a spring at C. The stiffness of the spring is 200 N/cm of extension. The end B carries a load of 10 kg. Before the load is applied the lever is horizontal. What are the downward displacements at C and B when the load is applied?

(12 mm; 28 mm)

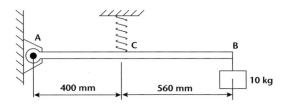

Fig. 2.17

2.8 A crane jib AB is shown in Fig. 2.18. The load rope applies a force of 4 kN at 20° to the jib at B and the tie rope applies a force of 5 kN at 25° to the jib at C. AB = 12 m, CB = 1·8 m. Find the resultant moment about A. What would be the pull in the tie rope for no resultant moment about A?

(5·14 kN m anticlockwise; 3·81 kN)

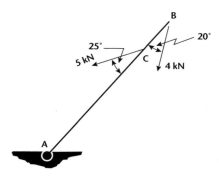

Fig. 2.18

2.9 The mechanism shown in Fig. 2.19 is part of the load-measuring device for a testing machine. Link CD is vertical and links AC and DF are horizontal. The links are pin-jointed at C, D and F. The top link rests on a knife edge at B. The vertical load at E is measured by the distance x from B of a counterpoise of mass 5 kg. If the load is 4 t, calculate the distance x.   (320 mm)

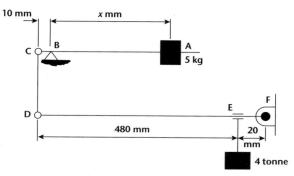

Fig. 2.19

2.10 Figure 2.20 shows two horizontal bars AC and DF supported by a spring at B and connected by a vertical tie CE. Bars AC and DF are uniform in section and each of mass 1·2 kg. At B there is a concentrated mass of 0·8 kg and at F a concentrated mass of 1 kg. The linkage is freely pinned at A, C, E, D. Find (i) the force in bar CE and (ii) the force in the spring. Neglect the mass of tie CE.

(22 N; 46·8 N)

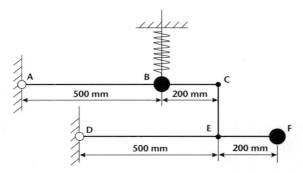

Fig. 2.20

2.11 A charge machine for a furnace, Fig. 2.21, consists of a trolley and counter-weight A, mounted on rails B, C and carrying a scoop D. The trolley, counterweight, scoop and handle together weigh 117·6 kN and their centre of gravity is on a vertical line midway between B and C. Calculate the greatest safe load in the scoop in tonnes.   (3 t)

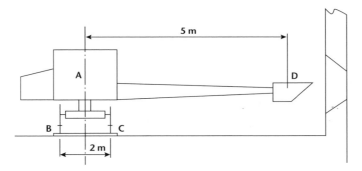

Fig. 2.21

2.12 Figure 2.22 shows a mobile grab. Calculate the load on the front and rear axles respectively when the grab carries a load of 1 tonne, given the following data: length of boom, 10 m at angle 60° to horizontal: distance between axles, 5 m: the boom is pivoted immediately above the rear axle: mass of boom, 0·6 t, centre of gravity at midpoint of boom: mass of grab, empty, 0·3 t: mass of engine driver's cab and chassis, 6 t; centre of gravity, 0·3 m behind front axle: mass of operator's cab and counterweight, 3 t; centre of gravity, 1·8 m behind front axle.   (front 58·4 kN; rear 48·3 kN)

2.13 A rotating flywheel is at a particular instant maintained in equilibrium by tangential forces F_1, F_2, radial forces R_1, R_2, and an anticlockwise couple M, as shown in Fig. 2.23. The weight W of the flywheel is 400 N. If $F_1 = 40$ N and $M = 2·25$ N m, find the magnitude of the other forces.

(F_2, 55 N; R_1, 374 N; R_2, 362 N)

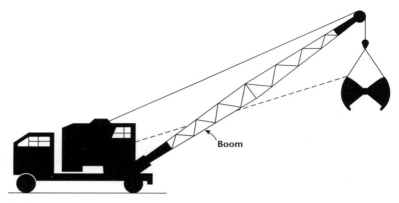

Fig. 2.22

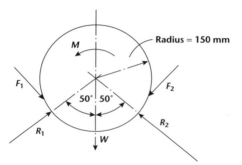

Fig. 2.23

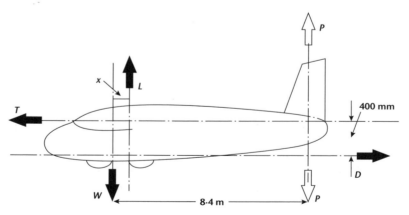

Fig. 2.24

2.14 When an aircraft cruises at steady speed in straight level flight it is maintained in equilibrium by the four main forces L, W, T and D, shown in Fig. 2.24. There is also a small tailplane load P which may act vertically upwards or downwards. For a medium-range transport plane $W = 93\,\text{kN}$,

$T = 12\,kN$ and $P = 2.5\,kN$ *downwards* at a particular instant. What is the distance x between the lines of action of L and W? Assuming the lines of action of W, T, and D remain the same, find the value of x corresponding with (i) a tailplane load of $2.5\,kN$, *upwards*, (ii) a zero tailplane load. What is the tailplane load when the lines of action of L and W coincide?

(170 mm; 285 mm, L in front of W; 52 mm, L in front of W; 571 N)

2.15 Calculate the magnitude and direction of the resultant of the force system shown in Fig. 2.25. Check by drawing the polygon of forces.

(586 N at 45° to, and below, AB)

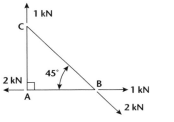

Fig. 2.25

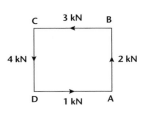

Fig. 2.26

2.16 Calculate the magnitude and direction of the resultant of the forces shown in Fig. 2.26. Where does the line of action of the resultant force cut line DA? The forces act round the sides of a square of 1 m side.

(2.83 kN at 45° to DA; 2.5 m to left of D on AD produced)

2.17 The equilateral triangle shown in Fig. 2.27 has a side of 100 mm. Find (a) the magnitude and direction of the resultant force and (b) the point at which the resultant cuts QR.

(5.3 N, 41° to QR; 150 mm to right of R on QR produced)

2.18 For the force system shown in Fig. 2.28 calculate: (a) the resultant force; (b) the angle which the line of action of the resultant makes with the Ox-axis; (c) the total moment about O; and (d) the point at which the resultant cuts the Ox-axis. The figure is marked off in 1 m squares.

(24.4 kN; 28.6°; 85.2 kN m, 7.3 m to right of O)

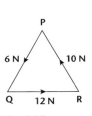

Fig. 2.27

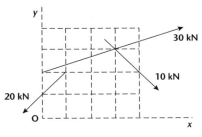

Fig. 2.28

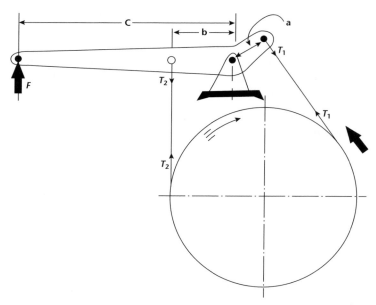

Fig. 2.29

2.19 Figure 2.29 shows a band-brake for a pulley of diameter 1200 mm. For clockwise rotation the tensions in the pulley belt are T_1 and T_2 when a braking force F is applied at the end of the lever. The lever is pivoted at a fixed fulcrum as shown. If $F = 100$ N, a = 30 mm, b = 120 mm, c = 720 mm and the ratio of the tensions is $\frac{T_1}{T_2} = 3$, find the braking torque.   (2880 Nm)

Chapter 3

Polygon of forces; frameworks

··

The work covered so far on conditions of equilibrium of a system of forces is extended here and applied, using the polygon of forces, to obtain the force diagrams for a statically determinate pin-jointed framework. Thus the nature and magnitude of the internal forces in the members of the frame and the reactions at the supports can be found.

3.1 Polygon of forces

We have already seen that if three forces acting on a body are in equilibrium, their lines of action must pass through the same point and the forces may be represented by the three sides of a *triangle of forces* (Ch. 1). Similarly, if four or more forces are in equilibrium they may be represented by a *polygon of forces* or *force polygon*. The forces do not necessarily have to meet in a single point.

Figure 3.1(a) shows four forces denoted **1**, **2**, **3**, **4** acting at one point. Such a diagram is called a *space diagram*.

The same four forces are represented in magnitude, direction and sense by the vectors **ab**, **bc**, **cd** and **dc**, in the *force diagram*, Fig. 3.1(b); this is the force polygon, and assuming that the forces are in equilibrium, the polygon will close as shown.

Note that, as in the triangle of forces, the vectors are 'taken in order', i.e. as we traverse polygon **abcd**, each vector arrow follows that of the force direction in the space diagram, and follows on *head to tail* from the preceding vector in the force diagram.

The four-sided polygon shown can be drawn in six different ways according to the order in which the forces are taken from the space diagram; all six polygons are correct. Figure 3.1(c) shows an alternative to the force polygon of Fig. 3.1(b). The student should sketch the remaining four alternative polygons.

The verification of the above polygon of forces construction is by experiment. However, we can also think of the polygon as being made up of *triangles* of forces. Thus in Fig. 3.1(b), if **ac** is joined we may consider the two triangles **abc** and **acd**. These triangles are redrawn separately in Fig. 3.1(d). If **ca** represents a force (which may be imaginary) then **ab**, **bc** and **ca** represent three forces, forming a triangle of forces in which the vector arrows are head to tail ('taken in order'). The three forces are, therefore, in equilibrium. Thus forces **1** and **2** are in equilibrium with

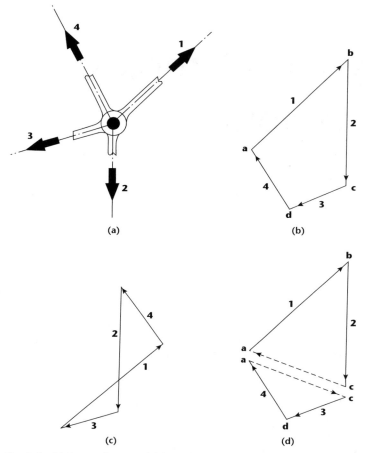

Fig. 3.1   (a) Space diagram; (b) force diagram; (c) alternative force diagram

some force represented by **ca**. Similarly, in triangle **acd**, the forces 3 and 4 (represented by **cd** and **da**) are in equilibrium with a third force represented by **ac**.

But force **ca** in triangle **abc** is equal and opposite to force **ac** in triangle **acd**.
If, therefore, we join the two triangles together to form the original polygon **abcd**,
the two imaginary forces **ac** and **ca**, required for equilibrium in each case, are
themselves in balance, and cancel.

Hence we may consider the polygon of forces as made up of several force
triangles. Each triangle represents three forces in equilibrium and the whole polygon represents all the forces in equilibrium.

3.2   Bow's notation

It is convenient, and in certain problems necessary, to have a systematic method
of describing the forces 1, 2, 3, etc. in the space diagram. Accordingly, we letter

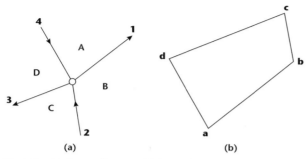

Fig. 3.2 (a) Space diagram; (b) force diagram

the *spaces* between the forces in *cyclic* order as shown in Fig. 3.2(a). Thus the spaces are lettered A, B, C, etc. in a clockwise direction and, for example, force **1** is referred to as force AB. It is represented in the force diagram of Fig. 3.2(b) by vector **ab**. Similarly, force **4** is called force DA and represented by **da**.

We may now drop the description **1, 2, 3** . . . for forces and follow *Bow's notation*: AB, BC, CD, . . .

Note that, since the spaces are lettered systematically clockwise, we may also drop the use of arrows on the force diagram to indicate directions. **ab** now represents a force AB taken in the direction **a** to **b**. This direction corresponds to the direction of the arrow on the force between spaces A and B.

When using Bow's notation the force diagram *must* be drawn by following the cyclic order A, B, C, . . . Otherwise, when drawing a force polygon without using Bow's notation, the forces may be picked in any order from the space diagram.

3.3) Equilibrant and resultant

We have already seen that, if four or more forces are in equilibrium, they may be represented by the sides of a *closed* polygon. If, when we draw the polygon of forces, *it does not close* then the forces are not in equilibrium, i.e. there is an unbalanced force. The vector required to complete a closed polygon is called the *equilibrant*. It represents, in magnitude and direction, the force required to balance the given system of forces.

Figure 3.3(a) shows four forces AB, BC, CD and DA, acting at the point O. The forces are represented in the force diagram, Fig. 3.3(b), by vectors **ab**, **bc**, **cd** and **de**. The last vector **de** does not close the polygon, therefore the forces are not in equilibrium.

The vector **ea**, representing the fifth force required to close the polygon and produce equilibrium, is therefore the *equilibrant* of the given system of forces. The equal and opposite vector **ae** represents the *resultant* force. This is the single force that has the same effect as the four given forces (see Ch. 1).

Figure 3.3(c) shows the space diagram with the equilibrant added. Note that the system of lettering does not apply to the equilibrant **ea**. The equilibrant actually lies in the space D.

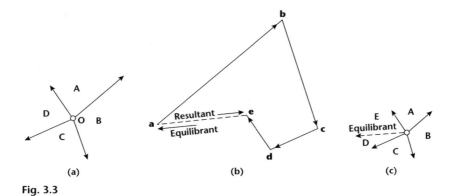

Fig. 3.3

Example

Figure 3.4(a) shows three members OX, OY, OZ of a roof truss connected together at a pin-joint O. The joint carries a load of 1000 N as shown. If the force in OX is 2000 N acting towards the joint, find the loads in members OY and OZ.

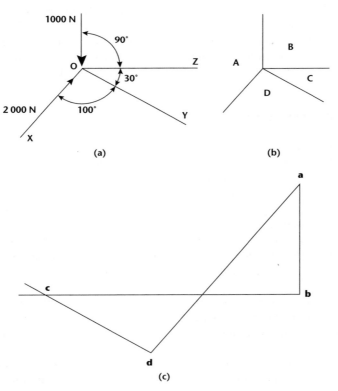

Fig. 3.4

Solution

First letter the spaces between the loads A, B, C and D in clockwise rotation as shown in Fig. 3.4(b).

- Draw **da** parallel to DA to represent to scale the 2000 N force in OX, Fig. 3.4(c).
- Through **a** draw **ab** parallel to force AB, to represent the load of 1000 N.
- Draw **bc** from **b** parallel to BC (the position of point **c** is unknown).
- From point **d** draw a line **cd** parallel to CD.

The intersection of **cd** and **bc** fixes the point **c**.

By construction,

load BC = **bc** = 2207 N from **b** to **c**

and

load CD = **cd** = 1064 N from **c** to **d**

Hence

force in OY is **1064 N** from O to Y

and

force in OZ is **2207 N** from Z to O

(Problems)

3.1 Four forces act at a point as follows: 1 kN due E; 4 kN at 45°N of E; 3 kN due N; 5 kN 50°S of W.

Find the magnitude and direction of the resultant force.

(2·09 kN at 73° N of E)

3.2 Figure 3.5 shows the loads on a gusset plate forming a joint in a roof truss. Verify graphically that the net vertical load on the plate is zero and find the horizontal force *P*.   (39·8 kN)

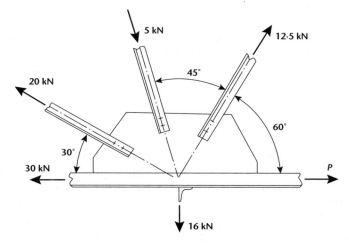

Fig. 3.5

3.3 Three horizontal wires are attached to the top of the vertical pillar of a jib crane. The directions and magnitudes of the pulls in the wires are as follows: 2·5 kN due W; 2·25 kN due N; 2 kN SW.

Find graphically the magnitude and direction of the pull in a horizontal cable attached to the same point required to balance these forces.

(4 kN, 12˚S of E)

3.4 Figure 3.6 shows a system of forces acting at point O. Determine graphically the forces *P* and *Q* if the system is in equilibrium.

(*P*, 7·7 kN away from O; *Q*, 3 kN towards O)

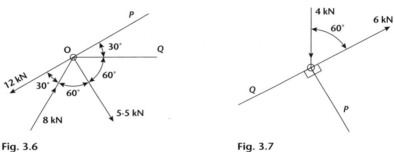

Fig. 3.6   Fig. 3.7

3.5 A heavy roller of mass 1000 kg is hauled steadily up a gradient whose sine is 0·5. The roller is pulled up by a horizontal force of 2·5 kN together with a pull *P* at 20˚ to, and above, the incline.

Neglecting friction, determine graphically the magnitude of pull *P* and the normal reaction of the incline on the roller.   (2·9 kN; 8·75 kN)

3.6 Figure 3.7 shows the forces acting at a pin-joint of a framework. Find graphically the magnitude of the unknown forces *P* and *Q*.   (*P*, 3·5 kN; *Q*, 4 kN)

(3.4) Frameworks

A *framework* is a structure built up of straight bars, or members, connected by pin-joints. The external loads on such a structure are applied at the pin-joints.

Figure 3.8 is a typical framework, a Warren girder. The frame is built up of triangles formed by three pin-jointed members enclosing a space. Such a structure is said to be *stiff* and can carry a load. Since the structure is in equilibrium the force polygon for the external loads and support reactions must close and the resultant moment about any point must be zero.

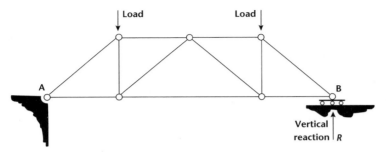

Fig. 3.8

3.5 Framework supports and reactions

The Warren girder of Fig. 3.8 is supported at B on rollers lying on a horizontal surface. It is usual to assume *perfectly smooth* surfaces and rollers; the reaction R at B is therefore *vertical*.

The left-hand support at A is a pin-joint; hence the direction of the reaction at A will depend on the loading. In the case shown the external loads are vertical and the reaction at B is vertical also. Thus the support reaction at A is vertical. The separate reactions at A and B may therefore be found by taking moments about A and B in turn.

The wall crane of Fig. 3.9 is fixed to the wall by pin-joints at A and B respectively. Since there is only one member attached to the wall at A the line of action of the reaction there must lie along the axis of the member 1.

Since there are two members, 2 and 3, attached to the pin at B the line of action of the wall reaction there is not yet known. However, the load W at C and the two reactions at A and B respectively must all pass through the same point Z, as shown (principle of concurrency for three forces in equilibrium, Ch. 1).

The line of action of the reaction at B lies therefore along the line BZ. Hence the triangle of forces may be drawn for the three *external* forces on the frame and the unknown reactions determined in magnitude and direction.

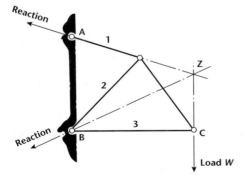

Fig. 3.9

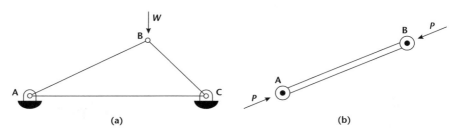

Fig. 3.10

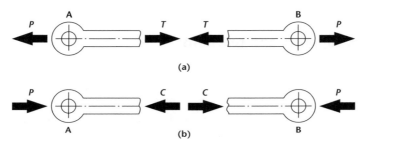

Fig. 3.11

3.6 Internal forces

Each bar of a frame must itself exert forces to balance the external loads. These forces exerted by the members are called *internal resisting forces*.

Consider the triangular frame shown in Fig. 3.10(a). The loads and reactions are applied at the pin-joints, which are assumed perfectly smooth and therefore frictionless. The only forces on each bar are therefore equal and opposite and act along the axis of the bar. Each bar is therefore subject to a push or a pull and is either in tension or compression. If in tension it is called a *tie*; if in compression it is a *strut*. In Fig. 3.10(b) the bar AB is in compression and is a strut.

In Fig. 3.11(a) the bar AB is a tie since the external forces P together exert a pull. Now imagine AB to be cut in two as shown. To maintain equilibrium there would have to be a pair of equal and opposite forces T at the cut. One force T is required to balance the pull at A, the other is required to balance the pull at B. The two equal and opposite forces T are the internal forces exerted by the bar and are each equal to the external load P. Note that the internal forces act *away* from the pin ends A and B respectively.

Similarly, when AB is subjected to compressive forces P, Fig. 3.11(b), the internal forces C in the strut are each equal in magnitude to the external load P and act *towards* the pin-joints A and B respectively.

The internal forces exerted by a tie act away from a pin-joint.
The internal forces exerted by a strut act towards a pin-joint.

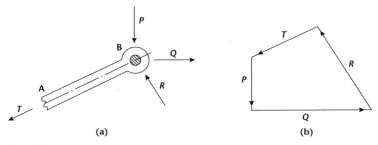

Fig. 3.12

3.7 Equilibrium of forces at a pin-joint

Consider now the forces at a pin-joint B, Fig. 3.12(a), which forms part of a framework. At such a joint there might be a load P, together with forces Q (tensile) and R (compressive) exerted *on* the pin by other members of the structure. In the figure a further bar AB is shown in which the force T is unknown.

For equilibrium we must consider the balance of forces acting at the pin B. Hence we must take into account the force exerted by the bar AB *on* the pin. This is the internal force T, say.

For a given structure the *directions* of all the forces P, Q, R and T are known. Hence if the magnitude and nature (tension or compression) of all except *two* of these forces are known, the magnitude and nature of these two unknown forces may be determined from a polygon of forces for the joint. For example, if, in Fig. 3.12(a) the forces P and Q are known then R and T may be found graphically. The directions of the forces R and T may be found from the force polygon and these directions determine whether the corresponding members are in tension or compression.

In Fig. 3.12(b) the polygon of forces shows that T acts away from the joint. Hence bar AB is in tension.

Note that if there are more than two unknown forces at a joint the polygon of forces cannot be completed. It is then necessary to consider some other joint of the frame first in order to reduce the number of unknown forces.

3.8 Force diagram for a frame structure

Consider the structure shown in Fig. 3.13(a). There is a vertical load W at joint X and the reactions at the supports are L and R vertically upwards.

To determine the forces in each bar we draw the force polygon for each joint. It is essential to use Bow's notation. Hence we letter the spaces between each bar and support reaction as shown. There is no 'magic' way of lettering the spaces but it should be done systematically.

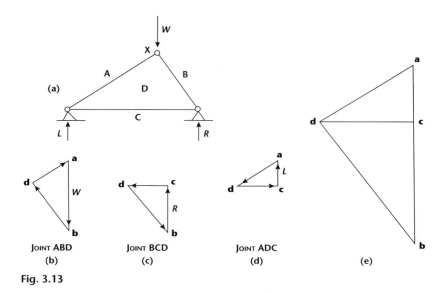

Fig. 3.13

We remember, however, that when drawing the force polygon *for a joint* the force vectors must be drawn in cyclic order (usually by going clockwise round the joint). Thus the load *W* is described by the letters AB (clockwise round the joint) and the support reaction *R* by BC. Similarly a joint may be described by the letters of the spaces meeting at the joint. For example, joint X is described as joint ABD.

The force polygon **abd** for joint ABD is shown in Fig. 3.13(b).

- **ab** represents the load AB (= *W*) in magnitude and direction.
- **bd** represents the load in bar BD and is parallel to BD.
- **da** represents the load in bar DA and is parallel to DA.

Since *W* acts vertically downwards, the sense of force **ab** is from **a** to **b**.

Following the arrowed vectors in order round **abd** we see that the force **bd** is in the sense **b** to **d** and **da** is in the sense **d** to **a**.

Since force **da** in DA acts towards the joint, then DA is in compression. Similarly BD is in compression.

Similarly, the force polygons for joints BCD and ADC are drawn in Fig. 3.13(c) and (d) respectively.

Note that, when drawing the separate force polygons, it is necessary to go round each joint in the same order as every other joint. *Once begun the order cannot be changed.*

In Fig. 3.13(d), **dc** represents the force in CD at joint ADC. The force is away from the joint, hence DC is in tension and is a tie. Also, the line **acb** represents the closed polygon for the external load and reactions. Thus

- **bc** represents the upward reaction BC (= *R*), and
- **ca** represents the upward reaction CA (= *L*).

The combined force polygon

Since

- **dc** is equal and opposite to **cd**,
- **ad** is equal and opposite to **da**, and
- **bd** is equal and opposite to **db**,

then it can be seen that the three separate force polygons of Fig. 3.13 may be combined into a single diagram as shown in Fig. 3.13(e).

In practice, therefore, it is not necessary to draw separate polygons for each joint. A single diagram will suffice for the whole framework. It is no longer possible, however, to use the vector arrows on the force polygon since they are oppositely directed for each joint. Hence, as the diagram is completed for each joint, a separate sketch is required of the original framework on which the force *directions* at a joint are inserted. From the directions of the two arrows on each bar, the nature of the force is determined. This is best illustrated by numerical examples.

Example

The triangular framework shown in Fig. 3.14(a) is loaded by a vertical force of 1 kN at joint ABD.

Determine graphically the magnitude and nature of the force in each member and the vertical support reactions at the pin-joints ADC and BCD.

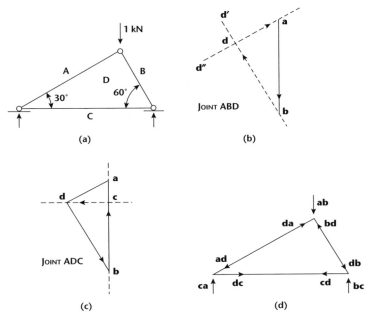

Fig. 3.14

Solution

Joint ABD, Fig. 3.14(b):

- Draw **ab** vertically downwards to represent the 1 kN load AB.
- Draw **bd'** from **b** parallel to BD.
- Draw **ad''** from **a** parallel to DA.

The intersection point **d** of **bd'** and **ad''** completes the force diagram for the joint.

The directions of the forces at the joint are fixed by the downward direction of the vector **ab** representing the known 1 kN load. Going round the triangle **abd** clockwise (i.e. round the joint ABD) we see that **bd** is in the sense **b** to **d** and **da** is in the sense **d** to **a**.

The force directions are taken from the force diagram and inserted in the sketch of the frame, Fig. 3.14(d).

The force in BD is **bd** *towards* the joint.
The force in DA is **da** *towards* the joint.

Joint BCD, Fig. 3.14(c): the *known* part of the force diagram **abd** is redrawn in full lines (for clarity only).

The only known force at this joint is DB represented by **db**, i.e. equal and opposite to the force **bd** at the top joint.

The force in BC is represented by a vertical line **bc** from **b**, but the point **c** is as yet unknown.

The force in CD is found by drawing a line **dc** from **d** parallel to CD.

The intersection of **dc** and **bc** fixes the point **c** and completes the force diagram for the joint.

The direction of the forces at the joint are fixed by the known direction of the vector **db**. Going round the triangle **dbc** shows that the force BC is vertically upwards (**bc**) and the force CD is *away* from the joint (vector **cd**).

These force directions are inserted in Fig. 3.14(d).

Joint ADC, Fig. 3.14(c): the force in AD is **ad**, equal and opposite to the force **da**.

The force in DC is **dc**, equal and opposite to the force **cd**. Traversing the triangle **adc** in order shows that the vertical support reaction CA is **ca** vertically upwards. (Note that if we went round the triangle – or the joint – in the wrong order we should find that the force CA was downwards, which is obviously incorrect.)

These force directions are inserted in the sketch, Fig. 3.14(d).

The bars BD and DA are in compression since the internal forces are towards the joints. Bar DC is in tension since the forces are away from the joints.

By construction, the following results are obtained:

Bar	Tension (kN)	Compression (kN)
AD	–	0·5
BD	–	0·87
CD	0·43	–

Reaction BC = **0·75 kN**
Reaction CA = **0·25 kN**

Note. In practice only *one* force diagram is required, i.e. Fig. 3.14(c), and arrows should not be drawn on the vectors. Also the force directions at the joints may be drawn on a scale diagram of the framework instead of on a separate sketch.

Example

The wall crane shown in Fig. 3.15(a) is loaded by a vertical force of 10 kN at the joint ABD. The bar AD is horizontal and the framework is pinned to the vertical wall at joints ADC and BC. Determine the magnitude and nature of the forces in the framework. What are the magnitude and direction of the reaction at the top wall joint?

Solution

Note that the vertical line of the wall is *not* a load or bar of the frame. Space C separates the reaction BC at the lower wall joint and the reaction CA at the upper wall joint.

The reaction CA is unknown in magnitude and direction. The reaction of the wall at the lower joint is equal and opposite to the force in bar BC.

Joint ABD, Fig. 3.15(b):

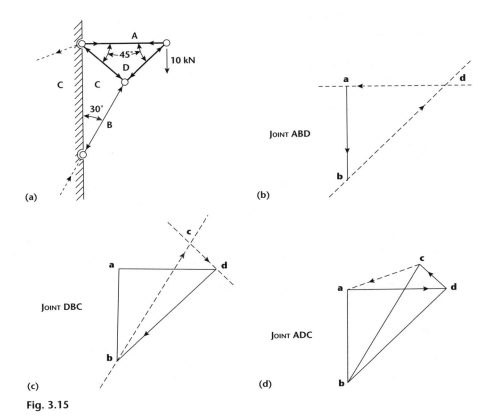

Fig. 3.15

● Draw **ab** vertically downwards from **a** to represent the 10 kN load AB.
● Draw **bd** parallel to bar BD, and **ad** parallel to bar DA.

The intersection **d** of **bd** and **ad** completes the force diagram for the joint.

The directions of the forces at the joint are fixed by the known downward vector **ab**. Following the vectors in order round triangle **abd** determines the direction **bd** (towards the joint) and the direction **da** (away from the joint). These directions are shown in Fig. 3.15(a).

Joint DBC, Fig. 3.15(c): the known force at this joint is **db**, equal and opposite to **bd**.

● From **b** draw **bc** parallel to BC.
● From **d** draw **dc** parallel to CD.

Thus point **c** is determined.

The force directions at the joint are fixed by the vector arrow from **d** to **b**, as shown.

Joint ADC, Fig. 3.15(d): the known force is **ad** in AD, equal and opposite to **da**, already found.

Join **ca**. This then represents the force in CA, the reaction at the joint. It acts away from the joint.

The polygon **abdc** is the complete force diagram for the framework.

By construction, the forces are found to be as follows:

Bar	Tension (kN)	Compression (kN)
BC	–	14·6
CD	–	3·8
DB	–	14·1
DA	10	–

Reaction at top wall joint = CA = **7·8 kN**, at 20° to, and below, the vertical.

Example

The Warren girder shown in Fig. 3.16(a) is loaded vertically by forces of 20 and 30 kN at the lower panel pin-joints. It is simply supported at the joints ABG and AED.

Determine the magnitude and nature of the forces in each bar and the magnitudes of the support reactions.

Solution

There are more than two unknown forces at every joint; hence we cannot begin the force diagram immediately. It is necessary first to calculate the vertical reactions at the supports by taking moments about each support in turn.

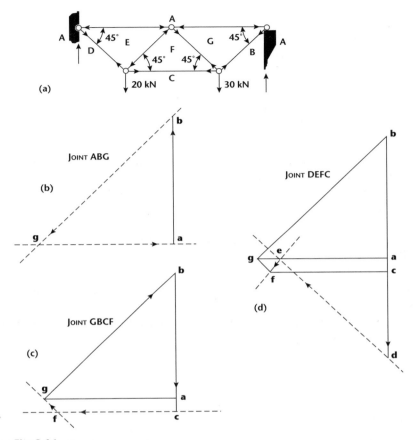

Fig. 3.16

Let

L = vertical reaction DA

R = vertical reaction AB

Assume each horizontal bar of the frame to be of length 2 units.

Taking moments about joint AED

$$4 \times R = (30 \times 3) + (20 \times 1)$$

therefore

$$R = 27 \cdot 5 \, \text{kN}$$

Taking moments about joint ABG

$$4 \times L = (30 \times 1) + (20 \times 3)$$

therefore

$$L = 22 \cdot 5 \, \text{kN}$$

Check:

$$L + R = 22 \cdot 5 + 27 \cdot 5 = 50 \, \text{kN}$$

$$= \text{total load}$$

Joint ABG, Fig. 3.16(b):

- Draw **ab** vertically upwards to represent the 27·5 kN reaction AB.
- Draw **bg** through **b** parallel to BG.
- Draw **ga** through **a** parallel to GA.

The force directions at the joint are fixed by the upward vector **ab**. The direction of the force in BG is away from the joint. The direction of the force in GA is towards the joint. The direction arrows are now added to Fig. 3.16(a).

Joint GBCF, Fig. 3.16(c):

- Draw **bc** to represent the downward load BC of 30 kN.
- Draw **fg** through **g** parallel to FG.
- Draw **cf** through **c** parallel to CF.

The intersection of **fg** and **cf** gives point **f** and completes the diagram for the joint.

The directions of the forces at the joint are fixed by the known direction **bc** of the load BC. Traversing the polygon **bcfg** in order determines the force directions.

Joint DEFC, Fig. 3.16(d):

- Draw **cd** vertically downward from **c** to represent the known load CD of 20 kN.
- Draw **de** parallel to DE.
- Draw **ef** parallel to EF.

If correctly drawn, the point **e** lies on the intersection of **de** and **ef**, and also on the horizontal line **ae**, parallel to AE.

The directions of the forces at the joint are fixed by the known direction of **cd**. The directions are found by following the arrows round the polygon **cdef** in order.

The forces in the frame are scaled off and tabulated as follows:

Bar	Tension (kN)	Compression (kN)
BG	39	–
GA	–	27·5
CF	25	–
FG	3·5	–
DE	32	–
EF	–	3·5
EA	–	22·5

Reaction AB = **27·5 kN**
Reaction DA = **22·5 kN**

Problems

3.7 Determine the magnitude and nature of the forces in the frameworks shown in Fig. 3.17 and state in each case the magnitude and direction of the reactions at the supports.

Answers (negative answers denote compression)

Fig. (a): AD, 11·5; BD, −5·8; DC, −11·5; CA, 11·5 kN.
　　　　　Reaction at X = 11·5 kN horizontal
　　　　　Reaction at Y = 15·3 kN at 48° to vertical

Fig. (b): CG = BF = −2·8 kN; FE = GE = 2·5 kN; FG = 0
　　　　　Reaction at X = Reaction at Y = 2·25 kN vertical

Fig. (c): AE, −28·9; BD, −57·7; CD, 28·9; DE, 0; GF, −57·7; FC, 28·9; EF, 0 kN
　　　　　Reaction at X = Reaction at Y = 50 kN vertical

Fig. (d): AE, 17·3; EB, −20; BD, −17·3; CD, 5; DE, −10 kN
　　　　　Reaction at X = 13·2 kN at 41° to vertical
　　　　　Reaction at Y = pull in cable = 8·7 kN

Fig. (e): AB = AF = −50; BC = FG = 70·7; ED = 50; BE = FE = 0 kN
　　　　　Both reactions 50 kN vertical

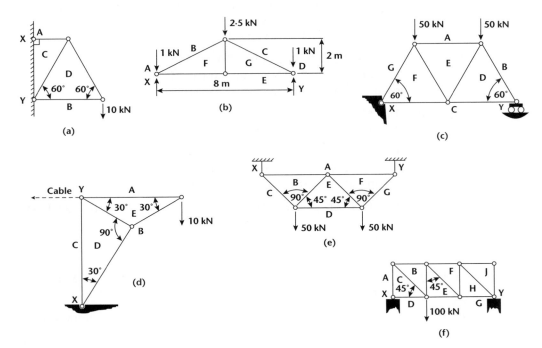

Fig. 3.17

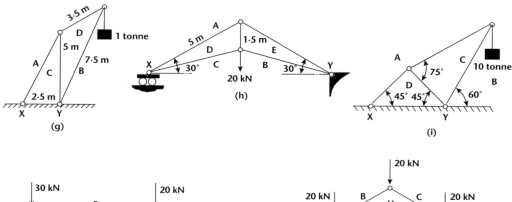

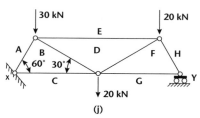

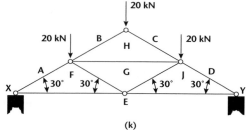

Fig. 3.17   *(cont.)*

Fig. (f): AB, −66·7; CD, 0; vertical AJ, 0; horizontal AJ, 0; BC, 94·3; BE, 33·3; EG, 66·7; HG, 33·3; FH, 33·3; EF, −47·1; HJ, −47·1; AF, −33·3; AC, −66·7
Reaction at X = 66·7 kN vertical
Reaction at Y = 33·3 kN vertical

Fig. (g)*: AC, 12·8; AD, 6·9; BD, −14·7; CD, −7·6 kN
Reaction at X = 12·8 kN parallel to bar AC
Reaction at Y = 21·9 kN at 15·3° to vertical

Fig. (h): AD = AE, −33·3; DC = BE, +29·6; DE, 33·3 kN
Both reactions 10 kN vertical

Fig. (i)*: AC, 98·1; BC, −170; CD, −25·4; AD, 94·8 kN
Reaction at X = 94·8 kN parallel to bar AD
Reaction at Y = 178 kN at 76·2° to horizontal

Fig. (j): AB, −44·8; ED, −37·6; FH, −36; FG, 1·8; BD, 17·5; BC, 22·4; DF, 22·4 kN
Reaction at X = 38·8 kN vertical
Reaction at Y = 31·3 kN vertical

Fig. (k): AF = DJ, −60; BH = CH, −20; HG, 35; FG = GJ, 0; EF = JE, 52 kN
Both reactions 30 kN vertical

* *Note* that the dead loads are in tonnes and must be converted to kilonewtons.

Chapter 4

Friction

••

Until now it has been assumed that the surfaces of bodies in contact have been smooth, but in practice there is always some degree of roughness, however small, so that friction is involved. The experimental laws of friction and their limitations are described in relation to contact between two dry and clean surfaces; friction in bearings and that between wheels and tracks are also explained. The problem of forces acting on a body on a rough inclined plane occurs frequently in engineering and scientific machines and is therefore given special attention. Finally the different aspects of friction as an aid or a hindrance and the methods of reducing its effects are discussed.

4.1 Friction force

Consider a load lying on a flat surface as in Fig. 4.1. Suppose that the surfaces are pressed together with a *normal* force N (this might be due to the weight of the load for example). Then, in order to move the load on the plane, a force tangential to the surfaces has to be overcome. This force is known as the *friction force*.

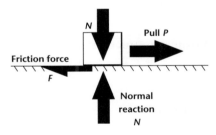

Fig. 4.1

4.2 Limiting friction force; laws of friction

Calculations on friction between two dry and clean surfaces in contact are based upon the following experimental facts:

1. If a force P is applied, tending to slide one body over another, the opposing friction force brought into play is tangential to the surfaces in contact and is just sufficient to balance the applied force.

2. There is a limit beyond which the friction force cannot increase; when this limit is reached sliding is about to start. This maximum value is termed the *limiting friction force*.
3. When the limit is reached the value of the limiting friction force F is proportional to the *normal* force N pressing the surfaces together and is independent of the area of contact.
4. The ratio of the limiting friction force F to the normal force N is a constant which depends only on the nature of the pair of surfaces in contact; thus

$$\frac{F}{N} = \mu, \text{ a constant}$$

μ is called the *coefficient of static friction*.

5. Before sliding starts, the pull P is equal to the friction force but is *less* than the limiting value μN.

When sliding is just about to start, the pull P is equal to the limiting friction force μN, i.e.

for limiting friction, $P = \mu N$

After sliding starts, the direction of the friction force is opposite to that of the *resultant motion*. The friction force is again given by $F = \mu N$, where μ is now called the *coefficient of sliding or kinetic friction*, and is usually slightly less than the coefficient of static friction. It is also roughly independent of the speed of sliding and of area of contact. Otherwise, sliding friction obeys the same laws as limiting friction.

When P remains equal to μN during sliding the body moves with a steady speed. When P is greater than μN the body is accelerated. (This particular case is dealt with in Chapter 9.)

These laws of friction are only approximately true but are sufficient for most engineering purposes.

4.3 Limiting friction on the horizontal plane

Figure 4.2(a) shows a block being pulled along a horizontal surface by force P inclined at angle α to the surface. The forces acting *on* the block are P, the weight W downwards, the normal reaction N of the plane upwards, and the friction force F opposing the motion.

When sliding is just about to start, or during steady motion, we have

$$F = \mu N$$

The force P has vertical and horizontal components, $P \sin \alpha$ and $P \cos \alpha$ respectively. To find reaction N, consider the vertical components of the forces acting. The weight W is balanced by N, together with the vertical component of P, hence

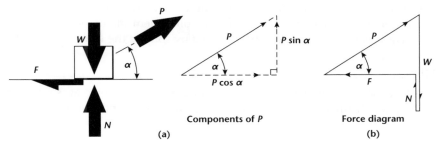

Fig. 4.2

$$W = N + P \sin \alpha$$

i.e. $N = W - P \sin \alpha$ [1]

and the normal reaction N is therefore less than the weight of the body.

Resolving horizontally, the friction force F balances the horizontal component of the pull P, i.e.

$$F = P \cos \alpha$$

or

$$\mu N = P \cos \alpha$$ [2]

From equations [1] and [2], P can be found in terms of μ, W and α.

The force diagram is shown in Fig. 4.2(b). It reduces to a triangle whose sides represent P, F, and $(W - N)$. The solution of this triangle also gives equations [1] and [2].

If the body is being *pushed* along the horizontal by an inclined force P as shown in Fig. 4.3(a), then resolving horizontally

$$P \cos \alpha = F = \mu N$$

and resolving vertically

$$N = W + P \sin \alpha$$

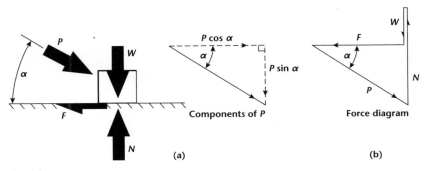

Fig. 4.3

In this case the normal force N pushing the two surfaces together is greater than the weight W, owing to the downward component of the push P.

These equations may also be obtained by solving the force diagram, Fig. 4.3(b).

Example

A load of 10 t is pulled along a horizontal track by a force at 30° to, and above, the track. If the coefficient of sliding friction is 0·2, find the friction force.

Solution

Let P be the unknown pull and F the friction force, Fig. 4.4.

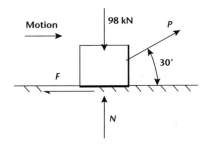

Fig. 4.4

For sliding friction,

$$F = 0·2\,N$$

Resolving horizontally,

$$F = P \cos 30°$$

i.e.

$$0·2\,N = 0·866P \quad \text{or} \quad N = 4·33P$$

Weight of 10 tonnes = $10 \times 9·8\,\text{kN} = 98\,\text{kN}$
Resolving vertically,

$$N + P \sin 30° = 98$$

i.e. $$4·33P + 0·5P = 98$$

Hence

$$P = 20·3\,\text{kN}$$

and

$$\text{Friction force, } F = P \cos 30°$$

$$= 20·3 \times 0·866$$

$$= \mathbf{17·6\,kN}$$

Example

Figure 4.5(a) shows the piston, crank, and connecting rod for a reciprocating engine. The force on the piston is 50 kN and the coefficient of sliding friction between piston and cylinder surfaces is 0·1. Calculate: (a) the force Q in the connecting rod and (b) the friction force resisting motion of the piston. The weight of the piston may be neglected.

Solution

The forces acting at the piston besides the 50 kN driving force are: Q the thrust in the rod, F the friction force and N the side thrust. These forces are shown in Fig. 4.5(b).

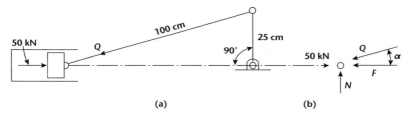

(a)   (b)

Fig. 4.5

$$F = \mu N$$

$$\sin \alpha = \tfrac{1}{4} \quad \therefore \ \alpha = 14°29'$$

Resolving vertically,

$$N = Q \sin \alpha = \frac{Q}{4}$$

Resolving horizontally,

$$50 = F + Q \cos 14°29'$$

$$= \mu N + 0·968Q$$

$$= 0·1\frac{Q}{4} + 0·968Q$$

Hence

$$Q = 50·3 \, \text{kN}$$

The friction force

$$F = 0·1 \, N = 0·1 \times \frac{50·3}{4}$$

$$= 1·26 \, \text{kN}$$

If the weight of the piston is taken into account the value of N is increased accordingly.

4.1 A girder is pulled along a horizontal floor by a rope at 45° to, and above, the floor. The pull in the rope is 1·5 kN. Calculate the coefficient of friction between girder and floor. The mass of the girder is 250 kg.   (0·76)

4.2 A casting of mass 50 kg is pushed along level ground by a force at an angle of 20° (downwards) to the horizontal. Calculate this force if the coefficient of friction between casting and floor is 0·4.   (244 N)

4.3 It requires a horizontal force of 375 N to drag a truck of mass 300 kg at uniform speed along a level track. Calculate the force required to move the truck if the force is inclined at 40° to the horizontal and is (a) a push into the track and (b) a pull away from the track.   (548 N; 442 N)

4.4 A load of 500 kg mass is pulled along level ground at constant speed by a pull of 1·5 kN at an angle of 69° to, and above, the ground. Calculate the coefficient of sliding friction.   (0·15)

4.5 The table of a planing machine weighs 2·5 kN and is traversed in its guides at a speed of 0·15 m/s. The coefficient of kinetic friction between table and guides is 0·1. Find the power employed to traverse the table. (Power, see Ch. 5.)   (37·5 W)

4.4 Friction on the inclined plane

Angle of repose

Consider a body of weight W on the point of slipping down a plane inclined at angle θ to the horizontal, Fig. 4.6.

The component of the weight down the slope is $W \sin \theta$ and that normal to the slope is $W \cos \theta$. The friction force F acts up the plane to oppose the motion.

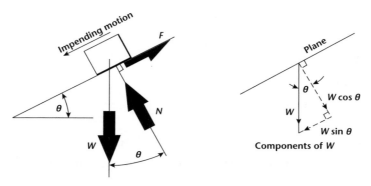

Fig. 4.6

Resolving parallel to the plane

$$F = W \sin \theta$$

Resolving normal to the plane

$$N = W \cos \theta$$

Therefore

$$\frac{F}{N} = \frac{W \sin \theta}{W \cos \theta} = \tan \theta$$

but

$$\frac{F}{N} = \mu$$

Hence

$\tan \theta = \mu$

where μ is the coefficient of static friction.

This particular angle of inclination of the plane is known as the *angle of repose*. If the plane is inclined at an angle greater than the angle of repose the body will slide down. If the inclination is less than the angle of repose the body will remain at rest. This gives a convenient method of obtaining μ by experiment.

Motion up the plane; pull *P* parallel to, and acting up, the plane

In Fig. 4.7, the friction force F opposes the motion of the body up the plane, N is the normal reaction of the plane on the body and the weight W acts vertically downwards. In the force diagram, F and N are at right angles, the angle between W and N is θ, the angle of the plane.

For limiting friction

$$F = \mu N$$

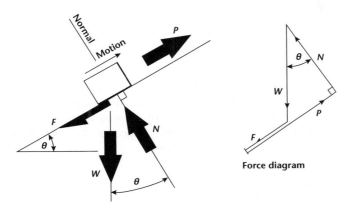

Fig. 4.7

Resolving perpendicular to the plane,

$N = W \cos \theta$

Resolving parallel to the plane,

$P = F + W \sin \theta$

Thus the pull must overcome both the friction force and the resolved component of the weight down the plane.

Motion down the plane; pull *P* parallel to the plane

1. When the body is *pulled* down the plane, Fig. 4.8, the component of the weight down the plane assists the pull. Hence, resolving parallel to the plane,

$P + W \sin \theta = F = \mu N$

and resolving normal to the plane,

$N = W \cos \theta$, as before

The force diagram is shown in Fig. 4.8.

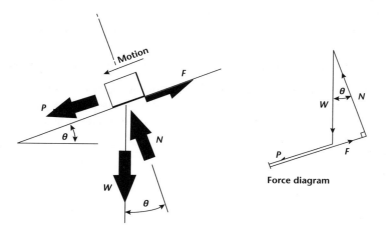

Force diagram

Fig. 4.8

2. When the weight component is sufficiently large, it may overcome the friction force without the assistance of a downward pull. However, if the body is just to be prevented from slipping down the incline or allowed to move down *steadily* then the force *P* must be applied *up* the plane, opposing the motion, Fig. 4.9. Then, in this case, resolving parallel to the plane,

$P + F = W \sin \theta$

and resolving normal to the plane,

$N = W \cos \theta$

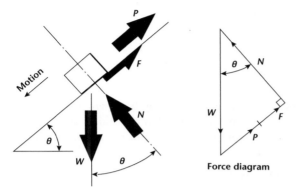

Fig. 4.9

and for limiting friction

$$F = \mu N$$

Example

A force of 540 N acting parallel to a plane inclined at 20° to the horizontal is required to just move a body of mass 61·3 kg up the plane. Find (a) the coefficient of friction between the surfaces and (b) the force parallel to the plane required to drag the body down the plane at a steady speed.

If the surface of the plane can be lubricated to alter the value of the coefficient of friction, what should this value be for the body to be just on the point of moving down under its own weight?

Solution

(a) Fig. 4.10(a) shows the forces acting on the body about to move up the plane. The weight of the mass of 61·3 kg is 61·3 × 9·8, i.e. 600 N. The normal force between the surfaces is 600 cos 20° N. The friction force F is therefore $\mu \times 600 \cos 20°$ and acts down the plane. Resolving forces parallel to the plane,

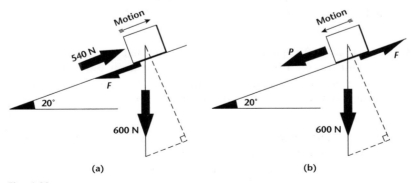

(a)   (b)

Fig. 4.10

$$540 = 600 \sin 20° + \mu \times 600 \cos 20°$$

$$= (600 \times 0.342) + (\mu \times 600 \times 0.9397)$$

Hence

$$\mu = 0.59$$

(b) Fig. 4.10(b) shows the forces acting on the body when it is moving down the plane at a steady speed. We may assume that the value of μ for static conditions will be true for sliding conditions. The friction force now acts up the plane. Resolving forces parallel to the plane,

$$P + 600 \sin 20° = \mu \times 600 \cos 20°$$

Therefore

$$P = 0.59 \times (600 \times 0.9397) - (600 \times 0.342)$$

$$= 128\text{N}$$

When the body is about to move down under its own weight, the component of the weight down the slope is just equal to the friction force acting up the slope. Therefore

$$\mu \times 600 \cos 20° = 600 \sin 20°$$

$$\mu = \tan 20°$$

$$= 0.36$$

Problems

4.6 A block is placed on an inclined plane. The plane is then raised until it makes an angle θ with the horizontal. If the coefficient of friction between block and plane is 0.3, find the value of θ to which the plane may be raised before the block begins to slide.   (16.7°)

4.7 A trolley of mass 50 kg just starts to move when placed on a plane inclined at 20° to the horizontal. Calculate the coefficient of static friction. Find also the horizontal force required to just start the trolley moving along the level if μ is unaltered.   (0.36; 179 N)

4.8 A force of 650 N parallel to a plane inclined at 30° to the horizontal is sufficient to pull a body up the plane at uniform speed. Calculate the force required to prevent it from moving down the plane if the body's mass is 100 kg. What is the coefficient of friction between body and plane?
(331 N; 0.2)

4.9 A force of 340 N parallel to an inclined plane is required to just move a body of mass 50 kg up the plane. Calculate the force parallel to the plane required to drag the body down the plane at a steady speed. The plane is inclined at 15° to the horizontal. What is the value of μ?   (86 N; 0.45)

4.10 A load of mass 25 kg is just on the point of sliding down a plane inclined at 35° to the horizontal when supported by a force of 90 N acting up the plane. What force parallel to the plane will just cause the load to move up the plane?

(191 N)

4.11 State the 'laws' of dry friction.

A winch hauls a casting of mass 800 kg up a slope of 1 in 15 (sine) at a uniform speed. Find the pull in the rope if the coefficient of friction is 0·5. The hauling cable may be assumed parallel to the slope.

(The laws state that the friction force is (i) independent of area of contact; (ii) proportional to the normal force; (iii) almost independent of the sliding velocity; 4·44 kN)

Motion up the plane; horizontal pull *P* assisting the motion

The body shown in Fig. 4.11 is about to move up the plane under the action of a horizontal pull *P*. The forces acting are: the horizontal pull *P*, the friction force *F* down the plane, the weight *W* vertically downwards, and the normal reaction *N* of the plane on the body. Resolving parallel to the plane,

$$P \cos \theta = F + W \sin \theta$$

$$= \mu N + W \sin \theta$$

Resolving normal to the plane,

$$N = W \cos \theta + P \sin \theta$$

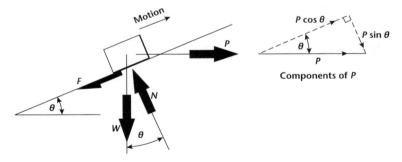

Fig. 4.11

Motion down the plane; horizontal pull *P*

Pull P *opposing motion*

The body shown in Fig. 4.12 is about to move down the slope or is moving down at a steady speed. The force *P* in this case is a restraining force. The friction force *F* now acts up the plane.

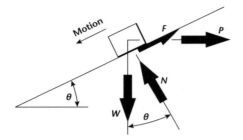

Fig. 4.12

Resolving parallel to the plane,

$$P \cos \theta + F = W \sin \theta$$

Resolving normal to the plane,

$$N = W \cos \theta + P \sin \theta$$

and

$$F = \mu N$$

Pull P *assisting motion*

The body shown in Fig. 4.13 is about to move down the plane *assisted* by a horizontal pull P. The friction force acts up the plane. Resolving parallel and normal to the plane, respectively,

$$F = W \sin \theta + P \cos \theta$$

and

$$N + P \sin \theta = W \cos \theta$$

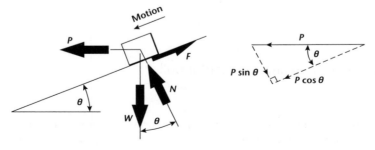

Fig. 4.13

A carriage of mass 1 tonne is to be pulled up a track inclined at 30° to the horizontal by a force P inclined at 20° to, and above, the track. Calculate the value of P if the effective coefficient of friction is 0·15.

Solution

In Fig. 4.14, the pull P has components $P \cos 20°$ and $P \sin 20°$ parallel to and normal to the track respectively. The weight of 1 tonne is $1 \times 9·8 = 9·8 \, \text{kN}$. Resolving forces parallel to the track,

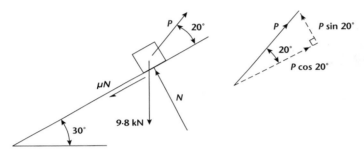

Fig. 4.14

$$P \cos 20° = 0·15 \, N + 9·8 \sin 30°$$

i.e. $P \times 0·9397 = 0·15 \, N + 9·8 \times 0·5$

Therefore

$$P = 0·16 \, N + 5·22 \qquad [1]$$

Resolving forces normal to the track,

$$N + P \sin 20° = 9·8 \cos 30°$$

Therefore

$$N = (9·8 \times 0·866) - (P \times 0·342)$$

$$= 8·5 - 0·342P$$

Substituting for N in equation [1],

$$P = 0·16(8·5 - 0·342P) + 5·22$$

Hence

$$P + (0·16 \times 0·342P) = (0·16 \times 8·5) + 5·22$$

i.e. $1·055P = 6·58$

Therefore

$$P = \mathbf{6·24 \, kN}$$

4.5 The angle of friction and total reaction

In all the cases considered so far, the body is in equilibrium under the action of *four* forces – *P*, *W*, *N* and *F*. The method of resolution of forces will solve all problems of this type but two equations are required for a solution. However, a simpler method of calculation exists for certain problems. This makes use of the resultant *R* of the friction force *F* and the normal reaction *N*, together with the angle ϕ between *R* and *N*. This angle ϕ is known as the *angle of friction*.

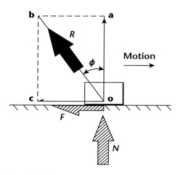

Fig. 4.15

Consider a body about to move to the right, as in Fig. 4.15. The force *R* is the resultant of *N* and *F*. Since these are at right angles, the angle ϕ between *R* and *N* is given by

$$\tan \phi = \frac{ab}{oa}$$

$$= \frac{F}{N}$$

$$= \frac{\mu N}{N} \text{ for limiting friction}$$

$$= \mu$$

The direction in which *R* must be drawn is determined by the fact that its tangential component *F* must oppose the motion, i.e. *R* is always drawn backwards to the direction of motion. If *R* lies inside the angle of friction the force of friction is less than the limiting value and slipping cannot take place.

When *N* and *F* are replaced by one force *R*, the forces *P*, *W* and *R* form *three* forces in equilibrium and a triangle of forces can be drawn, Fig. 4.16. Note that *in the absence of friction the only force possible between two surfaces is normal* to the surfaces.

Examples on the solution of problems on the inclined plane using the angle of friction are given below.

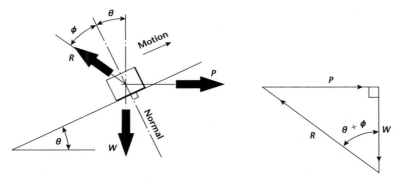

Fig. 4.16

Motion up plane; horizontal pull P

The force R, Fig. 4.16, is the resultant force exerted by the plane on the body. It acts at an angle ϕ to the normal to the plane and when slipping is just about to start, $\tan \phi = \mu$. Since the motion is up the plane, R must have a component down the plane and is therefore directed as shown. The pull P, the weight W and R form a triangle of forces as shown. Since P and W are at right angles,

$$P = W \tan (\theta + \phi)$$

where $\tan \phi = \mu$.

Motion down plane; horizontal pull P

Pull P opposing motion

In this case R has a component up the plane and is directed backwards to the direction of motion at an angle ϕ to the normal to the plane as shown in Fig. 4.17. From the triangle of forces,

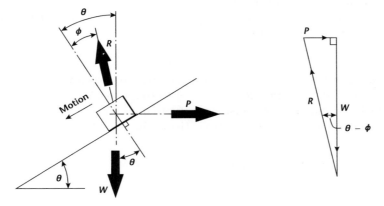

Fig. 4.17

$$P = W \tan (\theta - \phi)$$

This case corresponds to an angle of the plane θ greater than the angle of friction ϕ.

Pull P assisting motion

In Fig. 4.18, R is directed as in the previous case, while P is reversed in direction. From the triangle of forces,

$$P = W \tan (\phi - \theta)$$

This case corresponds to an angle of friction ϕ greater than the angle of the plane θ.

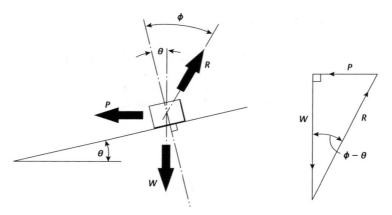

Fig. 4.18

(**Example**)

A casting of mass 2 tonnes is to be pulled up a slope inclined at 30° by a force at an angle to the slope. If the coefficient of friction is 0·3, find the least force required and its direction to the horizontal.

Solution

In the triangle of forces, Fig. 4.19(a), **oa** represents the known weight, and **ax** the direction of the total reaction R at $(30° + \phi)$ to the vertical. Pull P is represented by **ob** which, for equilibrium, must close the triangle. For P to be the least force required, **ob** must be perpendicular to **ax**. Hence the minimum value of P is given by

$$P = (2 \times 9\cdot8) \sin (30° + \phi)\,\text{kN}$$

where

$$\tan \phi = 0\cdot3 \quad \text{or} \quad \phi = 16°42'$$

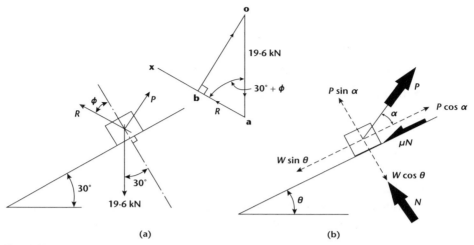

Fig. 4.19

Hence

$$P = 19{\cdot}6 \sin 46°42'$$

$$= \mathbf{14{\cdot}25\,kN}$$

From triangle **oba**, it can be seen that the minimum value of the force must be at an angle $(30° + \phi)$, i.e. 46°42', to the horizontal.

Alternative solution Fig. 4.19(b).

First consider the general case where θ is the angle of the slope to the horizontal, α is the angle the force P makes with the slope and W is the load. Resolving forces parallel and perpendicular to the slope gives the following equations:

$$P \cos \alpha = W \sin \theta + \mu N$$

$$N = W \cos \theta - P \sin \alpha$$

hence

$$P = \frac{W(\mu \cos \theta + \sin \theta)}{\cos \alpha + \mu \sin \alpha}$$

Since W, θ and μ are constant this may be written:

$$P = \frac{\text{constant}}{x}$$

where $x = \cos \alpha + \mu \sin \alpha$.

By drawing a graph of x against α for a range of values of α between 0° and 90°, the turning point of the graph will give the value of α for the least force P to pull the load up the slope. Using differential calculus* gives the mathematical solution which is $\alpha = \tan^{-1} \mu$ for x to be a maximum and P to have its *minimum* value. Thus

$$\alpha = \tan^{-1} 0.3 = 16°42'$$

and $\theta = 30°$, $W = 19·6\,\text{kN}$, hence $P = \mathbf{14·25\,kN}$.

* $dx/d\alpha$ is equated to zero, i.e.

$$\frac{d}{d\alpha} (\cos \alpha + \mu \sin \alpha) = 0$$

i.e. $-\sin \alpha + \mu \cos \alpha = 0$

therefore $\tan \alpha = \mu$ or $\alpha = \tan^{-1} \mu$.

(Example)

A body rests on an inclined plane. When the inclination of the plane is 25° to the horizontal, the body just slides down without acceleration. If it is pushed up the incline by a force parallel to the slope, find the mechanical efficiency of the arrangement.

Solution
When the body just slides down the plane under its own weight the angle of the plane and the angle of friction are equal. The angle of friction is therefore 25°.

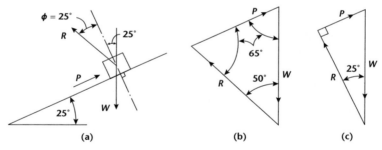

Fig. 4.20

When pushed up the incline the reaction R, Fig. 420(a), is directed at 25° to the normal so as to oppose the motion. The angle between R and the weight W in the triangle of forces, Fig. 4.20(b), is 50°; that between the weight W and the push P is $(90° - 25°) = 65°$. From the sine rule applied to the triangle of forces,

$$\frac{P}{\sin 50°} = \frac{W}{\sin 65°}$$

i.e. $P = \dfrac{W \sin 50°}{\sin 65°} = \dfrac{0·766\,W}{0·9063} = 0·845\,W$

When there is no friction, R is normal to the plane. The triangle of forces is now as in Fig. 4.20(c) and the force P is given by

$P = W \sin 25° = 0·4226\,W$

Mechanical efficiency (*see* Ch. 6) $= \dfrac{\text{pull without friction}}{\text{pull with friction}}$

$$= \frac{0·4226\,W}{0·845\,W} = 0·5 \quad \text{or} \quad \textbf{50 per cent}$$

Problems

4.12 A load is to be pulled down a slope inclined at 16° to the horizontal by a force *P* at an angle to the slope. The coefficient of friction is 0·4 and the pull *P* is to be the least possible. Calculate the angle to the horizontal which the line of action of force *P* must make.   (5·8°)

4.13 A load of 51 kg is pulled along a level floor by a force inclined at angle α to the floor. The coefficient of friction between load and floor is 0·3. If the pull is to be 300 N, find the angle α which the force must make with the horizontal. If the pull is to be the *least* possible, find its value.

(78° above horizontal, or 45° below; 144 N)

4.14 A 100 kg load is pulled up a plane inclined at 30° to the horizontal by a horizontal force *P*. Attached to the load is an anchored spring S which resists motion up the plane, Fig. 4.21. At the instant shown, the spring is horizontal and the tension in it is 500 N. If the coefficient of friction between the load and the plane is 0·15, calculate the force *P* at this instant.   (1·3 kN)

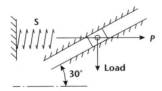

Fig. 4.21

4.15 A load of 200 kg rests on an inclined plane which makes 60° with the horizontal. The coefficient of friction is 0·1. It is prevented from sliding down the slope by a spring, the axis of which is parallel to the plane. Find the force in the spring.   (1·6 kN)

4.16 A load of 50 kg is being pulled up an inclined plane by a force *P* parallel to the plane. The angle of the plane is 45° to the horizontal and the coefficient of friction between load and plane is 0·1. Motion of the load is resisted by a spring, the axis of which is parallel to the plane. The strength of the spring is such that 10 mm extension requires a pull of 40 N. Calculate the force *P* on the load when the spring has been stretched 75 mm from its unstretched length.   (680 N)

4.17 A load is raised on an inclined plane by an effort acting parallel to the plane. If the coefficient of friction between load and plane is 0·1 and the angle of the plane is 45° to the horizontal, find (a) the ratio load:effort and (b) the efficiency of the inclined plane as a machine. (Efficiency, *see* Ch. 6.)

(1·3; 91 per cent)

4.18 A body of mass 10 kg is on the point of slipping down a plank 3 m long when one end of the plank is raised 900 mm. If this end of the plank is raised a further 450 mm, find the least force perpendicular to the plank that will maintain equilibrium. (53 N)

4.19 A load of 100 kg rests on an inclined plane. When the angle of the plane is 30° it moves with uniform velocity down the incline. If the body is pushed up the incline with uniform velocity by a horizontal force, find the value of the force. If the push is parallel to the incline, what is now the force required? (1·7 kN; 980 N)

4.20 A load of mass *M* is lowered from the drum D, Fig. 4.22. The load is prevented from running away by a vertical load applied to the brake drum E through the block B and the lever AC, hinged at C. The greatest vertical braking effort *P* at A is 200 N. What is the greatest load which can be lowered safely? The coefficient of friction between drum and block is 0·3. (137 kg)

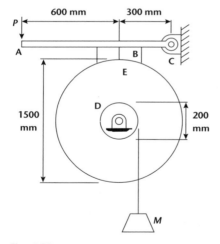

Fig. 4.22

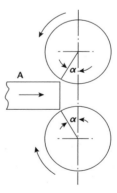

Fig. 4.23

4.21 A rolling mill, Fig. 4.23, consists of two rolls rotating as shown. In order that the billet shall be just pulled through by the rolls, the resultant force on the billet must have a horizontal component to the right. If the coefficient of friction between rolls and billet is 0·2, calculate the greatest angle *α* that will allow the billet to enter the rolls. (11·3°)

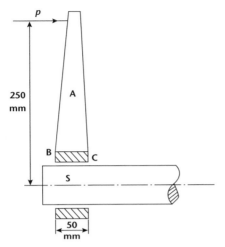

Fig. 4.24

4.22 The bush B, Fig. 4.24, just fits on the horizontal spindle S. The arm A is welded to the bush and carries a horizontal load *P*. The bush just starts to slip along the spindle when the load is applied at a point 250 mm from the axis of the spindle. Calculate the coefficient of friction between bush and spindle. (Assume that the bush tends to tilt about the point C.)   (0·1)

4.6 **Bearing friction**

Figure 4.25 shows a load shaft rotating anticlockwise in a plain bearing. The clearance between shaft and bearing is very often only about 0·002 mm for each millimetre of shaft diameter. Hence it may be assumed that the shaft diameter is very slightly less than the bearing diameter and that their centres coincide approximately. When stationary, contact takes place at the lowest point A. When rotating, and when flattening of the surfaces under load is neglected, the point of contact climbs to a point B at which slip commences. If *N* is the normal or radial reaction at B then the total reaction *R* is at angle ϕ to the direction of *N*, such that its tangential component is the friction force *F* opposing motion; ϕ is then the angle of friction. The forces on the shaft reduce to *W* (the bearing load) and *R*. These must be equal and opposite for balance; hence *R* = *W* and is vertically upwards.

The friction force $F = R \sin \phi$

$$= W \sin \phi$$

For a well-lubricated bearing, the coefficient of friction μ and the angle of friction ϕ are small. For example, if $\mu = \tan \phi = 0.1$, then $\phi = 5°43'$ and $\sin \phi = 0.0996$. Thus, for most practical purposes, $\sin \phi$ may be replaced by $\tan \phi = \mu$.

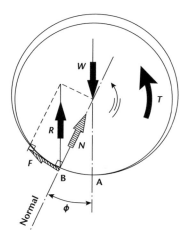

Fig. 4.25

Thus the friction force

$F = W \tan \phi$ approximately

$\quad = \mu W$

If the shaft diameter is d,

friction torque $= \frac{1}{2} Fd$

$\quad\quad\quad\quad\quad = \frac{1}{2} \mu Wd$ approximately

and is balanced by the equal and opposite driving torque T shown. This friction torque results in a loss of power in the bearing.

Example

The average load on each bearing of a four-cylinder engine amounts to 10 kN. The coefficient of sliding friction is 0·01 and the engine runs at 2000 rev/min. Find the power lost in friction if there are three bearings each of diameter 50 mm.

Solution
Friction force at each bearing $= \mu W$, approximately

$$= 0{\cdot}01 \times 10$$

$$= 0{\cdot}1\,\text{kN} = 100\,\text{N}$$

Friction torque $= \frac{1}{2} \mu Wd$

$$= \frac{1}{2} \times 100 \times 0{\cdot}05$$

$$= 2{\cdot}5\,\text{N m}$$

Then, using the formula developed in section 5.3 below,

$$\text{power lost in friction} = \frac{2\pi n T}{1000}$$

$$= \frac{2\pi \times 2000 \times 2 \cdot 5}{60 \times 1000}$$

$$= 0 \cdot 523 \, \text{kW}$$

Total friction loss for three bearings = **1·6 kW**

<u>**Problems**</u>

4.23 A turbine rotor of mass 1500 kg rotates in 100 mm diameter bearings at 1400 rev/min. If the coefficient of sliding friction is 0·02, calculate the power lost in friction. (Power, *see* Ch. 5.)   (2·2 kW)

4.24 In a test on a large journal bearing the power lost in friction at a speed of 10 rev/s was 722 W. The load on the bearing was 25 kN and the shaft diameter was 40 mm.

Calculate the coefficient of friction for the bearing.   (0·02)

4.7 Rolling friction

In order that a wheel shall roll steadily *without slip* there must be a friction force at the contact between wheel and track. This friction force acts so as to oppose slipping of the wheel on the track.

For example, if a trailer is hauled by means of a tow rope then, to cause the wheel to rotate clockwise, Fig. 4.26(a), the friction force F must act in the direction opposite to the motion of the axle, as shown.

On the other hand, if the wheel is *driven* by a torque T applied to the axle the friction force will act in the same direction as the motion, Fig. 4.26(b). If the wheel radius is r then, for balance of couples,

$$T = F \times r$$

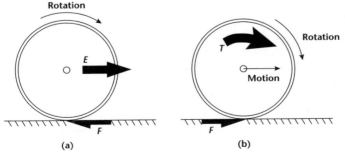

Fig. 4.26

Wheel slip occurs when F is equal to the limiting friction force.

Note that, in both the cases considered, friction between wheel and track acts merely to oppose slip. It is *not* a resistance to motion of the vehicle.

(4.8) Rolling resistance

In addition to the friction force which causes rotation of a rolling wheel, there is a resistance to motion of the vehicle owing to deformation of the wheel or track under load, Fig. 4.27. This resistance is called the *rolling resistance*. It acts so as to oppose the linear motion of the axle. The ratio of the rolling resistance to the normal force between wheel and track is called the *coefficient of rolling resistance* but this coefficent is little used.

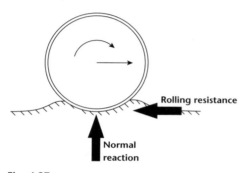

Fig. 4.27

The total force needed to maintain steady motion of a vehicle is that required to overcome rolling resistance at all wheels, sliding friction at the journal bearings, and air resistance. This total resistance is called the *tractive resistance*.

(4.9) Aspects of friction: tribology

We have discussed so far only the mechanics of the friction forces between two surfaces in contact. Friction has two aspects: without friction, traction (walking, motoring, skiing and so on) would not be possible, nor would the operation of many devices and machines such as wedges, screw threads, clutches, belt drives and brakes. Aircraft take advantage of air friction (drag) when landing and the Space Shuttle, without atmospheric drag, would require much more rocket power when descending from space. On occasion, it is necessary to simulate friction conditions, e.g. an artificial ski-run must reproduce the condition of skis moving over hard-packed ice. A further use of friction is in welding processes; a technique of 'friction stir welding', particularly applicable to ferrous materials but now being extended to joining other metals, uses the friction produced by a rotating probe and the workpiece at the joint line, to obtain the energy for the weld. On the other hand, friction is the engineer's enemy, where it has to be eliminated or reduced when bodies are sliding or rolling together. A particular field of industry where friction problems arise is the bulk handling of materials and viscous fluids.

The aspects associated with combating friction are indicated briefly in the following notes.

Lubrication is one of the most effective ways to reduce friction by separating the surfaces with a film of lubricant using some kind of lubrication system – manual, drip, pump, spray or bath; the lubricant may be a grease, mineral or synthetic oil, metallic or ceramic coating, or even a gas.

Wear is the loss of surface metal and this could decide the operating life of a machine component. The wear may be caused by friction between sliding surfaces, rolling friction or by grit causing abrasion. Such wear may be reduced by proper choice of contacting materials as to hardness and compatibility or by the use of surface coatings as well as surface-hardening methods including hard-facing by welding deposits.

Bearing friction must be kept to the absolute minimum to reduce surface wear, heat build-up and power loss; this is accomplished in sleeve bearings by lubrication and by the use of hardened steel balls and rollers to replace friction, although some rolling elements slide as well as roll. Lubricants are also used in ball and roller bearings and additionally enable heat to be removed from the bearing.

Fretting is the release of fine particles at the contact surfaces of two apparently hard, smooth materials. This does not cause a problem where the particles can be cleared away but in some situations it can lead to difficulties; for example, titanium alloys are used in medical implants because of their highly valued properties with regard to strength and hardness, but such implants may have several types of loading, together with complex chemical reactions between the mechanical and biological elements, so fretting may occur. In such cases the engineer has to take account of feedback from users and seek alternative materials such as cobalt, chrome and stainless steel.

Tribology is the name given to the range of topics mentioned above, encompassing the behaviour of interacting surfaces in contact under different types of load and conditions.

Chapter 5

Work and power

· ·

This chapter covers the work done and the rate at which it is done (power) by a force or forces. Initially, we find the work done on a body when displaced from its position by a force and the work done by a resisting force against the displacement of a body. The units of work and power are derived from the base units newton, metre and second, and examples show how the calculations are carried out. A final section deals particularly with the power demands, tractive effort and resistance of vehicles and trains moving at constant speed.

5.1 Work

Work is done when a force moves through a distance *in the direction* of the force. The distance must be measured along the line of action of the force.

If the force is *constant*, work is measured by the product

force × distance through which the force moves

Thus, if F is the force and s the distance moved,

work done = $F \times s$

If the force F is in newtons and the distance s in metres, then the units of work are newton metres, and the derived SI unit of work of $1\,\text{N}\,\text{m}$ is defined as the *joule*, denoted by **J**. Thus the unit of work is the joule defined as the *work done when the point of application of a force of one newton is displaced through a distance of one metre in the direction of the force*. Hence

work done = $F \times s$ joules

Work may also be expressed in multiples of the joule. Thus

$$1 \text{ gigajoule (GJ)} \quad = 10^9\,\text{J}$$

$$1 \text{ megajoule (MJ)} = 10^6\,\text{J}$$

$$1 \text{ kilojoule (kJ)} \quad = 10^3\,\text{J}$$

If there is no motion in the direction of the force then there is no work done. For example, in lifting the centre of gravity of a load of weight W through a height h, the work done against the force of gravity is Wh. But, since weight acts vertically downwards, no work is done *against gravity* in displacing the load horizontally.

When a body is moved against a resisting force, work is said to be done against the resistance.

Example

The load shown in Fig. 5.1 is pulled 20 m along the horizontal by a force of 0·4 kN at 60° to the horizontal. Calculate the work done.

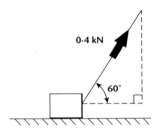

Fig. 5.1

Solution

Component of pull *in direction of motion*

$$= 0.4 \times \cos 60°$$

$$= 0.2 \, \text{kN}$$

Therefore

Work done = force × distance

$$= 0.2 \times 20$$

$$= 4 \, \text{kJ}$$

The vertical component of the force does not move vertically and therefore does no work.

Example

The block shown in Fig. 5.2 is constrained to move along fixed guides. A horizontal force of 950 N is required to move the block upwards along the slope a distance of 0·3 m. Calculate the work done.

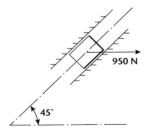

Fig. 5.2

If the block's mass is 60 kg and there is a frictional force opposing the motion of the block, find the magnitude of the friction force.

Solution

Horizontal distance moved by force

$$= 0 \cdot 3 \cos 45° = 0 \cdot 212 \, \text{m}$$

Therefore

Work done $= 950 \times 0 \cdot 212$

$$= 202 \, \text{J}$$

This work done is made up of (a) the work done against gravity and (b) the work done against the friction force, F.

Work done against gravity $=$ weight $\times$ height block is raised

$$= (60 \times 9 \cdot 8) \times 0 \cdot 3 \sin 45°$$

$$= 125 \, \text{J}$$

Therefore, work done against friction force (μF)

$$= 202 - 124 \cdot 5 = 77 \cdot 5 \, \text{J}$$

Thus, friction force $\times$ distance moved by block along the slope $= 77 \cdot 5 \, \text{J}$, so that

friction force $= \dfrac{77 \cdot 5}{0 \cdot 3}$

$$= 258 \, \text{N}$$

5.2 Power

Power is the rate of doing work. *Average* power is measured by the ratio of work done/time taken.

The unit of work is the joule (J) and the unit of time the second (s). The unit of power is therefore the *joule per second* (**J/s**). The **SI** unit of power is the watt (**W**) $= 1 \, \text{J/s}$. This is rather small and power is more often expressed in the higher multiples of the watt; thus

1 gigawatt (GW) $= 10^9 \, \text{W}$

1 megawatt (MW) $= 10^6 \, \text{W}$

1 kilowatt (kW) $= 10^3 \, \text{W}$

Note that $1 \, \text{kW} = 1 \, \text{kJ/s}$.

If a body moves at a speed v metres per second under the action of a force F newtons which acts in the direction of motion, then rate of working = work done per second

$$= F \times v \, \text{J/s}$$

or

power = Fv watts = $\dfrac{Fv}{1000}$ kW

(**Example**)

An electrically driven conveyor belt carries 60 000 packages per hour a distance of 80 m up an incline of 1 in 12. Each package weighs 48 N and the power absorbed by friction in the drive is 2 kW. What is the total power output of the motor?

Solution

Total weight moved per second $= \dfrac{60\,000 \times 48}{3600}$

$$= 800 \, \text{N}$$

The force being overcome in the direction of motion is the component of the weight along the slope, i.e. $800 \times 1/12 = 66 \cdot 7 \, \text{N}$.

Work done per second $= 66 \cdot 7 \times 80$

$$= 5336 \, \text{W}$$

i.e.   power $= 5 \cdot 34 \, \text{kW}$

Since 2 kW is absorbed by friction,

total power output $= 5 \cdot 34 + 2$

$$= \mathbf{7 \cdot 34 \, kW}$$

(**Example**)

The constant oil pressure on one side of the piston of a servo-motor is $6 \, \text{MN/m}^2$. The piston area is $20 \, \text{cm}^2$. Calculate the work done in moving the piston 25 mm. If the average power developed is $1 \cdot 9 \, \text{kW}$, calculate the time taken for the piston to traverse 25 mm.

Solution

$$\text{Area of piston} = 20 \, \text{cm}^2 = 0 \cdot 002 \, \text{m}^2$$

$$\text{Force on piston} = 6 \times 10^6 \times 0 \cdot 002 = 12\,000 \, \text{N}$$

Work done in moving piston 25 mm $= 12\,000 \times 0 \cdot 025$

$$= \mathbf{300 \, J}$$

If t seconds is the time taken to move the piston 25 mm, then

work done per second $= \dfrac{300}{t} \, \text{J/s}$

Now,

$$\text{average power} = \text{work done per second}$$

i.e. $$1000 \times 1.9 = \frac{300}{t}$$

Therefore

$$t = \frac{0.3}{1.9} = 0.16\,\text{s}$$

(Example)

Find the power required to pump 50 t of water to a height of 30 m in one second.

Solution

Weight of $50\,\text{t} = 50 \times 1000 \times 9.8 = 490\,000\,\text{N}$

Work done against gravity $= Wh$

$$= 490\,000 \times 30\,\text{J/s}$$

$$= 14.7 \times 10^6\,\text{W}$$

i.e.   power $= \mathbf{14.7\,MW}$

(Example)

The force on the cutting tool of a planing machine is 8 kN and the table of the machine makes a total of 360 forward and return strokes per hour of length 2 m. The friction resistance on the forward (cutting) stroke is 700 N. Calculate the power required, neglecting friction on the return stroke. If the table is driven by hydraulic pressure on a piston of 75 mm diameter, find the pressure.

Solution

Number of cutting strokes per second $= \frac{1}{2} \times \frac{360}{3600} = 0.05$

Speed of table $= 2 \times 0.05 = 0.1\,\text{m/s}$

Work done against friction $=$ friction force $\times$ speed

$$= 700 \times 0.1$$

$$= 70\,\text{J/s}$$

Work done in cutting $=$ cutting force $\times$ speed

$$= 8000 \times 0.1$$

$$= 800\,\text{J/s}$$

Total work done $= 800 + 70 = 870\,\text{J/s} = 870\,\text{W}$

i.e.   Power required = **0·87 kW**

$$\text{Area of piston} = \tfrac{\pi}{4} \times 0.075^2 = 0.004\,42\,\text{m}^2$$

Total force exerted in moving table = 8000 + 700 = 8700 N. Therefore

$$\text{pressure} = \frac{\text{force}}{\text{area of piston}} = \frac{8700}{0.004\,42}$$

$$= 1\,970\,000\,\text{N/m}^2$$

$$= \mathbf{1{\cdot}97\,MN/m^2}$$

(*Note*: for units of pressure *see* Ch. 16.)

Problems

5.1   An aluminium billet is extruded through a steel die by a piston 150 mm in diameter. If the piston stroke is 1·2 m and the average pressure on the face of the piston is 100 MN/m², calculate the work done in extrusion.

(2·12 MJ)

5.2   The table of a planing machine is driven horizontally by hydraulic pressure on a piston contained in a cylinder 80 mm in diameter. The force on the cutting tool is 5 kN, the mass of the moving parts is 1000 kg and the coefficient of friction is 0·05.

Calculate the pressure required in the cylinder. What is the work done by the hydraulic fluid in moving the table 2 m during the cutting stroke?

(1·1 MN/m²; 11 kJ)

5.3   A casting, of mass 300 kg, is hauled 12 m up an incline of 1 in 12. The friction force between the casting and slope is 150 N.

Calculate the work done against friction and the work done against gravity.

(1·8; 2·94 kJ)

5.4   A rocket motor exerts a thrust of 2·6 MN at a speed of 400 m/s. Find the power developed.   (1·04 GW)

5.5   A lift cage of mass 1 tonne is hauled up a shaft 300 m deep by a single cable, of mass 40 kg per metre length. Calculate the work done.

If the single cable is replaced by a tackle having three cables, what is then the work done? Neglect the weight of the lower pulley.

(20·6 MJ; 55·9 MJ)

5.6   A man of mass 100 kg stands on a platform of mass 30 kg. One end of a cable is attached to the platform, the other end passes over a wall pulley and the platform is hauled up by a pull on the free end.

How much work is done by the man in lifting himself and the platform a height of 4 m?   (5·1 kJ)

5.7   The piston of a steam operated hammer has a diameter of 200 mm and a stroke of 400 mm. The hammer delivers 18 working strokes per minute. Calculate the power developed by the hammer if the mean pressure on the piston during each working stroke is 700 kN/m². (2·64 kW)

5.8   A uniform ladder, of mass 30 kg, lying flat on the ground, is raised so that it lies flat against a vertical wall. Calculate the work done if the ladder is 6 m long.

If the ladder finally makes 30° with the wall, what is then the work done? (883 J; 764 J)

5.9   A 'Goliath' crane lifts a pressure vessel of mass 400 t through a height of 4 m in one minute. Calculate the average power required. (262 kW)

5.10 The force on the cutting tool of a planing machine is 20 kN and the cutting speed 0·3 m/s. If the power required to run the machine idle is one-quarter of the cutting power, what is the total power required at the driving motor? (7·5 kW)

5.11 1000 litres of water per minute are pumped to a height of 60 m. Calculate the power output of the pump. (Density of water = 1 Mg/m³.) (9·8 kW)

5.12 An endless conveyor belt transports 60 000 packages per hour a distance of 12 m along an incline of 1 in 12. If each package has a mass of 4 kg, calculate the power required. Neglect friction. (655 W)

(**5.3**) ## Work done by a torque

In Fig. 5.3(a) the arm OA, of length r, rotates about the fixed axis O. We consider the work done by a force F at A; the line of action of F is always perpendicular to the radius OA, i.e. tangent to the circle traced out by the point A.

When the point A moves to A' through a distance AA', measured round the circumference of the circle of centre O, the work done by the force F is

$F \times$ AA'

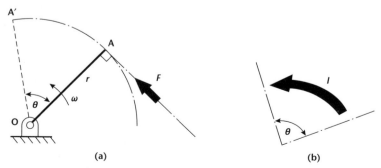

(a)   (b)

Fig. 5.3

But the distance AA' is equal to $r \times \theta$ where θ (*radian*) is the angle turned through by the arm OA. Thus

work done by $F = F \times r\theta$

But $F \times r$ is the moment of F about the centre O, i.e. the applied torque T. Hence

work done = $T \times \theta$

Note that it is no longer necessary to know either the force F or the radius r. The work done may be obtained directly from a knowledge of the applied torque and the rotation, Fig. 5.3(b). Thus the work done by a torque T in turning through an angle θ is $T\theta$.

If the torque is constant, and the time taken for OA to sweep out the angle θ is t, then

$$\text{work done per unit time} = \frac{T\theta}{t}$$

$$= T\omega$$

where $\theta/t = \omega$, *the angular velocity of rotation* (angular velocity, *see* Ch. 10).

If the torque T is in newton metres and the angular velocity in radians per second (rad/s), then

$$\text{rate of working (power)} = T\omega \, \text{N m/s}$$

$$= T\omega \text{ joules per second, or watts}$$

If n is the speed of rotation of the arm in *revolutions per second*, then

$\omega = 2\pi n$ radians per second

and

$$\textbf{power required} = \frac{T\omega}{1000} = \frac{2\pi n T}{1000} \textbf{ kW}$$

Example

The power output from an engine on test is absorbed by a rope brake at a pulley 1·2 m in diameter, Fig. 5.4. The rope is 30 mm in diameter. The lower end carries a load m of mass 50 kg, the upper end is wrapped round the drum and attached to a spring balance S which reads 80 N. The power is absorbed in friction between pulley and rope. If the speed of rotation of the pulley is 9 rev/s, calculate the power absorbed.

Solution

Allowing for the diameter of the rope, the effective radius R at which load and spring balance force act is given by

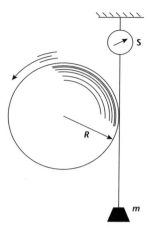

Fig. 5.4

$R = 600 + (\frac{1}{2} \times 30) = 615\,\text{mm}$

torque due to weight $= 50 \times 9\cdot8 \times 0\cdot615 = 301\,\text{N m}$, clockwise

torque due to spring $= 80 \times 0\cdot615 = 49\cdot2\,\text{N m}$, anticlockwise

net torque $T = 301 - 49\cdot2 = 251\cdot8\,\text{N m}$

power absorbed in friction $= \dfrac{2\pi nT}{1000}\,\text{kW}$

$$= \frac{2\pi \times 9 \times 251\cdot8}{1000}$$

$$= \mathbf{14\cdot3\,kW}$$

Problems

5.13 The power output of an engine under test is measured by a rope brake. The brake diameter is 1·5 m. When the engine is running at 720 rev/min and the load hung from the free end of the rope is 34 kg, the spring balance reading is 60 N. Calculate the power dissipated in friction at the brake.   (15·5 kW)

5.14 An electric motor drive for a rolling mill exerts a torque of 18 000 N m at a speed of 1000 rev/min. Calculate the power developed.   (1·89 MW)

5.15 In a test on a lathe cutting tool, the motor power required to cut a 150 mm diameter bar at 240 rev/min is found to be 5 kW. If the friction at the lathe bearings amounts to 1 kW, calculate the force on the cutting tool.

(2·13 kN)

5.16 A radio-telescope of mass 2000 t turns on rollers mounted on a horizontal circular track of 20 m diameter. If the coefficient of rolling friction is 0·05 and the telescope is turned through a right angle in five minutes, calculate the average power required. Neglect air resistance.   (51·3 kW)

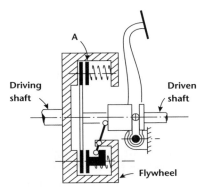

Fig. 5.5

5.17 In an overhead travelling crane the winding drum is driven by an electric motor through gearing. If the crane lifts a load of 15 t at a constant speed of 0·1 m/s, calculate the power required at the motor. Allow 6 kW for losses between motor and load. If the motor exerts a constant torque of 750 N m, what should be the motor speed in rev/min?   (20·7 kW; 264 rev/min)

5.18 The disc clutch shown in Fig. 5.5 has eight springs, each at a radius of 10 cm from the axis of the shaft. When engaged by releasing the foot pedal the force exerted by each spring on the pressure plate A is 45 N. If the coefficient of friction between disc and flywheel is 0·7, estimate the maximum torque which may be transmitted by the flywheel to the driven shaft and the corresponding power at a speed of 25 rev/s.   (25 N m; 4 kW)

5.4 Tractive effort and tractive resistance

When a vehicle moves at *constant speed* on the level the only force needed is that required to *maintain* the motion against the various resistances which would otherwise tend to bring the vehicle to rest. These resistances are due to bearing friction, track friction and wind forces; these are all included in the term *tractive resistance* and this quantity is often expressed in newtons per tonne mass of the vehicle, particularly for trains. To find the tractive resistance, it is only necessary to determine the force needed to *maintain* the constant speed.

Rolling resistance and drag

Rolling resistance (R) arises mainly from axle friction and deformation of wheels and tyres. It is sometimes expressed as a *coefficient of rolling resistance* given by the ratio of the *rolling resistance to the normal force between wheel and ground*; this coefficient, however, is not often used. The rolling resistance increases with speed and can be allowed for by expressing it in the form:

rolling resistance = **a** + **b**v

where v is the road speed and **a** and **b** are constants.

The resistance to motion produced by the air in front of the vehicle is called *drag* and depends on the density of the air, the shape of the vehicle and its projected frontal area (*form* drag) as well as the amount of disturbance of the air flow around the skin of the car (*skin* drag). The drag force is roughly proportional to the square of the road speed and the total resistance may be expressed as

$$\text{total resistance} = \mathbf{a} + \mathbf{b}v + \mathbf{c}v^2$$

where **c** is another constant. The resisting forces act mainly parallel to the road but at high speeds there is also a component force at right angles to the road surface, tending to lift the vehicle and reduce the normal reaction between tyres and road. Racing cars employ aerodynamic devices to produce a downforce to counteract this effect when cornering.

Tractive effort

The driving force exerted by a vehicle or train is called the *tractive effort* and it can be taken as equivalent to the pull in a tow-rope or coupling. A vehicle is driven by a force produced at the road surface usually by an engine torque transmitted to the driving axles; this is the force with which the ground *pushes on the vehicle*. The vehicle will be driven forward by this torque only if there is a friction force at the road surface to prevent slipping of the road wheels. On the level at constant speed the tractive effort is equal to the tractive resistance. On an incline the tractive effort must also overcome the effect of the component of the weight of the vehicle down the slope. In Chapter 8 we discuss the relation between tractive effort and changes in speed of the vehicle.

(Example)

A car of mass 2 tonnes is driven up a hill of gradient 1 in 10 (sine) at a constant speed of 36 km/h. The road resistance to motion amounts to 75 N per tonne mass of the vehicle. Calculate the tractive effort employed and the power developed.

Solution

The forces acting on the car are shown in Fig. 5.6. E is the tractive effort up the incline, R the road resistance acting down the incline, W the weight of the car acting vertically downwards and N the normal reaction of the road on the car.

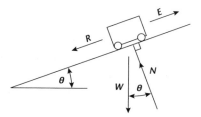

Fig. 5.6

Resolving parallel to the incline,

$$E = R + W \sin \theta$$

$$= (2 \times 75) + (2 \times 1000 \times 9\cdot8) \times \frac{1}{10}$$

$$= \mathbf{2110\,N}$$

$$36\,km/h = 36 \times \frac{1000}{3600} = 10\,m/s$$

At a speed v m/s up the slope the power developed

$$= \frac{Ev}{1000}\ kW$$

$$= \frac{2110 \times 10}{1000}$$

$$= \mathbf{21\,kW}$$

(**Example**)

During a test, a 6 t lorry is driven up an incline of 1 in 15 (sine) at a constant speed of 45 km/h. Instruments indicate that the engine is then working at 65 kW. Calculate the power required to overcome the road resistance.

Solution

$$Power = \frac{Ev}{1000}\ kW$$

and

$$v = 45\,km/h = 12\cdot5\,m/s$$

Therefore,

$$tractive\ effort\ E = \frac{power\ in\ kW \times 1000}{v}$$

$$= \frac{65 \times 1000}{12\cdot5}$$

$$= 5200\,N$$

Tractive effort = component of weight down slope + tractive resistance

$$5200 = \left(6 \times 1000 \times 9\cdot8 \times \frac{1}{15}\right) + R$$

Hence

$$R = 1280\,N$$

Therefore power required to overcome road resistance

$$= \frac{Rv}{1000} \text{ kW}$$

$$= \frac{1280 \times 12 \cdot 5}{1000}$$

$$= 16 \text{ kW}$$

(Example)

A motor car travels at a constant speed of 72 km/h in still air. The resistance to motion is given in newtons by the formula $R = 180 + 0.5v + 0.06v^2$, where v is the road speed in km/h relative to the air. What is the power output?

Solution
At 72 km/h the resistance to motion is

$$R = 180 + 0.5 \times 72 + 0.06 \times 72^2$$

$$= 527 \text{ N}$$

At constant speed the tractive effort is equal to the resistance, hence

power output = tractive effort × road speed

$$= 527 \times \frac{72}{3 \cdot 6} \times 10^{-3} \text{ kW} = 10 \cdot 5 \text{ kW}$$

This is the power developed by the engine when all losses are neglected. Note that v is the speed relative to the air which is at rest in this problem.

(Example)

A train, of total mass 800 t, is hauled by a locomotive along a level track at a constant speed of 30 km/h. The power developed at this speed is 600 kW. Calculate the tractive resistance in newtons per tonne mass of train.

If the train ascends an incline of 1 in 75, what power must be developed if the speed is to be maintained constant at 30 km/h? The tractive resistance may be assumed to be the same as on the level track.

Solution

$$v = 30 \text{ km/h} = 8 \cdot 33 \text{ m/s}$$

$$\text{Tractive effort } E = \frac{\text{power in kW} \times 1000}{v}$$

$$= \frac{600 \times 1000}{8 \cdot 33}$$

$$= 72\,000 \text{ N}$$

At constant speed on the level, tractive resistance

= tractive effort

= 72 000 N

$$= \frac{72\,000}{800} \text{ N/tonne}$$

= 90 N/tonne mass of train

On the incline, the tractive effort must overcome the component of the weight down the slope ($W \sin \theta$) together with the tractive resistance (72 000 N); therefore

$$\text{tractive effort } E = 72\,000 + \left(800 \times 1000 \times 9.8 \times \frac{1}{75}\right)$$

$$= 176\,500 \text{ N}$$

$$\text{Power} = \frac{Ev}{1000} = \frac{176\,500 \times 8.33}{1000} = \textbf{1470 kW} \quad \text{or} \quad \textbf{1.47 MW}$$

Problems

5.19 A train of mass 200 t is pulled up an incline of 1 in 100. In 60 s it is drawn 800 m along the incline. Calculate the average rate of working. Find also the power developed when the train travels up the incline at a steady speed of 100 km/h. Resistance to motion may be neglected.   (262 kW; 545 kW)

5.20 A diesel-electric locomotive of mass 300 t travels at 60 km/h. The train's mass is 700 t and the tractive resistance is 120 N/t for the locomotive and 50 N/t for the train. If the train is steadily ascending an incline of 1 in 100, calculate the power developed at the driving wheels.   (2.82 MW)

5.21 Calculate the tractive effort required to haul a train of mass 300 t at uniform speed up an incline of 1 in 100 if the resistance to motion amounts to 120 N/t mass of the train. If the speed is 30 km/h, find also the power exerted.   (65.4 kN; 546 kW)

5.22 The resistance to motion of a motor car under test on a track is given in newtons by $R = 200 + 0.5v + 0.05v^2$, where v is the road speed in km/h. What is the tractive effort and the output power when the car is travelling at a constant speed of 60 km/h?   (410 N; 6.83 kW)

5.23 A locomotive of mass 60 t pulls 12 wagons each of mass 25 t. The tractive resistance is 100 N/t for the locomotive and 60 N/t for the wagons. If the locomotive pulls the wagons along the level at a constant speed of 90 km/h what power is developed? What *additional* power must be developed if the train is to ascend an incline of 1 in 100 at the same speed?   (600 kW; 882 kW)

5.24 A motor vehicle has a maximum speed on the level road of 220 km/h at which speed the engine develops a torque of 210 N m at 6000 rev/min. Calculate the road resistance. Neglect any losses in the transmission.   (2·2 kN)

5.25 A car of mass 1500 kg rolls at a steady speed down an incline of 1 in 12. Calculate the road resistance. What tractive effort would be required to pull the car up the incline at a steady speed?   (1·23 kN; 2·45 kN)

5.26 A supercharged road racing automobile has an engine capable of developing a maximum torque of 700 N m at an engine speed of 5000 rev/min. If the vehicle travels on the level at a steady speed of 210 km/h, the engine speed being then 5000 rev/min, calculate the tractive resistance. If the car now ascends an incline of 1 in 10 what steady speed can be maintained if all other conditions remain unchanged? Mass of vehicle = 1100 kg. Neglect losses in transmission and gearing.   (6·3 kN; 179 km/h)

5.27 The power at the output (brake) shaft of a naval ship's engine is 110 MW when travelling at a steady 25 knots. Of this power 35 per cent is lost due to mechanical and propeller inefficiencies. Find the resistance to the ship's motion. 1 knot = 0·514 m/s.   (5·6 MN)

5.28 A car on test travels at a constant speed of 130 km/h in still air and the engine develops 55 kW. The rolling and air resistance to motion is given in newtons by $R = 155 + 0·5v + bv^2$, where v is the road speed in km/h relative to the air. Find the constant b.   (0·08)

5.29 A locomotive pulls a train at constant speed on a horizontal track against a track resistance given by $R = 4 + 0.28v$ where R is in kN and v, the velocity, is in km/h. What is the power exerted at 30 km/h? If the train climbs a gradient of 1 in 200 at a constant speed of 20 km/h, find the power required, assuming the tractive resistance formula to be the same as a level track. The total mass of locomotive and train is 450 t.   (103 kW; 176 kW)

5.30 A vehicle requires 14·5 kW to maintain a constant speed of 72 km/h on a level track. The resistance to motion is given by $R = a + bv^2$ where R is in newtons, **a** and **b** are constants and v is the road speed in m/s. Find the value of the constant **b** if **a** = 250. Assuming the same formula for R find the power required at a constant speed of 36 km/h on an up-gradient of 1 in 10. The mass of the vehicle is 1200 kg.   (1·19; 15.5 kW)

Chapter 6

Machines

··

Brief descriptions of basic machine elements are given in this chapter including wheel-and-axle, screw jacks, gear trains, worm and worm-wheel, and belt drives. The treatment is sufficient for an understanding of the law of a machine and how to measure its mechanical advantage, velocity ratio and efficiency.

6.1 The simple machine

A *machine* is an assembly of rigid bodies for transmitting motion, transmitting force or doing work. An example of a simple machine is the wheel-and-axle, Fig. 6.1. A load of weight W, carried by a cord from an axle of diameter d, is raised by an *effort E* at a cord attached to, and wrapped round, a wheel of diameter D. The wheel and axle rotate together.

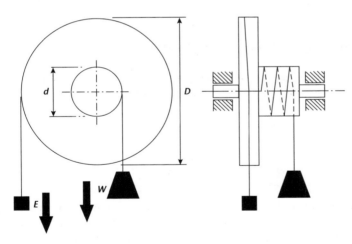

Fig. 6.1

6.2 Mechanical advantage

A machine is usually designed so that the load overcome is greater than the effort applied. The ratio of load to effort is known as the *mechanical advantage* of the machine. Thus

$$\text{mechanical advantage} = \frac{\text{load}}{\text{effort}}$$

$$= \frac{W}{E}$$

Since both E and W have the same units, mechanical advantage has no units and is only a number which varies with the load. The mechanical advantage is usually greater than unity. For example, in the simple wheel-and-axle of Fig. 6.1, taking moments about the axis of the shaft and neglecting friction,

$$W \times \frac{d}{2} = E \times \frac{D}{2}$$

or

$$\frac{W}{E} = \frac{D}{d}$$

i.e.   mechanical advantage $= \dfrac{\text{diameter of effort drum}}{\text{diameter of load drum}}$

In an ideal machine there is no friction and the ratio load/ideal effort is then the *ideal mechanical advantage*. Thus, for the wheel-and-axle the ratio D/d is the ideal mechanical advantage. In practice, the actual mechanical advantage is always less than the ideal and is not usually calculated but obtained by experiment.

6.3   The load–effort graph; law of a machine

If an experiment is carried out on a machine to determine the effort E required to lift a load W for a range of values of the load the following table might be obtained:

W (kN)	0	1	2	4	6	8	10
E (kN)	0·11	0·33	0·59	1·11	1·61	2·07	2·62
W/E	0	3·0	3·39	3·6	3·73	3·86	3·82

If E is plotted against W and the best straight line drawn through the points, the load–effort graph, Fig. 6.2, is obtained. Since the graph is a straight line the relation between E and W may be represented by the equation

$$E = aW + b$$

where a and b are constants.

This equation is known as the *law of the machine*.

The constant b is determined by setting $W = 0$, so that $E = b$; b is represented by the intercept on the effort axis and is the effort required to just set the machine in motion by overcoming the friction at no load.

From the graph $b = 0·1$.

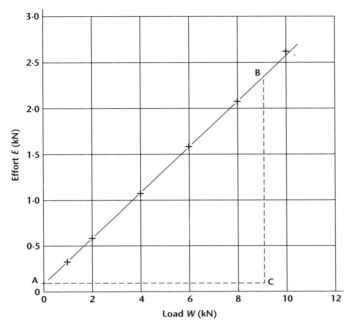

Fig. 6.2

The constant a is the gradient of the straight line and is found by taking two points A and B on the graph as far apart as possible. Lines AC and BC are drawn parallel to the load and effort axes respectively as shown.

From the graph, AC = 9·0; BC = 2·26.

Therefore gradient a is given by

$$a = \frac{BC}{AC} = \frac{2\cdot26}{9\cdot0} = 0\cdot25$$

The law for this machine is therefore

$$E = 0\cdot25W + 0\cdot1 \, kN$$

The values of a and b should not be obtained using only two experimental points since these are always bound to involve error.

However, in examinations we are often given only two sets of values for load and effort and we have to use these values to find the law of the machine. For example, suppose we are given two values for load and effort, thus

load = 2 kN, effort = 0·59 kN

load = 10 kN, effort = 2·62 kN

Then, substituting in the equation

$$E = aW + b$$

we obtain two equations:

$$0.59 = (a \times 2) + b \qquad\qquad [1]$$

$$2.62 = (a \times 10) + b \qquad\qquad [2]$$

Subtracting equation [1] from equation [2],

$$2.03 = 8a$$

therefore

$$a = 0.25$$

Hence, from equation [1],

$$b = 0.59 - 0.254 \times 2$$

$$= 0.08$$

therefore

$$E = 0.25W + 0.08 \,\text{kN}$$

6.4 The load–mechanical advantage graph

From the table in section 6.3 values of the mechanical advantage W/E may be plotted against load W to obtain the graph shown in Fig. 6.3. From such a graph the mechanical advantage at any load may be obtained.

The mechanical advantage increases with load but there is not a straight line relationship between mechanical advantage and load, i.e. the increase in mechanical advantage is not proportional to the corresponding increase in load.

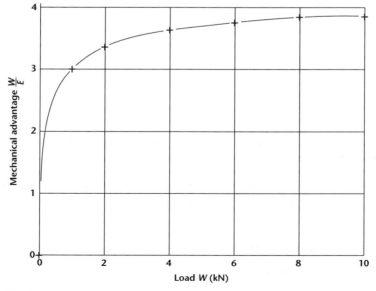

Fig. 6.3

6.5 Friction effort

The effort required at any particular load to overcome friction alone is called the *friction effort.*

The actual effort E required at any load is made up of two parts:

1. The effort to move the load if the machine were ideal, i.e. no friction.
2. The effort to overcome friction.

Hence

friction effort = actual effort − ideal effort

$$= E - \frac{W}{\text{ideal mechanical advantage}}$$

For the test results of section 6.3, suppose that the ideal mechanical advantage of the machine is 5; then the friction effort for each of the various loads is calculated as in the following table:

W (kN) (load)	0	2	4	6	8	10
$W/5$ (kN) (ideal effort)	0	0·4	0·8	1·2	1·6	2
E (kN) (actual effort)	0·11	0·59	1·11	1·61	2·07	2·62
$E - W/5$ (kN) (friction effort)	0·11	0·19	0·31	0·41	0·47	0·62

In Fig. 6.4 the friction effort is shown plotted against the load W. The graph shows that the friction effort increases linearly with load, but the increase in

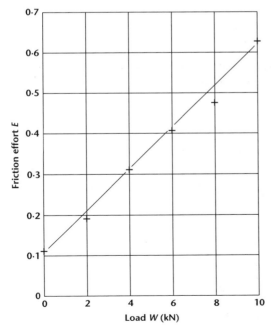

Fig. 6.4

friction effort is not as great as the corresponding increase in load. Hence the mechanical advantage increases with load as already shown in Fig. 6.3.

6.6 Velocity ratio

In practice a mechanical advantage greater than unity is obtained only by allowing the effort to move a greater distance than that through which the load moves. Since the distances moved by load and effort take place in the same time, the ratio of these distances is the same as the ratio of the velocities.

The *velocity ratio V* of a machine is therefore defined as

$$V = \frac{\text{distance moved by the effort}}{\text{distance moved by the load}}$$

Velocity ratio, like mechanical advantage, has no units. Unlike mechanical advantage, however, the velocity ratio usually remains constant for all loads for a given machine.

For example, in the simple wheel-and-axle, Fig. 6.5, the velocity ratio is found as follows:

In one revolution of the wheel and axle,

distance moved by effort $= \pi D$

distance moved by load $= \pi d$

Therefore

$$\text{velocity ratio} = \frac{\pi D}{\pi d} = \frac{D}{d}$$

For an ideal frictionless wheel-and-axle, it was shown (section 6.2) that D/d was the ideal mechanical advantage, so that

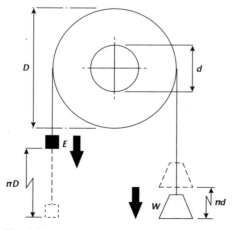

Fig. 6.5

velocity ratio = *ideal* mechanical advantage

$$= \frac{\text{load}}{\textit{ideal} \text{ effort}}$$

This statement is true for all ideal machines and will be proved in the following section.

The determination of velocity ratios for various common types of machines is dealt with in section 6.12.

Note: for a real machine, the velocity ratio is not the same as the mechanical advantage since there is always some friction present and the actual mechanical advantage is always less than the ideal.

6.7 The principle of work

The *principle of work* states that, for an ideal frictionless machine, the work output from the machine is equal to the work input. It is therefore assumed that there is no loss of work due to friction, slipping of surfaces, or stretching of ropes or cables.

Now,

work input = work done by effort

= effort × distance moved by effort

work output = work done on load

= load × distance moved by load

For an ideal machine,

work input = work output

Therefore effort × distance moved by effort = load × distance moved by load, i.e.

$$\frac{\text{distance moved by effort}}{\text{distance moved by load}} = \frac{\text{load}}{\text{effort}}$$

Thus

velocity ratio = *ideal* **mechanical advantage**

since for an ideal machine the effort is the ideal value.

In practice the work output is always less than the work input because of friction and other losses.

6.8 Efficiency

The efficiency of a machine is defined as the ratio of the useful work done by the machine to the actual work put into the machine. Thus

$$\text{efficiency} = \frac{\text{work output}}{\text{work input}}$$

$$= \frac{\text{load} \times \text{distance moved by load}}{\text{actual effort} \times \text{distance moved by effort}}$$

$$= \frac{\text{mechanical advantage}}{\text{velocity ratio}}$$

From the experimental results given in section 6.5, where the ideal mechanical advantage was 5 (and thus $V = 5$), the following results can be obtained:

W (kN) (load)	0	1	2	4	6	8	10
E (kN) (actual effort)	0.11	0.33	0.59	1.11	1.61	2.07	2.62
W/E (actual mech. adv.)	0	3.0	3.39	3.6	3.73	3.86	3.82
$\eta = \dfrac{W/V}{E}$ (efficiency)	0	0.6	0.68	0.72	0.75	0.77	0.76
η (%)	0	60	68	72	75	77	76

The efficiency–load graph for this machine is shown in Fig. 6.6. The efficiency is seen to increase with load in the same manner as the mechanical advantage.

For small loads, the proportion of the effort taken up in overcoming friction is large. The efficiency at small loads is therefore low. As the load is increased, the friction effort also increases but the proportion of friction work to actual work done decreases. The efficiency therefore increases with load.

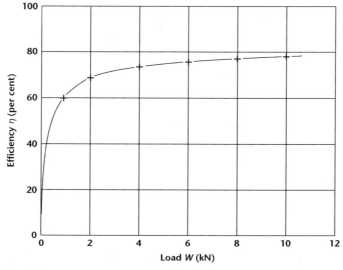

Fig. 6.6

6.9 Limiting efficiency

As the load on a machine is increased, the efficiency increases as shown in Fig. 6.6. However, the efficiency does not continue to increase indefinitely with load. There is a limiting value to the efficiency of a particular machine. This limiting value is always less than 100 per cent since there is always friction present. The *limiting efficiency* is found as follows.

Since

$$\text{efficiency } \eta = \frac{\text{mechanical advantage}}{\text{velocity ratio}}$$

$$= \frac{\text{load}}{\text{effort} \times \text{velocity ratio}}$$

$$= \frac{W}{E \times V}$$

and the law of a machine is

$$E = aW + b$$

then

$$\eta = \frac{W}{(aW + b) \times V}$$

$$= \frac{1}{aV + (bV/W)}$$

The greater the value of W, the smaller becomes the value of the quantity bV/W in the denominator, and since the other term aV of the denominator is independent of the load W, it can be seen that as W increases, the efficiency increases. When W is made infinitely large, the denominator reaches the limiting value aV and the limiting efficiency is therefore $1/aV$.

For test results of section 6.3, $a = 0.25$ and $V = 5$, therefore

$$\text{limiting efficiency} = \frac{1}{aV}$$

$$= \frac{1}{0.25 \times 5}$$

$$= 0.8 \quad \text{or} \quad 80 \text{ per cent}$$

From the efficiency curve of Fig. 6.6, it can be seen that the limiting efficiency is approximately 80 per cent, as calculated.

Example

A lifting machine has a velocity ratio of 4 and can lift a load of 100 kg when the effort applied is 327 N. Find (a) the efficiency, (b) the effort required to overcome friction at this load and (c) the work done against friction when the load is lifted 2 m.

Solution

(a) Mechanical advantage = $\dfrac{\text{load}}{\text{effort}}$ = $\dfrac{980}{327}$ = 3

Since

load, $W = mg = 100 \times 9{\cdot}8 = 980\,\text{N}$

efficiency = $\dfrac{\text{mechanical advantage}}{\text{velocity ratio}}$

= $\dfrac{3}{4}$ = 0·75   or   **75 per cent**

(b) Ideal effort = $\dfrac{\text{load}}{\text{velocity ratio}}$

= $\dfrac{980}{4}$ = 245 N

Friction effort = actual effort − ideal effort

= 327 − 245 = **82 N**

(c) Distance moved by effort = 2 × 4 = 8 m
Work done by effort against friction

= 82 × 8 = 656 N m = **656 J**

Example

The law of a small lifting machine is given by E = 0·22 W + 18, where E is the effort and W the load, in kilonewtons. The velocity ratio for the machine is 185. Calculate the power required to raise a load of 500 N at a speed of 3 m/min.

Solution
Effort required to lift a load of 500 N,

$E = 0{\cdot}22 \times 500 + 18$

= 128 N

Speed at which effort moves = $\dfrac{3 \times 185}{60}$

= 9·25 m/s

Power required = $E(N) \times v(\text{m/s})$ watts

= 128 × 9·25

= 1184 W

= **1·2 kW**

Problems

6.1 The following table gives values of load and effort from an experiment on a machine. Determine the law of the machine and draw the load–friction effort graph. The ideal mechanical advantage is 5. What is the effort required to overcome friction at a load of 4·5 kN?

Load W (kN)	1	2	3	4	5	6
Effort E (kN)	1·35	1·7	2·1	2·3	2·65	2·95

$(E = 0·32\,W + 1·04\,\text{kN};\ 1·58\,\text{kN})$

6.2 A rope tackle has an efficiency of 67 per cent and a velocity ratio of 7. Calculate the effort required to lift a load of 1000 N.   (213 N)

6.3 A machine raises a load of 900 N by means of an effort of 120 N. The distance moved by the load is 30 cm and the effort moves through 270 cm in the same time. Calculate the efficiency of the machine at this load.

(83·3 per cent)

6.4 The velocity ratio of a machine is 12 and it is found that an effort of 118 N is needed to lift a load of 65 kg. What is the efficiency at this load? If a load of 120 kg requires an effort of 216 N, find the effort to lift a load of 250 kg assuming a straight line graph for load and effort.   (45 per cent; 447 N)

6.5 The law of a lifting machine is given by $E = 0·25W + 1·3\,\text{kN}$. Find the efficiency at a load of 40 kN if the velocity ratio is 5. What is the power required to raise a load of 40 kN at 5 m/min?   (71 per cent; 4·7 kW)

6.6 A wheel-and-axle has a wheel diameter of 250 mm and an axle diameter of 100 mm. Find the velocity ratio. If a load of 120 kg is raised 0·3 m by an effort of 510 N, find the efficiency and the work done against friction. What is the effort required to overcome the friction?

(2·5; 92·3 per cent; 29·7 J; 40 N)

6.7 Find the effort required to lift a load of 810 N in a machine having a velocity ratio of 50 and an efficiency of 45 per cent. What is the work done against friction if the load rises 2 m?   (36 N; 2 kJ)

6.8 A lifting gear is used to haul a structure up a gradient. The effort–load law for the gear is $E = 0·18P + 60$ newton, where E is the effort and P the pull in the cable attached to the structure. The velocity ratio of the gear is 8. The structure's mass is 1000 kg and the coefficient of friction for the sliding surfaces is 0·2. The gradient is at 10° to the horizontal and the lifting cable is at 20° to the incline. Find, for this position, the pull P in the lifting cable, the effort required, and the efficiency of the gear.

(3·64 kN; 714 N; 63·7 per cent)

6.9 The effort E required to raise a load of weight W kN by means of a winch is given by $E = 1·2W + 0·4$. If the velocity ratio is 8, find the limiting efficiency.

(10·4 per cent)

6.10 A machine is operated by an effort applied to a wire wrapped round the circumference of a pulley of diameter 28 cm. When the pulley rotates once, the load moves a distance of 4 mm. The gradient of the load–effort graph is 0·05. Find the limiting efficiency of the machine.   (9·1 per cent)

6.11 A hoisting gear raises a load of 1800 kg with an efficiency of 45 per cent. The velocity ratio is 56 and an effort of 60 N is required to overcome friction at no load. Find the probable effort required to raise a load of 2400 kg. Find also the limiting efficiency at high loads.   (913 N; 49 per cent)

6.12 A lifting machine has a velocity ratio of 35. An effort of 36 N is required to raise a load of 550 N and an effort of 100 N is required for a load of 2·4 kN. Assuming the two sets of load–effort values to give a straight line graph, determine:

(a) the law of the machine;
(b) the mechanical advantage and efficiency when the load is 1·5 kN;
(c) the limiting efficiency of the machine.

($E = 0\cdot0346W + 17\,\text{N}$; 21·8, 62·3 per cent; 82·6 per cent)

6.13 A machine when tested gave the following results:

Velocity ratio = 8

Load (kN)	10	20	30	40	50	60	70
Effort (kN)	2·2	4·19	6·21	8·2	10·2	12·2	14·2

Draw the load–effort graph. Using the corrected values of the effort from the straight line graph obtained, draw the load–efficiency curve. State the law of the machine and the limiting efficiency.

($E = 0\cdot2W + 0\cdot2\,\text{kN}$; 62·5 per cent)

6.10 Overhauling

A machine is said to *overhaul* if, when the effort is removed, the load overcomes the friction forces and the operation of the machine is reversed. This is possible only when the efficiency of the machine is *greater than* 50 *per cent*. In other words, only when the friction is great enough to reduce the efficiency to below 50 per cent will the load be prevented from 'running away' when the effort is removed.

This statement is proved as follows. For a frictionless machine,

velocity ratio = ideal mechanical advantage

$$= \frac{\text{load}}{\text{ideal effort}}$$

$$\text{ideal effort} = \frac{\text{load}}{\text{velocity ratio}}$$

$$= \frac{W}{V}$$

In an actual machine the effect of friction can be considered as an 'additional load', say F. The additional effort required to overcome this friction load is F/V. Hence

actual effort = effort to overcome load + effort to overcome friction

i.e.
$$E = \frac{W}{V} + \frac{F}{V}$$

Efficiency $\eta = \dfrac{\text{mechanical advantage}}{\text{velocity ratio}}$

$$= \frac{W}{EV}$$

$$= \frac{W}{\left(\dfrac{W}{V} + \dfrac{F}{V}\right) \times V}$$

$$= \frac{W}{W + F}$$

When the effort is removed ($E = 0$), the load tries to reverse the machine and the friction load F now opposes the load. If the applied load is just sufficient to overcome the friction load F, the machine is reversed and the load overhauls, i.e. for overhauling, F is equal to or less than W.

When $F = W$, the limiting case, the efficiency for normal (forward) operation becomes

$$\eta = \frac{W}{W + F} = \frac{W}{2W} = \frac{1}{2} \text{ or 50 per cent}$$

If the efficiency is greater than 50 per cent $(W + F)$ must be less than 2, i.e. F must be less than W, so that the load does not overcome the friction load and does not overhaul. It is usually assumed that the efficiency is the same in the forward and reverse directions.

Problems

6.14 The load on a lifting machine is gradually increased up to 400 kg. The machine then starts to overhaul. If the velocity ratio is 8, calculate the effort required to lift this load.   (980 N)

6.15 A load of 600 kg is carried from a cable wrapped round the wheel of a simple wheel-and-axle. The effort is applied to a cable attached to the wheel. The axle diameter is 30 cm and the wheel diameter 90 cm. The efficiency of the machine at this load is 75 per cent.

Calculate (a) the effort required to lift the load and (b) the effort required to prevent the load from falling.   (2·62 kN; 1·3 kN)

6.16 A casting, of mass 50 kg, just begins to slip down an incline when the slope is 30°. What is the force parallel to the incline which will just pull the casting up the slope? What is the efficiency of this inclined plane considered as a machine?   (490 N; 50 per cent)

6.11 Calculation of velocity ratio

The velocity ratio V of a machine may be calculated from the dimensions alone, and methods of determining V for various machines will now be shown. It is assumed that there is no slipping between the parts of the machine and no stretch in the ropes or chains supporting the loads.

In the design of the following machines, the aim has been to produce as high a velocity ratio as possible using a minimum number of pulleys and gears.

The greater the velocity ratio of a machine the greater is the mechanical advantage for a given efficiency. Thus for a given effort an increase in velocity ratio means that a heavier load can be raised. However, it usually happens that the greater the velocity ratio of a machine, the lower the efficiency for a given load.

6.12 The differential wheel and axle

The differential wheel and axle, Fig. 6.7, consists of a pair of axles A, B of diameters d_1, d_2 respectively, rotating with a wheel C of diameter D. The effort E is applied to a rope wrapped round the circumference of C. The load is carried by a snatch block, a pulley suspended by a rope attached to the axles A and B as shown. The rope from the load is wound on to axles A and B in opposite directions. This arrangement gives a much higher velocity ratio than the simple wheel and axle.

As the effort E turns the wheel and axle the rope is unwrapped from the wheel and the load rope is wrapped on to axle B and unwrapped from axle A.

Thus, in one revolution of the effort wheel C,

length of rope wound on axle B $= \pi d_2$

length of rope unwound from axle A $= \pi d_1$

Since d_2 is greater than d_1, more rope wraps on to B than leaves A so that the rope carrying the load shortens.

Net shortening of load rope $= \pi d_2 - \pi d_1$

This shortening is shared equally between the two parts of the rope; hence

distance load W is raised $= \frac{1}{2}\pi(d_2 - d_1)$

Therefore

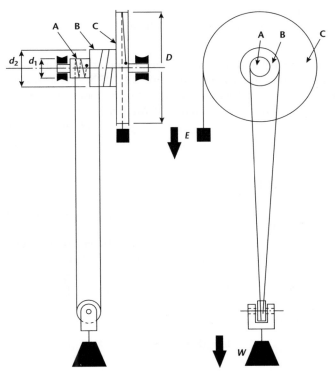

Fig. 6.7

$$\text{velocity ratio} = \frac{\text{distance moved by effort}}{\text{distance moved by load}}$$

$$= \frac{\pi D}{\frac{1}{2}\pi(d_2 - d_1)}$$

$$= \frac{2D}{d_2 - d_1}$$

By making d_1 nearly equal to d_2 a large velocity ratio, and therefore large ideal mechanical advantage, may be obtained, using a convenient size of wheel and axle.

Problems

6.17 A differential wheel and axle has axles 30 cm and 15 cm in diameter. The load is 700 N and the effort required at an efficiency of 45 per cent is 64 N. Calculate the wheel diameter.   (183 cm)

6.18 A differential wheel and axle is used to raise a load of 4 t. It has an efficiency at this load of 52 per cent. The wheel diameter is 200 mm and the axles have diameters 100 mm and 84 mm respectively.

(a) Calculate the effort required.

(b) What is the work done by the effort if the load is raised 2 m?

(3 kN; 151 kJ)

6.19 In a differential wheel and axle, the wheel diameter is 650 mm and the axle diameters are 50 mm and 75 mm. An effort of 735 N lifts a load of 1200 kg. Find the velocity ratio of the machine and the efficiency at a load of 1200 kg. What is the work done against friction if the load is raised 300 mm?

(52; 30·8 per cent; 7·9 kJ)

(6.13) Weston differential pulley block

This pulley block, Fig. 6.8, uses three pulleys. It gives a higher velocity ratio than the simple three-pulley block of section 6.12.

An endless chain passes over pulley 1 of the top block, then over the lower snatch block pulley 3 and finally over the smaller pulley 2 of the top block. The two pulleys of the top block rotate together. Rotation of the top block by an effort E, applied at the rope leaving pulley 1, causes shortening of the load chain A and lengthening of chain B.

In one revolution of the compound pulley block,

distance x moved by E at C = length of chain moved off pulley 1

Therefore

$$x = \pi D_1$$

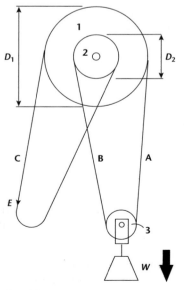

Fig. 6.8

Length of chain wound on from A = πD_1

Length of chain wound off to B = πD_2

Net shortening of load chain AB = $\pi D_1 - \pi D_2$

$$= \pi(D_1 - D_2)$$

This is shared between portions A and B, hence the load is raised a distance

$$y = \tfrac{1}{2}\pi(D_1 - D_2)$$

$$\text{velocity ratio} = \frac{x}{y} = \frac{\pi D_1}{\tfrac{1}{2}\pi(D_1 - D_2)}$$

$$= \frac{2D_1}{D_1 - D_2}$$

By making $(D_1 - D_2)$ small the velocity ratio can be made large. Ropes or chains may be used in this machine. For heavy loads link chains are normally used. In a chain drive the chain may engage with either teeth or 'flats' cut to fit the shape of the links. The number of flats or teeth on a pulley is proportional to the circumference and thus to the diameter. Hence, if T_1 and T_2 are the number of flats or teeth on the pulleys 1 and 2 respectively,

$$\text{Velocity ratio} = \frac{2T_1}{T_1 - T_2}$$

Problems

6.20 The top block of a Weston pulley has pulleys 21 cm and 18 cm in diameter. Find the effort required to lift a load of 750 kg if the efficiency at this load is 35 per cent.   (1·5 kN)

6.21 The differential pulleys of a Weston block have 18 and 15 flats respectively. If an effort of 50 N raises a load of 450 N, find the efficiency.   (75 per cent)

6.22 The fixed pulleys of a Weston pulley block have 13 and 12 teeth. Calculate the velocity ratio. If the efficiency of the block is 30 per cent, what is the work done against friction per metre rise of load when the load raised is 400 kg?   (26; 9·13 kJ)

6.14 The screw thread

A screw thread is shown diagrammatically in Fig. 6.9(a). It is a single-start, right-hand thread. Figure 6.9(b) shows a double-start, right-hand thread and in this case there are two parallel threads.

The *pitch p* is the axial distance between corresponding points on the thread, such as A and B.

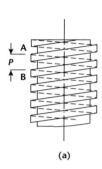 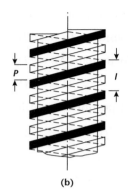

(a)   (b)

Fig. 6.9

The *lead l* is the distance moved axially by the screw in one revolution, Fig. 6.9(b).

For a single-start thread:

lead = pitch

For a double-start thread:

lead = 2 × pitch

For *n* starts:

lead = *n* × pitch

(6.15) The screw-jack

The screw-jack consists of a table A carrying the load of weight *W*, which is raised by a screw S turning in the body B, Fig. 6.10. The screw is rotated by a handle of radius *R*, at the end of which a tangential force *E* (the effort) is applied.

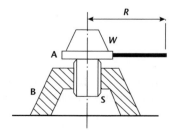

Fig. 6.10

In one revolution of the screw, the effort E moves a distance $x = 2\pi R$, and the load is raised a distance $y =$ lead, l. Therefore

velocity ratio $= \dfrac{x}{y}$

$= \dfrac{2\pi R}{l}$

The torque applied to the screw is $E \times R$ and this torque must overcome the friction between the screw and the body B. In Fig. 6.10, the load W is shown to rotate with the screw. When the load does not rotate with the screw, the top of the screw bears on a collar. Additional friction torque therefore has to be overcome between the collar and the screw.

(**Example**)

A screw-jack has a double-start thread with a pitch of 6 mm. If the load to be raised is 1 tonne, and the efficiency at this load is 18 per cent, calculate the torque required at the jack handle to raise the load.

Solution
Load $= 1$ tonne $= 1 \times 1000 \times 9{\cdot}8 = 9800$ N

Lead, $l = 2 \times$ pitch $= 2 \times 6 = 12$ mm $= 0{\cdot}012$ m

Velocity ratio $V = \dfrac{2\pi R}{l}$

$= \dfrac{2\pi R}{0{\cdot}012}$

$= 524R$

where $R =$ radius at which effort is applied (metres).

Mechanical advantage $= V \times$ efficiency

$= 524R \times 0{\cdot}18$

$= 94R$

Effort $E = \dfrac{\text{load}}{\text{mechanical advantage}}$

$= \dfrac{9800}{94R}$

$= \dfrac{104}{R}$ N

Torque $= E \times R = \dfrac{104}{R} \times R$

$= \mathbf{104\,N\,m}$

(**Example**)

A table carrying a cutting tool is operated by a single-start screw of 15 mm pitch. The table's mass is 312 kg and the coefficient of friction between the table and its guides is 0·08. The force of the cut along the axis of the screw is 350 N when 1 kW is supplied at 600 rev/min. What are the speed of cutting and the efficiency of the operation?

Solution

The screw rotates at 600 rev/min, i.e. $n = 10$ rev/s.

In one revolution the screw moves forward a distance equal to the pitch, i.e. lead = pitch = 15 mm. Therefore in 10 revolutions, distance moved = $10 \times 15 = 150$ mm. Hence

speed of cutting = **150 mm/s**

$$\text{Power} = \frac{2\pi n T}{1000} \text{ kW}$$

Therefore

$$\text{torque on screw, } T = \frac{\text{power (kW)} \times 1000}{2\pi n}$$

$$= \frac{1 \times 1000}{2\pi \times 10} = 15 \cdot 9 \text{ N m}$$

Total force to be overcome by screw,

$$W = \text{force of cut + friction force}$$

$$= 350 + (0 \cdot 08 \times 312 \times 9 \cdot 8)$$

$$= 595 \text{ N}$$

$$\text{Efficiency} = \frac{\text{work done on } W \text{ per revolution}}{\text{work done by } T \text{ per revolution}}$$

$$= \frac{W \times \text{lead}}{T \times 2\pi}$$

$$= \frac{595 \times 0 \cdot 015}{15 \cdot 9 \times 2\pi}$$

$$= 0 \cdot 089 \quad \text{or} \quad \textbf{8·9 per cent}$$

(**Problems**)

6.23 A screw-jack is used to raise a load of 14 t with an effort of 4·5 kN at the end of an arm 38 cm long. The thread is single start with a pitch of 25 mm. Calculate the efficiency of the arrangement.   (32 per cent)

6.24 A screw press exerts a force of 12 kN by means of a single-start thread having a pitch of 3 mm. If the efficiency is 20 per cent, calculate the force required to operate the press. The force is to be applied at the end of a lever 50 cm long.   (57 N)

6.25 A screw-jack is used to raise a load of 2 t. The screw is single start, 12 mm pitch. The load does not turn with the screw. If the efficiency of the jack at this load is 12 per cent and the torque to overcome the friction torque at the bearing collar is 100 N m, calculate the torque to overcome thread friction.
                                                                                     (212 N m)

6.26 A table carrying a light machine tool is traversed by a three-start screw of 3 mm pitch. The table's mass is 20 kg and the coefficient of friction between the table and its guides is 0·1. The screw is driven by a small electric motor at 12 rev/s. What are the speed of operation of the tool and power required, assuming an efficiency of 20 per cent?   (108 mm/s; 10·6 W)

6.27 A screw-jack is used horizontally in sliding a casting into position on its foundation. The screw is single start, 12 mm pitch. The mass of the casting is 8 t and the coefficient of friction between casting and foundation is 0·3. If the torque applied to the jack is 107 N m, what is the efficiency of the operation?   (42 per cent)

(6.16) Simple gear train

Figure 6.11 shows two gear wheels A and B in mesh. The smaller of the two wheels is called the *pinion*, the larger is called the *spur* or *wheel*.

For the teeth to engage correctly the spacing or *pitch* of the teeth on the two wheels must be the same. Hence the number of teeth in each wheel is proportional to its circumference (and therefore its diameter). Thus

$$\frac{\text{circumference of A}}{\text{circumference of B}} = \frac{\text{number of teeth on A}}{\text{number of teeth on B}}$$

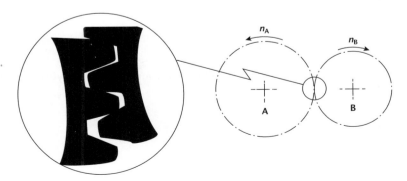

Fig. 6.11

Suppose there are 100 teeth on A and 50 teeth on B. Then the circumference of A is twice that of B. When A turns through one revolution B therefore turns through two revolutions, i.e. the speed of B is twice that of A, or in general, *the speeds of two wheels in gear are inversely proportional to the numbers of teeth in the wheels.*

Thus

$$\frac{\text{speed of A}}{\text{speed of B}} = \frac{\text{number of teeth on B}}{\text{number of teeth on A}}$$

or

$$\frac{n_A}{n_B} = \frac{T_B}{T_A}$$

where n_A, n_B are the speeds, and T_A, T_B the numbers of teeth respectively.

6.17 Idler wheels

Figure 6.12 shows a train of three wheels A, B and C. The speed ratio between A and C is independent of the number of teeth on wheel B.

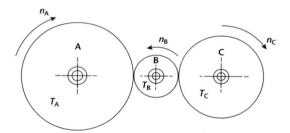

Fig. 6.12

This is shown as follows:

for wheels A and B, $\dfrac{n_A}{n_B} = \dfrac{T_B}{T_A}$ [1]

for wheels B and C, $\dfrac{n_B}{n_C} = \dfrac{T_C}{T_B}$ [2]

The speed of C in terms of the speed of A is found by multiplying equation [1] by equation [2]:

$$\frac{n_A}{n_B} \times \frac{n_B}{n_C} = \frac{T_B}{T_A} \times \frac{T_C}{T_B}$$

Thus

$$\frac{n_A}{n_C} = \frac{T_C}{T_A}$$

That is, the gear ratio of the train ABC is *independent* of the number of teeth in the intermediate wheel B. The only functions of this wheel are to make C rotate in the same direction as A and to increase the centre distance of the shafts of A and C.

Wheel B is called an *idler*. Several idlers may be used and the final direction of rotation of C is determined by the number of idlers.

6.18 Compound gear train

Figure 6.13 shows a compound gear train A, B, C, D. Gears B and C are rigidly fixed together and form a *compound wheel*.

Let n_A, n_{BC} and n_D be the speeds, and T_A, T_B, T_C and T_D the numbers of teeth. For wheels A and B:

$$\frac{n_{BC}}{n_A} = \frac{T_A}{T_B} \qquad\qquad [1]$$

For wheels C and D:

$$\frac{n_D}{n_{BC}} = \frac{T_C}{T_D} \qquad\qquad [2]$$

$$\text{Gear ratio of train} = \frac{\text{speed of D}}{\text{speed of A}}$$

$$= \frac{n_D}{n_A}$$

$$= \frac{n_D}{n_{BC}} \times \frac{n_{BC}}{n_A}$$

$$= \frac{T_C}{T_D} \times \frac{T_A}{T_B} \text{ from equations [1] and [2]}$$

$$= \frac{T_C \times T_A}{T_D \times T_B}$$

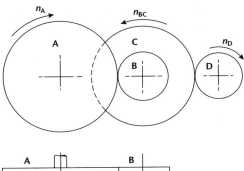

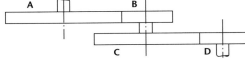

Fig. 6.13

Problems

6.28 In the double-reduction gear train shown, Fig. 6.14, B and C form a compound wheel free to rotate on the layshaft Z. The speed of D is to be one-tenth the speed of A. The numbers of teeth are A 24, C 24, and D 60. Find a suitable number of teeth for wheel B.   (96)

6.29 In the double-reduction gear train shown, Fig. 6.15, wheel A has 20 teeth and wheel C 16 teeth. B and C form a compound wheel free to rotate on shaft Z. The speed reduction from A to D is to be 9 and the speed reduction from A to B is to be the same as that from C to D. Find suitable tooth numbers for wheels B and D.   ($T_B = 60$; $T_D = 48$)

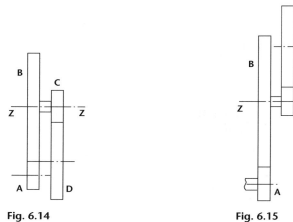

Fig. 6.14   Fig. 6.15

6.19 The crab winch

The crab winch shown in Fig. 6.16 consists of a double-reduction compound gear train A, B, C and D, an effort drum rotating with the first gear wheel A and a load drum rotating with the last wheel D.

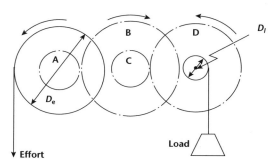

Fig. 6.16

Let T_A, T_B, T_C and T_D represent the number of teeth in gear wheels A, B, C and D respectively. Let D_e be the diameter of the effort drum and D_1 the diameter of the load drum.

The speed ratio of the gear train is (section 6.18):

$$\frac{n_D}{n_A} = \frac{T_A \times T_C}{T_B \times T_D}$$

For one revolution of the effort drum, therefore,

distance x moved by the effort = πD_e

distance y moved by load = $\pi D_1 \times \dfrac{n_D}{n_A}$

Hence the velocity ratio V for the winch is

$$V = \frac{x}{y}$$

$$= \frac{\pi D_e}{\pi D_1 \times \dfrac{n_D}{n_A}}$$

$$= \frac{D_e}{D_1} \times \frac{n_A}{n_D}$$

i.e. the velocity ratio is

$$\frac{\text{diameter of effort drum}}{\text{diameter of load drum}} \times \frac{\text{speed of effort drum shaft}}{\text{speed of load drum shaft}}$$

Example

The crab winch of Fig. 6.17 has an effort drum 80 cm in diameter and a load drum 20 cm in diameter. The number of teeth on each gear are as follows: $T_A = 12$, $T_B = 60$, $T_C = 18$, $T_D = 72$.

If the efficiency of the gear train is 90 per cent, calculate the effort required to lift a load of 1200 kg.

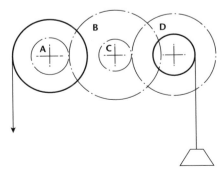

Fig. 6.17

Solution

Velocity ratio of winch $= \dfrac{D_e}{D_1} \times \dfrac{n_A}{n_D}$

$$= \frac{80}{20} \times \frac{60 \times 72}{12 \times 18} \left(\text{since } \frac{n_A}{n_D} = \frac{T_B \times T_D}{T_A \times T_C} \right)$$

$$= 80$$

Mechanical advantage = efficiency × velocity ratio

$$= 0 \cdot 9 \times 80$$

$$= 72$$

Effort $= \dfrac{\text{load}}{\text{mechanical advantage}}$

$$= \frac{1200 \times 9 \cdot 8}{72}$$

$$= \mathbf{163 \cdot 3\,N}$$

Problems

6.30 In the gear drive shown in Fig. 6.17, a drum 45 cm in diameter rotates with the gear A and a drum is keyed to the gear D. A load of 1 tonne is to be raised by a rope wrapped round drum D and the effort is applied to a rope wrapped round drum A. The number of teeth on each gear is as follows: A 30, B 120, C 20, D 100.

(a) If the velocity ratio is to be $37\frac{1}{2}$, what is the diameter of the drum D?

(b) If the efficiency of the arrangement is 65 per cent, what is the effort required?   (24 cm; 402 N)

6.31 Calculate the velocity ratio of the crab winch shown in Fig. 6.18. The effort is applied by a rope attached to the drum A, 36 cm in diameter, and the load is suspended from the drum B, 27 cm in diameter. The number of teeth in each wheel is as shown.

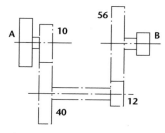

Fig. 6.18

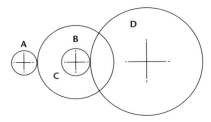

Fig. 6.19

If an effort of 1075 N raises a load of 2000 kg, calculate the efficiency of the winch at this load. (25; 73 per cent)

6.32 A winch has the following wheel train, Fig. 6.19: A 32 teeth, B 64, C 48, D 128. B and C form a compound wheel rotating together. The load drum is keyed to the shaft of D and is 30 cm in diameter. The effort is applied by an arm 80 cm long to wheel A. Find the velocity ratio of the winch. (28·4)

6.20 Worm and worm-wheel

A *worm* may be considered as one or more gear teeth wrapped around a cylinder, as in a screw. The number of teeth on the worm is equal to the number of starts of the worm. The *lead* of a worm is defined in the same way as that of a screw, i.e.

lead = number of starts × pitch of thread

If the worm rotates once, a point on the circumference of the wheel which meshes with the worm is moved round its circumference a distance equal to the lead of the worm. Hence, for one revolution of the worm,

$$\text{revolutions of wheel} = \frac{\text{lead of worm}}{\text{circumference of wheel}}$$

$$= \frac{\text{number of starts on worm} \times \text{pitch}}{\text{number of teeth on wheel} \times \text{pitch}}$$

$$= \frac{\text{number of starts } (n) \text{ on worm}}{\text{number of teech } (T) \text{ on wheel}}$$

$$= \frac{n}{T}$$

The velocity ratio of the worm and wheel is therefore equal to that of an equivalent simple gear train having n and T teeth respectively,

i.e. $\dfrac{\text{speed of wheel}}{\text{speed of worm}} = \dfrac{n}{T}$

Figure 6.20 shows a lifting gear consisting of a worm and wheel, the load drum being keyed to the wheel shaft and the effort being applied to a drum on the

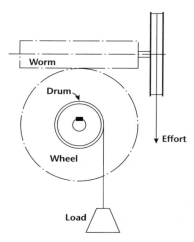

Fig. 6.20

worm shaft. If D_e = effort drum diameter and D_l = load drum diameter then, as in section 6.19, the velocity ratio of the lifting gear

$$= \frac{\text{diameter of effort drum}}{\text{diameter of load drum}} \times \frac{\text{speed of effort drum shaft}}{\text{speed of load drum shaft}}$$

$$= \frac{D_e}{D_l} \times \frac{T}{n}$$

(**Example**)

A lifting gear consists of a load drum 32 cm in diameter keyed to a worm wheel having 180 teeth. The meshing worm has a double-start thread and is keyed to an arm of 96 cm radius, at the end of which an effort may be applied. If the effort required to lift a load of 8000 kg is 314 N, calculate the efficiency of the apparatus.

Solution
The 96 cm radius arm is equivalent to a 192 cm diameter effort drum. The worm and wheel mechanism is equivalent to a simple gear train of velocity ratio 180/2.

$$\text{Velocity ratio} = \frac{192}{32} \times \frac{180}{2}$$

$$= 540$$

$$\text{Mechanical advantage} = \frac{\text{load}}{\text{effort}}$$

$$= \frac{8000 \times 9 \cdot 8}{314}$$

$$= 250$$

$$\text{Efficiency} = \frac{\text{mechanical advantage}}{\text{velocity ratio}}$$

$$= \frac{250}{540}$$

$$= 0\cdot46 \quad \text{or} \quad \textbf{46 per cent}$$

Problems

6.33 In a worm and worm-wheel mechanism, the wheel has 40 teeth and the worm is single start. The load wheel is 18 cm in diameter and the effort is applied at a wheel 27 cm in diameter. Find the velocity ratio.   (60)

6.34 In a worm and worm-wheel lifting gear, the wheel has 30 teeth and the worm has two starts. The load wheel is 36 cm in diameter and the effort is applied to the worm shaft at the end of a lever 42 cm long. An effort of 100 N is needed to raise a load of 204 kg. What is the efficiency of the machine?   (57 per cent)

6.35 In a worm and wheel lifting mechanism, the wheel has 72 teeth and a double-start worm is used. The load is lifted by a rope wrapped around a drum 72 cm in diameter mounted on the worm shaft. A load of 1400 kg is to be lifted by an effort of 250 N at the end of an arm keyed to the wheel shaft. Assuming an efficiency of 50 per cent for the machine, calculate the length of the lever arm required. Neglect the rope thickness.   (1·1 m)

6.36 In a testing machine used for twisting steel bars, a worm and worm-wheel mechanism is used to produce a large torque. A torque of 100 N m is applied to the worm, which has a single-start thread. The worm-wheel has 120 teeth. Assuming an efficiency of 40 per cent, find the torque exerted at the wheel shaft.   (4800 N m)

6.21 Belt drives

Figure 6.21(a) shows an open belt drive, Fig. 6.21(b) a crossed belt drive. In the open drive the pulleys rotate in the same direction, whereas in the crossed drive they rotate in opposite directions.

The angle θ shown is the *angle of lap* or *contact*. For the crossed belt drive θ is the same for both pulleys. This is not necessarily the case for the open drive.

Velocity ratio

Consider pulley A of diameter D_1. In one revolution the length of belt unwrapped from one side is πD_1. Hence, in n_1 revolutions a length $n_1 \times \pi D_1$ is unwrapped. If pulley B, of diameter D_2, makes n_2 revolutions *in the same time* then the length of belt going on to it is $n_2 \times \pi D_2$.

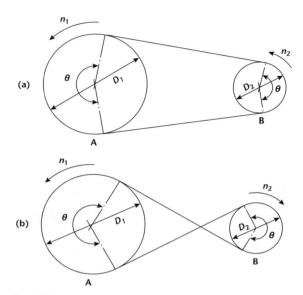

Fig. 6.21

If there is no slip or stretching of the belt the length unwrapped from A must equal the length wrapped on to B; thus

$$n_1 \pi D_1 = n_2 \pi D_2$$

or

$$n_1 D_1 = n_2 D_2$$

i.e. $\dfrac{n_1}{n_2} = \dfrac{D_2}{D_1}$

Thus *the speeds of two pulleys connected by a belt are inversely proportional to their diameters*.

Power transmitted

If a driving torque is applied to pulley A the belt coming on to the pulley tightens, and the belt leaving the pulley slackens. The belt is mounted on the two pulleys with an initial tension so that, when driving, the tension on the tight side is greater than that on the slack side. The ratio of the tight and slack side tensions depends on the angle of lap and the coefficient of friction between belt and pulley.

If, when transmitting power, the tension on the tight side is T_1 and the tension on the slack side is T_2, Fig. 6.22, then

Net clockwise torque, $T = T_1 r - T_2 r$

$$= (T_1 - T_2)r$$

where r is the radius of pulley.

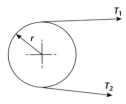

Fig. 6.22

The *effective* tension is therefore $(T_1 - T_2)$.

At a pulley speed n revs/s the power is given by

$$\text{power} = \frac{2\pi n T}{1000} \text{ kW}$$

where the torque T is in newton metres.

Alternatively, since the effective tension in the belt is $(T_1 - T_2)$, the power transmitted may be written in terms of the belt speed v, thus:

$$\text{power} = \frac{\text{effective tension (N)} \times \text{speed (m/s)}}{1000}$$

$$= \frac{(T_1 - T_2)v}{1000} \text{ kW}$$

It is shown in Chapter 10 that the relation between belt speed v and the speed of rotation n is

$$v = \omega r$$

where $\omega = 2\pi n$ radians per second and n is in revolutions per second.

Example

The ratio of the tensions in the two sides of a belt drive is 3:1. Find the tensions when 10 kW is being transmitted at 6 rev/s. The diameter of the driving pulley is 750 mm.

Solution

$$\text{Power} = \frac{2\pi n T}{1000} \text{ kW}$$

Therefore

$$\text{torque } T = \frac{10 \times 1000}{2\pi \times 6} = 265 \text{ N m}$$

so that

$$(T_1 - T_2)r = 265 \text{ N m}$$

and

$$T_1 - T_2 = \frac{265}{0.375} = 707 \text{ N}$$

Also

$$\frac{T_1}{T_2} = 3$$

Hence the tensions are

$$T_1 = \mathbf{1062\,N} \text{ and } T_2 = \mathbf{354\,N}$$

Example

A compound belt and countershaft drive is shown in Fig. 6.23. The pulley diameters are: A, 72 cm; B, 27 cm; C, 54 cm; D, 24 cm. The power input to pulley A is 10 kW at a speed of 100 rev/min.

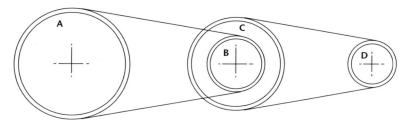

Fig. 6.23

If the belt drive is 85 per cent efficient, calculate:

(a) the output power (kW) at D,
(b) the speed of rotation of D,
(c) the torque exerted at D.

Solution
In this drive pulleys B and C form a compound pulley, rotating as one pulley on the countershaft. Since the diameters of the pulleys B and C are different, the belt speeds are different.

(a)   Efficiency $= \dfrac{\text{output}}{\text{input}}$

Therefore,

output kW at D = efficiency × input kW at A

$$= 0{\cdot}85 \times 10$$

$$= \mathbf{8{\cdot}5\,kW}$$

(b)   Speed of A = 100 rev/min

speed of B, C $= 100 \times \dfrac{72}{27}$ rev/min

speed of D $= \left(100 \times \dfrac{72}{27}\right) \times \dfrac{54}{24}$

$$= \mathbf{600\,rev/min}$$

(c) Power $= \dfrac{2\pi nT}{1000}$ kW

At D, $n = \dfrac{600}{60} = 10$ rev/s and power $= 8{\cdot}5$ kW

Therefore,

torque $T = \dfrac{\text{power (kW)} \times 1000}{2\pi n}$

$= \dfrac{8{\cdot}5 \times 1000}{2\pi \times 10}$

$= \mathbf{136\,N\,m}$

Notes on belt drives

The tension ratio T_1/T_2 for a leather or solid woven belt is often about 5, and should not be less than 3. The efficiency of a good drive is usually about 97 per cent, power losses being due to belt slip, bearing friction and bending of the belt as it passes over the pulley.

The pulley face for a flat belt drive is usually cambered, Fig. 6.24. If the belt tends to wander from the centre-line joining two pulleys the camber ensures that there is a component of the belt tension acting towards the crown of the camber, thus tending to restore the belt to the centre.

The efficient transmission of large powers at high speeds is best performed by the Vee rope drive, Fig. 6.25. The vee-grooves cut in the pulley face ensure a higher friction force between belt and pulley and therefore higher driving torque. A number of ropes can be used on one drive with a tension ratio of up to 20.

Cambered face

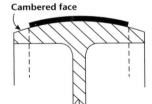

Fig. 6.24

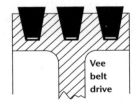

Vee belt drive

Fig. 6.25

Belt drives are often used when it is required to prevent sudden changes of load and shocks from damaging the driving motor or engine. Sudden overloads are absorbed in belt slip or stretching of the belt.

Problems

6.37 A belt passes over a pulley at 16 m/s when transmitting 20 kW. What is the ratio of the belt tensions if the greatest tension in the belt is 1·8 kN?

(3·3:1)

6.38 Two pulleys are connected by a crossed belt drive. The angle of contact of the belt on each pulley is 200° and the shafts are 2 m apart. One of the pulleys is 46 cm diameter and rotates at 450 rev/min. What is the speed of the other pulley?   (885 rev/min)

6.39 The tensions on the tight and slack sides of a belt passing over a pulley are 1 kN and 0·5 kN respectively. The pulley diameter is 30 cm and the speed of rotation 300 rev/min. Calculate the power transmitted.   (2·36 kW)

6.40 The maximum permissible tension in a belt transmitting power is 1·5 kN. Find the belt speed when 12 kW is transmitted if the ratio of the belt tensions is 1·8 to 1.   (18 m/s)

6.41 A compound belt drive transmits 80 kW at 600 rev/min at the output shaft. The pulley mounted on the output shaft is 24 cm diameter and it is driven by a 48 cm diameter pulley on the intermediate shaft. The drive is taken from a 96 cm diameter engine pulley to a pulley of diameter D on the intermediate shaft. If the engine makes 50 rev/min, find the value of D.

What power and torque are exerted by the engine if the drive is 85 per cent efficient?   (16 cm; 94·2 kW; 18 kN m)

6.42 A compound belt drive for papermaking machinery consists of a 24 cm diameter pulley on a motor shaft, which drives a 72 cm diameter pulley on the intermediate shaft. The latter pulley is keyed to the same shaft as an 18 cm diameter pulley which in turn drives a 96 cm diameter pulley on the driven shaft.

Calculate the velocity ratio of the arrangement.

If the power required at the driven shaft is 30 kW and the efficiency of the drive is 96 per cent, calculate the power lost in the drive.   (16; 1·25 kW)

6.43 An ore-crusher is driven through a compound belt drive by an electric motor running at 1000 rev/min. The pulley on the motor shaft is 16 cm diameter. The driven pulley on the intermediate shaft is 72 cm in diameter and the driving pulley 24 cm in diameter. The pulley on the crusher line shaft is 84 cm in diameter.

If the power required at the crusher is 400 kW and the efficiency of the drive is 90 per cent, calculate:

(a) the power at the motor;
(b) the motor torque;
(c) the speed of the line shaft at the crusher.

(444 kW; 4·23 kN m; 63·5 rev/min)

6.44 A pump operates at 10 rev/s when delivering 1000 litres of water per minute against a head of 40 m. The overall efficiency of the pump is 65 per cent. The pump is driven by a motor through a belt drive which is a single belt on a pulley of 40 cm diameter on the pump shaft. If the maximum permissible tension in the belt is 16 N/mm width and the ratio of the belt tensions is 3:1, find the width of belt required to transmit the power.   (75 mm)

6.45 The output from a motor is 10 kW at 12 rev/s. The motor drives a machine through a compound gear train and belt drive. The drive from the motor to the intermediate shaft is by a pinion and wheel, and a belt drive connects pulleys on the intermediate and machine shafts. The pinion has 18 teeth and the wheel on the intermediate shaft has 54 teeth. The pulley diameters are 12 and 48 cm respectively on the intermediate and machine shafts. The overall efficiency of the drive is 90 per cent. If the tension in the tight side of the belt is 15 kN, find the tension in the slack side.   (9 kN)

Chapter 7

Motion; velocity and acceleration

....

We have already mentioned speed and velocity in respect of bodies in steady motion, but now the linear motion of a body that has a change in its speed and direction is considered, irrespective of the forces involved. The equations of linear motion are derived, connecting time with the vector quantities of displacement, velocity and acceleration: these quantities can be treated in the same manner as forces in terms of resolution and combination. The speed–time graph, relative velocity, triangle of velocities and the equations of motion are applied to problems involving the motion of ships and aircraft, etc., freely falling bodies and the flight of projectiles.

7.1 Displacement and motion

When a body moves in a straight path in one plane, its displacement alters in magnitude but its direction remains constant. The body is said to have *plane rectilinear motion*. If the body moves in a curved path, both the magnitude and direction of its displacement alter and the motion is said to be *curvilinear*.

7.2 Speed

The *average speed* of a body, a car for example, is defined as the distance travelled along the path of its motion divided by the time taken. Thus if the distance travelled along any curved path PQ, Fig. 7.1, is denoted by s, and the time taken is denoted by t, the average speed v_{av} is given by

$$v_{av} = \frac{s}{t}$$

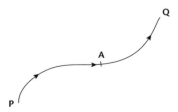

Fig. 7.1

Speed is a derived quantity since it is defined in terms of the two fundamental quantities, length and time. In metre second units the unit of speed is the *metre per second* (m/s). This is the derived SI unit. Other units used are the kilometre per second and the kilometre per hour. A non-SI unit in use for navigation work is the *knot*, equal to one nautical mile per hour and equivalent to 0·514 m/s or 1·85 km/h. (*See* section 7.14.)

If the distance travelled is the same in successive intervals of time then the speed is said to be *constant*. For constant speed *v*, we have

$$v = \frac{s}{t} \qquad\qquad\qquad [1]$$

In practice the speed of a car or other body is not always constant but may change from instant to instant. As shown, the average speed may always be calculated if both the total distance and the time taken are known. This tells us nothing, however, of the *speed at a point*. Suppose therefore that at some point A on the path PQ the speed is said to be of amount *v*. This merely means that if that speed were to be maintained from the point A onwards the body would travel a distance *v* in unit time. The speed at a point is often described as a *rate of change* of distance with time. A precise definition of this idea requires the use of the calculus and is given in the authors' book *Applied Mechanics*.

(7.3) ## Acceleration

If the speed of a body increases from one instant to the next it is said to *accelerate*, and the change of motion is termed *acceleration*. If the speed decreases the body is said to *decelerate* or *retard*, and the change of motion is said to be *deceleration* or *retardation*.

The *average* acceleration is defined as the increase in speed divided by the time taken. Thus, if *t* is the time taken for the speed to increase from an amount *u* to an amount *v*, the average acceleration is

$$a_{av} = \frac{\text{increase in speed } (v - u)}{\text{time taken } (t)}$$

$$= \frac{v - u}{t}$$

A deceleration may be considered as a negative acceleration. For example, if the speed of a vehicle decreases from $u = 60$ m/s, to $v = 36$ m/s in a time $t = 12$ s, the average acceleration is

$$a_{av} = \frac{36 - 60}{12}\left(\frac{\text{metres per second}}{\text{seconds}}\right)$$

$$= -2\,\text{m/s}^2 \text{ (metres per second per second)}$$

This is negative; hence the change in speed is a deceleration. If the deceleration is constant, this means that the speed of the body falls by an amount 2 m/s in every second.

If the speed of a body increases by equal amounts in equal times the acceleration is said to be *constant*. Thus a constant acceleration of $+2\,\text{m/s}^2$ means that the speed of the body increases by an amount $2\,\text{m/s}$ in every second. The term *uniform acceleration* is sometimes used to mean constant acceleration.

Note that the only recommended unit of acceleration is the *metre per second per second*. Multiples and sub-multiples should not be used.

The formula for constant acceleration is identical with that for average acceleration. Thus constant acceleration a is given by

$$a = \frac{v - u}{t}$$

or

$$at = v - u$$

i.e. $v = u + at$ [2]

7.4 The speed–time graph

The relation between speed and time expressed by the equation $v = u + at$ is shown in the speed–time graph, Fig. 7.2. For a constant speed u, the graph is a horizontal line CB and OC = u. For constant acceleration a, the graph is a straight line CD and AD = v, OA = t. Hence

$$a = \frac{v - u}{t}$$

$$= \frac{AD - OC}{OA}$$

$$= \frac{AD - AB}{OA}$$

$$= \frac{BD}{CB}$$

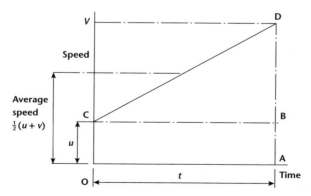

Fig. 7.2

Thus the acceleration a is given by the gradient BD/CB of the graph.

When the acceleration is *constant*, the average speed is given by the mean height, i.e.

$$\text{average speed} = \tfrac{1}{2}(u + v)$$

But average speed = s/t, hence

$$v_{av} = \frac{s}{t} = \tfrac{1}{2}(u + v)$$

Therefore

$$s = \tfrac{1}{2}(u + v)t \qquad\qquad [3]$$

Also, we have

$$v = u + at$$

Hence, substituting for v,

$$s = \tfrac{1}{2}[u + (u + at)]t$$

or

$$s = ut + \tfrac{1}{2}at^2 \qquad\qquad [4]$$

This gives the distance s travelled in time t.

Alternatively the distance can be expressed by an equation which does not involve the time. Thus, from equation [3],

$$\text{distance, } s = \tfrac{1}{2}(u + v)t$$

Also,

$$\text{time, } t = \frac{v - u}{a}$$

Therefore

$$s = \tfrac{1}{2}(v + u)\frac{(v - u)}{a}$$

$$= \frac{v^2 - u^2}{2a}$$

or

$$2as = v^2 - u^2$$

Thus

$$v^2 = u^2 + 2as \qquad\qquad [5]$$

7.5 Summary of equations for constant acceleration

$$v = u + at$$

$$s = \tfrac{1}{2}(u + v)t$$

$$s = ut + \tfrac{1}{2}at^2$$

$$v^2 = u^2 + 2as$$

Note that these equations are *not* valid if the acceleration is not constant. The first equation only may be used, however, for a varying acceleration if a is replaced by the average acceleration a_{av}. *Finally, it should be remembered that* a *is negative for a retardation.*

Example

Calculate the time taken for a train to travel 1000 m if its initial speed is 30 m/s and it has a constant acceleration of 0·1 m/s².

Solution
Since

$$s = 1000\,\text{m}$$

$$u = 30\,\text{m/s}$$

$$a = 0\cdot1\,\text{m/s}^2$$

and

$$s = ut + \tfrac{1}{2}at^2$$

then

$$1000 = 30t + \tfrac{1}{2} \times 0\cdot1 \times t^2$$

i.e. $0\cdot05t^2 + 30t - 1000 = 0$

$$t^2 + 600t - 20\,000 = 0$$

Hence

$$t = -300 \pm 332$$

Since the negative answer has no meaning here,

$t = -300 + 332$

$t = 32\,\text{s}$

(**Example**)

Find the deceleration of a car that is brought to rest in 60 m from a speed of 45 km/h. What is the time taken?

Solution

The following factor for converting kilometres per hour to metres per second is useful and should be memorized:

$$1 \, km/h = \frac{1 \times 1000}{3600} = \frac{1}{3 \cdot 6} \, m/s$$

Hence, **to convert km/h to m/s divide by 3·6.** Thus

$$u = 45 \, km/h = \frac{45}{3 \cdot 6} = 12 \cdot 5 \, m/s$$

$$v = 0$$

$$s = 60 \, m$$

Since

$$v^2 = u^2 + 2as$$

$$0 = 12 \cdot 5^2 + (2 \times a \times 60)$$

$$a = -1 \cdot 3 \, m/s^2$$

This is negative since the car is being retarded.

Again, to find the time t, from

$$v = u + at$$

$$0 = 12 \cdot 5 - 1 \cdot 3t$$

Therefore

$$t = 9 \cdot 6 \, s$$

(**Problems**)

7.1   A car travelling at 80 km/h is brought to rest with constant retardation in a distance of 60 m. Calculate the retardation.   ($4 \cdot 12 \, m/s^2$)

7.2   A piston moves from rest to a speed of 5 m/s in one twenty-fifth of a second. What are the average acceleration and the distance travelled?
                                                                ($125 \, m/s^2$; 100 mm)

7.3   The constant acceleration of a train is $1 \, m/s^2$. Calculate the time taken to increase its speed from 10 to 40 km/h and the distance travelled in this time.
                                                                ($8 \cdot 3$ s; $57 \cdot 8$ m)

7.4   Calculate the time taken for a car to cover a distance of 100 m if the initial speed is 10 km/h and it has a constant acceleration of $1 \cdot 5 \, m/s^2$.   ($9 \cdot 9$ s)

7.5 The maximum retardation of a lift is $3\,\text{m/s}^2$. Calculate the shortest time required to bring it to rest from a speed of $9\,\text{m/s}$. What is the distance travelled in this time?   (3 s; 13·5 m)

7.6 A planing machine table comes to rest from a speed of $27\,\text{m/min}$ in a distance of $8\,\text{cm}$. If the retardation is constant, calculate its value.

$(1·27\,\text{m/s}^2)$

7.7 A train moves with constant acceleration from a speed of $20\,\text{km/h}$ to a final speed of $150\,\text{km/h}$. What is its average speed? If the time taken is $6\,\text{min}$, calculate the distance travelled in this period.   (85 km/h; 8·5 km)

7.8 An aircraft increases its speed from $160\,\text{km/h}$ to $960\,\text{km/h}$ in one minute. If the acceleration is constant, calculate its value and the distance travelled in this time. $(3·7\,\text{m/s}^2; 9·34\,\text{km})$

7.9 A motor boat is travelling at 16 knots in calm water when the engine cuts out. The boat slows to 10 knots in $15\,\text{s}$. Assuming constant deceleration, find its value and the distance travelled in kilometres in this time. What further time elapses before the boat comes to rest? 1 knot = $0·514\,\text{m/s}$.
$(0·21\,\text{m/s}^2; 0·1\,\text{km}; 25\,\text{s})$

7.10 An airliner is required to have a speed of $370\,\text{km/h}$ along the ground on a level runway before it can take off. The initial acceleration is limited to $3\,\text{m/s}^2$. Assuming constant acceleration during the run-up, find the minimum length of runway needed.   (1·76 km)

7.11 An aeroplane lands on the deck of a stationary aircraft carrier at a horizontal speed of $30\,\text{m/s}$ and is brought to rest by the arrestor gear in a distance of $60\,\text{m}$. Find the deceleration of the plane, assuming it to be constant, and the time taken to halt the plane. $(7·5\,\text{m/s}^2; 4\,\text{s})$

7.6 Use of the speed–time graph

Many problems are conveniently solved by making use of the fact that *the area under the speed–time graph is equal to the distance travelled*. This is proved as follows.

In Fig. 7.3 the line CD represents the *v–t* graph for motion with constant acceleration.

Area under *v–t* graph = area OADC

$$= \text{area OABC} + \text{area CBD}$$

$$= \text{OC} \times \text{OA} + \tfrac{1}{2}\text{CB} \times \text{BD}$$

$$= ut + \tfrac{1}{2}t(v - u)$$

But

$$v - u = at$$

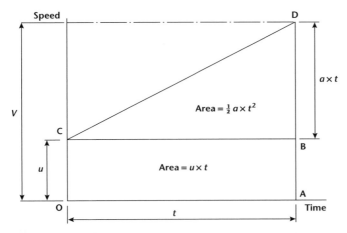

Fig. 7.3

Hence

$$\text{area} = ut + \tfrac{1}{2}at \times t$$

$$= ut + \tfrac{1}{2}at^2$$

$$= s \quad \text{from equation [4], p. 143}$$

Thus

$$\text{average speed} = \frac{s}{t}$$

$$= \frac{\text{area under } v\text{–}t \text{ graph}}{\text{total time taken}}$$

<hr>

Example

A cycle accelerates with a constant acceleration of $0.2\,\text{m/s}^2$ from rest for a period of 7 s and continues at this speed for 30 s before coming to rest in 5 s. The deceleration is constant. Sketch the v–t diagram and determine the time taken to travel the first 15 m and the total distance travelled.

Solution

In the v–t diagram, Fig. 7.4, OA represents the period of constant acceleration, AB the period of constant speed and BC the retardation. Since at O the velocity $u = 0$, the velocity at the end of the acceleration is

$$v = at$$

$$= 0.2 \times 7 = 1.4\,\text{m/s}$$

Distance travelled during acceleration

$$= \text{area OAE}$$

$$= \tfrac{1}{2} \times 1.4 \times 7$$

$$= 4.9\,\text{m}$$

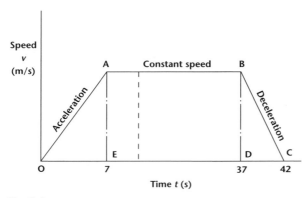

Fig. 7.4

In order to cover a distance of 15 m, the additional distance to be travelled at a constant speed of 1·4 m/s

$$= 15 - 4·9$$

$$= 10·1 \text{ m}$$

Time taken to travel this additional distance at constant speed

$$= \frac{\text{distance}}{\text{speed}}$$

$$= \frac{10·1}{1·4}$$

$$= 7·2 \text{ s}$$

Total time to travel 15 m

$$= 7 + 7·2 = \mathbf{14·2 \, s}$$

Total distance travelled at constant speed

$$= \text{area ABDE}$$

$$= 1·4 \times 30$$

$$= 42 \text{ m}$$

Distance travelled in coming to rest

$$= \text{area BCD}$$

$$= \tfrac{1}{2} \times 1·4 \times 5$$

$$= 3·5 \text{ m}$$

Total distance travelled in coming to rest

$$= 4·9 + 42 + 3·5$$

$$= \mathbf{50·4 \, m}$$

(**Example**)

An electric train starting from rest is uniformly accelerated during the first 0·4 km, runs 1·2 km at the maximum speed attained and is afterwards brought to rest in 0·2 km.

If the time for the whole journey is 5 min, find the uniform acceleration at the start.

Solution

The v–t diagram is shown in Fig. 7.5. Let v = maximum speed attained (km/h), x (h) the time for the acceleration, y the time at constant speed, z the time to come to rest.

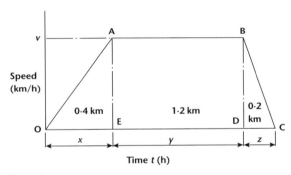

Fig. 7.5

Area OAE = 0·4 km

i.e. $\tfrac{1}{2}vx = 0\cdot4$

or

$$vx = 0\cdot8$$

Area ABDE = 1·2 km

therefore

$$vy = 1\cdot2$$

Area BDC = 0·2 km

therefore

$$\tfrac{1}{2}vz = 0\cdot2$$

$$vz = 0\cdot4$$

Hence

$$v = \frac{0\cdot8}{x} = \frac{1\cdot2}{y} = \frac{0\cdot4}{z} \text{ km/h}$$

therefore

$$y = \frac{3x}{2} \text{ h and } z = \tfrac{1}{2}x\,\text{h}$$

Also

$$x + y + z = \text{total time taken} = 5\,\text{min} = \tfrac{1}{12}\,\text{h}$$

i.e. $x + \tfrac{3}{2}x + \tfrac{1}{2}x = \tfrac{1}{12}$ h

Therefore

$$x = \tfrac{1}{36}\text{ h} = 100\,\text{s}$$

For the first stage of the journey,

$$s = 0{\cdot}4\,\text{km} = 400\,\text{m}$$

$$t = 100\,\text{s}$$

From $s = \tfrac{1}{2}at^2$, we have

$$a = \frac{2s}{t^2} = \frac{2 \times 400}{100^2} = 0{\cdot}08\,\text{m/s}^2$$

Problems

7.12 A diesel train accelerates uniformly from rest to 70 km/h for 30 s, continues at this speed for 60 s and decelerates uniformly to rest in 60 s. Calculate the total distance travelled.   (2·04 km)

7.13 A car accelerates uniformly from rest at 0·6 m/s² for 11 s and continues at the maximum speed attained. Calculate the time taken to travel the first 75 m.   (16·9 s)

7.14 An electric train is uniformly accelerated from rest for 700 m, runs 1000 m at the maximum speed attained and is then brought to rest at the next station. The distance between stations is 3·2 km and the total time taken is 3 min. Calculate the maximum speed attained.   (30 m/s)

7.15 A hoist is accelerated uniformly from rest to 4·5 m/s and then retarded uniformly to rest. The distance travelled in the accelerated portion of its motion is double that in the second part. Calculate the total time taken if the initial acceleration is 1·5 m/s².   (4·5 s)

7.16 A planing machine accelerates from rest to its maximum speed of 80 mm/s over a distance of 8 cm, continuing at this speed for a distance of 48 cm and then comes to rest in 8 cm. Assuming the acceleration and retardation each to be constant, calculate the total time taken for the stroke.   (10 s)

7.7) Variable acceleration

When the acceleration is not constant, the increase or decrease in speed varies from instant to instant; the *v–t* diagram is no longer made up of straight lines, but may be as in Fig. 7.6. For example, if the accelerator pedal of a car is held at a constant position then, all other things being equal, the acceleration is constant. When the pedal is being moved, the acceleration is variable.

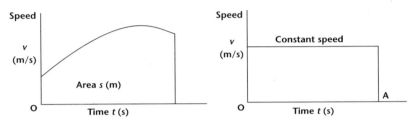

Fig. 7.6

It may be proved that the distance travelled is given by the area under the *v–t* diagram, as before. This area is found by *Simpson's rule*, or any other suitable rule for irregular areas.

The *average speed* is that constant speed at which the same distance would be covered in the same time. Thus

$$\text{Average speed} = \frac{s}{t} = \frac{\text{area of } v\text{–}t \text{ graph}}{\text{total time taken}}$$

Example

The following table gives the speed *v* of a car at various times *t*. Calculate the distance travelled and the average speed.

t (s)	0	1	2	3	4	5	6	7	8
v (km/h)	17	19	20	21	21	20	19	15	10

Solution

The distance travelled in 8 s is the area under the *v–t* diagram. There are an odd number of equally spaced ordinates, hence Simpson's rule may be used. If

s = area under graph

F = first ordinate (17)

L = last ordinate (10)

E = sum of all even ordinates (75)

O = sum of all odd ordinates, excepting the first and last (60)

D = spacing between ordinates = $1 \times \dfrac{1}{60} \times \dfrac{1}{60} = \dfrac{1}{3600}$ h then Simpson's rule states that

$$s = \frac{D}{3} (F + L + 4E + 2O)$$

Therefore $\quad s = \dfrac{1}{3 \times 3600} [17 + 10 + (4 \times 75) + (2 \times 60)]$

$$= 0{\cdot}0414 \, \text{km} = \textbf{41·4 m}$$

Average speed,

$$\frac{s}{t} = \frac{41{\cdot}4}{8} = 5{\cdot}17 \, \text{m/s} = \textbf{18·6 km/h}$$

Problems

7.17 The following table gives the speed of a train at intervals of 1 min. Sketch the *v–t* diagram and calculate the total distance travelled.

Speed (km/h)	0	15	45	60	90	100	100	80	62	20	0
Time (min)	0	1	2	3	4	5	6	7	8	9	10

(9·4 km)

7.18 The following table gives the speed of a car at intervals of 1 s when braking. Calculate the distance travelled in coming to rest and the average speed.

Speed (m/s)	30	29	25	20	15	10	5	2	0
Time (s)	0	1	2	3	4	5	6	7	8

(121 m; 54·6 km/h)

7.8 Displacement

Consider the translation of a body from its initial position at the point P, Fig. 7.7(a), to a final position at Q. The distance travelled, *s*, is measured along the curved path actually traversed. The *displacement*, on the other hand, is measured by the straight line joining the initial and final positions P and Q respectively.

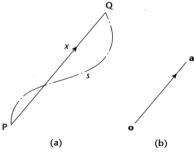

(a)   (b)

Fig. 7.7

The *magnitude* of the displacement is the length x = PQ; the *direction* is along the line PQ, and the *sense* is from P to Q. Displacement is a quantity which possesses magnitude, direction and sense. It is therefore a *vector quantity*.

7.9   Velocity

Suppose a body moves from P to Q in time t, Fig. 7.7(a). The distance travelled, measured along the curved path, is s. Hence the average speed is of amount $v_{av} = s/t$. The displacement PQ, on the other hand, is of amount x, measured along the straight line PQ.

We define the *average velocity* of the motion between P and Q as a vector of length x/t, direction along PQ and sense from P to Q. The *velocity vector* may be represented in a velocity diagram, Fig. 7.7(b), by a line **oa** drawn from the point **o** representing zero velocity, parallel to PQ, and of length x/t.

The velocity is said to be *uniform* if the motion is along the straight line PQ and the speed is constant. The magnitude (length) of the velocity vector then represents the speed.

7.10   Vector change of velocity

Consider the motion of a body at constant speed along the curved path PQ, Fig. 7.8(a). Although the speed may be constant, the velocity is changing continuously, for the direction of motion, and therefore the direction of the velocity vector, is everywhere tangential to the path of the motion.

If at some instant the body is at point A then its velocity may be represented to some scale by the vector **oa** in the velocity diagram Fig. 7.8(b). Its direction is tangential to the path of motion at A. At some later instant the body will be at A′ and its speed may have changed. The velocity at A′ may be represented by the vector **oa′**, parallel to the tangent to the path at A′. (The point **o** is the point of zero velocity in the velocity diagram and is therefore the same for both vectors.)

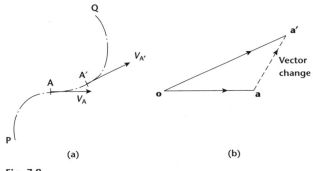

(a)   (b)

Fig. 7.8

The *vector change in velocity* in going from A to A′ is defined as the vector **aa′**. The vector change is itself a vector, of magnitude equal to the length of **aa′**, direction along **aa′**, and sense from **a** to **a′**.

7.11 Acceleration vector

The average acceleration between points A and A′, Fig. 7.8(a), may now be defined as a vector of magnitude (length) given by

$$\frac{\text{magnitude of vector change of velocity } (\mathbf{aa'})}{\text{time taken } (t)} = \frac{\mathbf{aa'}}{t}$$

where t is the time taken for the body to move from A to A′. The direction of the acceleration vector is along **aa′** and its sense from **a** to **a′**.

Uniform acceleration is motion along a straight line with constant acceleration.

The *acceleration at a point* is identical with the vector representing the average acceleration if the two points A and A′ are very close together. The acceleration due to motion in a circle is a special case of acceleration arising from a vector change in velocity (a change in direction) and this case is dealt with in Chapter 10.

7.12 Resultant displacement

Consider the points **o**, **a** and **b**, Fig. 7.9. They may be regarded as points on a map. The straight lines **oa**, **ab** and **ob** are distances or vector displacements on the map. Evidently it is possible to go from **o** to **b** either

- directly along the straight line **ob**, or
- indirectly by going first from **o** to **a** and then from **a** to **b**.

The displacement **ob** is equivalent to the two displacements **oa** and **ab** *taken in turn*; **ob** is called the *resultant* of **oa** and **ab**.

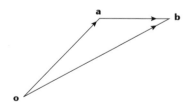

Fig. 7.9

7.13 Resultant velocity

A body may have several different velocities simultaneously, as, for example, when a person walks across a moving ship or an aircraft flies in a crosswind. Suppose that in Fig. 7.9 the three displacements **ob**, **oa** and **ab** take place *simultaneously* in a given time. The three vectors now represent *velocities*, each to the same scale. The vector **ob** represents the *resultant* of the two velocities **oa** and **ab**.

Velocities can therefore be compounded to give a resultant, in the same way as forces are compounded.

Example

A car travelling forwards at 50 km/h starts to skid sideways at 20 km/h. How far will it travel in the direction of the resultant motion in 3 s?

Solution

Figure 7.10 shows the velocity triangle for the car's motion. The car has two speeds – its forward speed, represented by vector **oa**, and its sideways speed, represented by **ab**; these two vectors are drawn head to tail. The resultant speed is

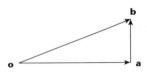

Fig. 7.10

$$\mathbf{ob} = \sqrt{(50^2 + 20^2)}$$

$$= 53\cdot85 \text{ km/h}$$

$$= 15 \text{ m/s}$$

In 3 s the car travels $3 \times 15 = $ **45 m**, in the direction of its resultant motion.

Problems

7.19 A lathe tool is traversed at 1·5 mm/s and its rate of feed into the work is 0·55 mm/s. What is its resultant velocity?   (1·6 mm/s)

7.20 A train travels due S at 60 km/h in a 45 km/h wind which is blowing due E. In what direction does the smoke travel?   (37°E of S)

7.21 A pin on the mechanism of a shaping machine is moving vertically at 1·8 m/s and horizontally at 1·2 m/s. What are its resultant velocity and direction of motion?   (2·17 m/s; 33·7° to vertical)

7.22 A sleeve is moved axially at 0·8 m/s along a rotating shaft. If the surface speed of the shaft is 1 m/s, what is the resultant rubbing velocity between the shaft and sleeve surfaces?   (1·3 m/s)

7.14 Resultant velocity: motion of ships and aircraft

In navigation work the metre is now used for heights and depths but distances are expressed in *nautical* or *sea-miles* and speed in *knots*. **The knot is defined as one nautical mile per hour.** The nautical mile is equivalent to one minute of latitude and varies slightly between poles and equator. For navigation purposes it is taken as 1852 m, making the knot equal to 0·514 m/s or 1·85 km/h.

Any vehicle moving through water or air has two speeds:

1. Its speed over the ground or *groundspeed*, required when calculating distance travelled and time taken for a journey. For an aircraft it is also important when taking off and landing.
2. Its speed through, or relative to, the water or air.

Ships

The speed through the water is in the direction the bow is pointed, i.e. its *heading*, given in degrees by compass from due north. In *still* water the speed through the water and the groundspeed are the same. When a ship encounters a current, the speed through the water is not affected but the groundspeed alters. The ship is steered in one direction (the *water track*) but moves in a different direction relative to the ground. The ship's groundspeed, i.e. its absolute velocity, is the resultant of its speed through the water and the speed of the current relative to the ground ('current' here includes effect of wind and tidal streams). The triangle of velocities applied to a ship's motion is shown in Fig. 7.11(a). The current is represented by vector **ab**, the ship's speed through the water by **oa**, and the groundspeed by **ob**. The course over the ground is from **o** to **b** and the heading is in direction **o** to **a**. The resultant **ob** is the vector sum of **oa** and **ab** and to add two vectors the tail of one is joined to the head of the other. In navigation work, a single arrowhead is used for the speed through the water, a double for the groundspeed, and a treble for current, tidal stream or wind vector.

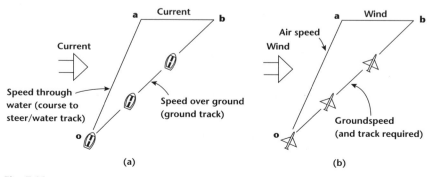

(a)   (b)

Fig. 7.11

Aircraft

As for a ship, a pilot maintains a course along a 'track required', i.e. a predetermined flight path, by steering the plane on a heading calculated to bring it to a desired destination, allowing for the effect of wind velocity. The speed through the air is called the *indicated airspeed** and it registers in knots or km/h on the instrument panel. The pilot uses this airspeed to control and manoeuvre the plane in the air.

Suppose a plane is flying in still air at airspeed 200 knots, then its groundspeed is also 200 knots. The wind velocity does not affect the airspeed but alters the groundspeed. Thus if there is a 40 knot headwind the groundspeed falls to 160 knots, whereas with a tailwind it rises to 240 knots. A wind blowing at an angle slows down or speeds up the plane and moves it off the course being steered.

Figure 7.11(b) shows the triangle of velocities for an aeroplane heading slightly W of N (**oa**) and due to a wind blowing from W to E (**ab**) having a resultant course over the ground roughly NW (**ob**).

Note

1. By convention, a wind blows *from* a direction but a current or tidal stream flows *towards* a direction. Thus an 'easterly' wind blows from the east, whereas an 'easterly' or 'eastgoing' current flows to the east. A current or tidal stream is said to *set* in the direction towards which it flows.
2. The time taken for a journey at steady speed is the distance travelled over the ground divided by the groundspeed.

(Example)

A ship is heading due S at a steady speed of 15 knots through the water in a tidal stream setting NE, at the rate of 5 knots. Find its groundspeed and true course over the ground. If the ship travels 8·4 nautical miles over the ground, how long does the journey take?

Solution

Draw **oa** to represent the speed through the water, 15 knots due S, Fig. 7.12. From **a** draw the tidal stream vector **ab** representing 5 knots NE (the flow is **a** to **b** and vectors **oa** and **ab** are head to tail at **a**). Then **ob** is the speed over the ground, i.e. groundspeed. Results may be obtained from an accurate scale drawing. Thus

$$ob = 12 \text{ knots}$$

and

$$\theta = 17°$$

---

* An aircraft is subject to 'dynamic' as well as 'static' air pressure. A measurement of dynamic pressure at a forward-facing intake by an instrument such as a pitot tube gives an indication of the speed of the plane relative to the air. Such an instrument is subject to errors depending on altitude, density of the air, etc. When this indicated airspeed reading is corrected for such errors a 'true airspeed' is obtained. The true airspeed corrected for wind velocity gives the groundspeed.

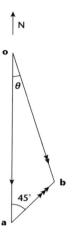

Fig. 7.12

To calculate **ob**, use the cosine rule for the triangle. Thus

$$\mathbf{ob}^2 = \mathbf{ab}^2 + \mathbf{oa}^2 - 2 \times \mathbf{oa} \times \mathbf{ab} \cos 45°$$

$$= 5^2 + 15^2 - 2 \times 15 \times 5 \times 0.707$$

$$= 144$$

Therefore

$$\mathbf{ob} = 12 \text{ knots.}$$

Using the sine rule,

$$\frac{\mathbf{ab}}{\sin \theta} = \frac{\mathbf{ob}}{\sin 45°}$$

i.e. $\sin \theta = \dfrac{0.707 \times 5}{12} = 0.295$

Therefore

$$\theta = 17.1°$$

The true course (over the ground) is **163°E of N**, and the groundspeed is **12 knots**.

The knot is one nautical mile per hour, hence

$$\text{time taken} = \frac{\text{distance over the ground}}{\text{groundspeed}}$$

$$= \frac{8.4 \text{ (n.m.)}}{12 \text{ (knots)}}$$

$$= 0.7 \text{ h}$$

$$= \mathbf{42 \text{ min}}$$

A boat crosses a river 3 km wide running at the rate of 4 knots to reach a point on the opposite bank 5 km upstream. The boat's speed in still water is 12 knots. Find the direction in which the boat must be headed, the speed over the ground, and the time for the journey (1 knot = 0·514 m/s).

Solution

The course over the ground is AB, Fig. 7.13(a). By drawing or calculation, $\alpha = 31°$, AB = 5·83 km. The speed through still water is the speed through the water when the boat moves in the current. Draw **oa** to represent the 4 knot current, Fig. 7.13(b). Through **o** draw a line parallel to AB to represent the direction of the groundspeed vector. With centre **a** and radius **ab** representing the speed through the water, 12 knots, draw an arc cutting the groundspeed line in **b**. Then **ob** is the resultant velocity of the boat, i.e. the groundspeed. From a scale drawing,

ob = 8·4 knots

$\theta = 21°$

Hence the heading required is at **21°** to the bank, upstream and speed over the ground is **8·4 knots**.

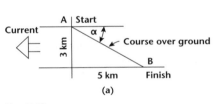

Fig. 7.13

The time taken is

$$\frac{\text{distance over ground}}{\text{groundspeed}} = \frac{AB}{\text{ob}} = \frac{5·83 \times 10^3}{8·4 \times 0·514}$$

$$= 1350\,\text{s}$$

$$= 22·5\,\text{min}$$

The student should confirm these results by calculation, using the cosine and sine rules for triangle **oab** noting that angle **aob** = 149°.

An aeroplane flies at 300 knots through the air on a heading due N and encounters an easterly wind of 80 knots. The distance travelled over the ground is 200 nautical miles. Find the groundspeed, time taken for the journey and the actual direction travelled by the aeroplane.

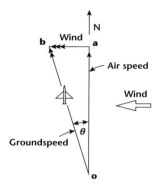

Fig. 7.14

Solution

Figure 7.14 shows the velocity triangle. The airspeed 300 knots due N is represented by vector **oa**, the windspeed 80 knots from E to W by **ab**, and the resultant **ob** is the groundspeed.

$$\textbf{ob} = \sqrt{(80^2 + 300^2)} = 310 \text{ knots}$$

$$\tan \theta = \frac{80}{300} = 0.267$$

i.e. $\theta = 14.9°$

The groundspeed is **310 knots** and the actual route flown by the plane is **14.9°W of N**. The plane travels 200 nautical miles over the ground at 310 nautical miles per hour, hence

$$\text{time taken} = \frac{\text{distance over ground}}{\text{groundspeed}}$$

$$= \frac{200}{310} \text{ h}$$

$$= 39 \text{ min}$$

(**Example**)

A pilot is informed that the wind velocity is 60 knots blowing due N when embarking on a journey to a destination 60°E of N. Estimated groundspeed is 200 knots. Find the required airspeed and heading.

Solution

Draw **ob** to represent the groundspeed, 200 knots 60°E of N, Fig. 7.15. Then **ob** is the resultant of the airspeed and windspeed vectors. Draw **oa** due N to represent the windspeed of 60 knots. Then **ab** is the airspeed in direction **a** to **b** (vectors **ab** and **ob** must be tail to head) and the plane must be headed in direction **a** to **b**. From a scale drawing

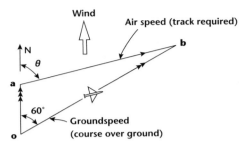

Fig. 7.15

ab = 178 knots

$\theta = 77°$

The airspeed therefore is **178 knots** and the heading required is **77°E of N**. The student should confirm these results by calculation using the cosine and sine rules applied to triangle **oab**.

Problems

(1 knot = 0·514 m/s or 1·85 km/h; 1 nautical mile = 1852 m.)

7.23 What course must a ship steer when steaming at 10 knots to have a resultant velocity due E if the tidal stream sets NE at 3 knots? How long does the ship take to travel 5 nautical miles?   (102°E of N; 25 min)

7.24 A boat's compass indicates it is steering due E in a tidal stream which sets SW at a rate of 4 knots. The boat's speed through the water is 10 knots. Find the actual course and speed over the ground. If the boat arrives at its destination in 12 minutes, what is the distance travelled in nautical miles?
(112°E of N; 7·7 knots; 1·54 n.m.)

7.25 A ship is headed 70°W of N at 17 knots when it encounters a 7 knot current running from N34°E. What is the actual course followed by the ship in speed and direction?   (20 knots due W)

7.26 To cross a river 1 km wide a motorboat is headed at a speed of 16 knots through the water with the bow pointing at right angles to the bank. The river flows at the rate of 5 knots. How far downstream does the boat meet the opposite bank and how long does the journey take?   (313 m; 122 s)

7.27 A boat can travel at 8 m/s in still water. It sets off at right angles to the bank of a river which flows at 6 m/s. Find the resultant speed and direction of motion of the boat.   (10 m/s; 53° to bank, in direction of flow)

7.28 A boat crosses a river flowing at 4 km/h. At what speed should the boat attempt to move in a direction normal to the bank in order to arrive at a point 30 m downstream, if the width of the river is 40 m?   (5·33 km/h)

7.29 A ship has a speed through the water of 25 knots when heading due N. Its speed and course over the ground are 22 knots and 15°E of N respectively. What is the magnitude and direction of the current?

(6·82 knots, sets 123°E of N)

7.30 An aeroplane flies in a straight line a distance of 216 nautical miles against a headwind of 32 knots and its airspeed indicator shows that it is travelling at 243 knots relative to the surrounding air. What is its groundspeed and how long does the journey take?   (211 knots; 61·4 min)

7.31 On a journey of 190 km a plane has an indicated airspeed of 175 knots with a tailwind of 40 knots. How long does the journey take?   (28·7 min)

7.32 A plane has an airspeed of 314 knots when flying against a cross-wind. The groundspeed is 320 knots. Find the wind velocity.   (61·7 knots)

7.33 An aircraft attempts to fly due N at an indicated airspeed of 200 km/h in an easterly wind of 50 km/h. Calculate its resultant velocity.

(206 km/h, 14°W of N)

7.34 In what direction should an aircraft fly at an airspeed of 400 km/h in a wind blowing due N at 120 km/h, in order to travel towards the NW?

(57·3°W of N)

7.35 An aeroplane flies in a straight line a distance of 400 km against a headwind of 60 km/h. Its airspeed indicator shows that it is travelling at 450 km/h relative to the surrounding air. What is its groundspeed and how long does the journey take?   (390 km/h; 61·5 min)

7.36 An aeroplane flies on a heading due E with an airspeed of 400 km/h. If the windspeed is 70 km/h due S, what is the groundspeed of the plane and the direction of the track over the ground?   (406 km/h; 100°E of N)

7.37 An aircraft carrier is heading into a wind of 15 knots at a speed of 25 knots when an aircraft touches down on the deck against the wind with an airspeed of 95 knots. What is the speed of the aircraft relative to the ship?

(55 knots)

7.38 An aircraft has a maximum flying speed of 380 knots in still air. The pilot wishes to travel at maximum speed to a point 150 nautical miles due N in a wind of 60 knots blowing from N 60°E. Find the aircraft's required heading, groundspeed, and time for the journey.   (7·9°E of N; 346 knots; 26 min)

7.39 A pilot sets off for a destination due N, a distance of 300 km. The plane encounters an easterly wind of 60 km/h and the journey takes 80 mins. Find the velocity of the plane in still air and the heading required.

(233 km/h; 14·9°E of N)

7.40 An aeroplane maintains 200 knots through the air to reach a destination due N, a distance of 500 n.m. How long does the journey take (i) in still air and (ii) in a 70-knot cross-wind? If the wind blows from due W, what is the heading required in the cross-wind?   (150 min; 160 min; 20·5°W of N)

7.41 A pilot requires to arrive at a destination 650 km due S of his starting point in a time of 100 min. The airspeed is 450 km/h, and allowing for windspeed the plane is steered 165°W of N. What is the magnitude and direction of the windspeed?   (125 km/h, from 69°W of N)

(7.15) Resolution of velocities

Consider the velocity diagram of Fig. 7.16(a). The velocity $v = $ **ob** is the resultant of the velocities **oa, ab**. Alternatively, the velocity **ob** can be considered as being replaced by the two separate parts **oa** and **ab**. These are called the *resolved parts* or the *components* of **ob** *in the directions* **oa** and **ab** respectively.

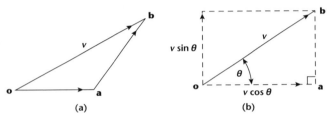

Fig. 7.16

It is usually convenient, but not necessary, to resolve **ob** into two components at right angles to one another, as shown in Fig. 7.16(b). Thus

horizontal component of $v = $ **oa** $= v \cos \theta$

vertical component of $v = $ **ab** $= v \sin \theta$

Example

A rocket is given a velocity of 300 km/h at 60° to the horizontal. Calculate the horizontal and vertical components of its velocity. Find the time for the rocket to travel a horizontal distance of 1000 m.

Solution
In Fig. 7.17 **ob** represents the velocity of 300 km/h at 60° to the horizontal.

Horizontal component = **oa** = **ob** cos 60°

$$= 300 \times 0 \cdot 5$$

$$= 150 \, \text{km/h} = 41 \cdot 8 \, \text{m/s}$$

Vertical component = **ab** = **ob** sin 60°

$$= 300 \times 0 \cdot 866$$

$$= 260 \, \text{km/h}$$

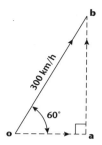

Fig. 7.17

It is only the horizontal component of the velocity which affects the horizontal travel of the rocket. Thus, assuming the velocity remains constant,

$$s = vt$$

$$1000 = 41 \cdot 8 \times t$$

Therefore

$$t = 24\,\text{s}$$

Problems

7.42 The speed of an aircraft flying due N is 500 km/h. What is its component of velocity in a north-easterly direction?   (354 km/h)

7.43 A train travels up an incline of 1 in 100 at 60 km/h. How fast is it ascending vertically?   (0·6 km/h)

7.44 A shell is given a velocity of 1000 m/s at 40° to the horizontal. Calculate the horizontal and vertical components of its velocity. How far will it have travelled in a horizontal direction in 3 s?   (766 m/s, 643 m/s; 2300 m)

7.45 An aircraft touches down on a level runway at 100 knots and comes to rest in 1·2 km. The approach is at 5° to the ground. What is its vertical component of velocity, in km/h, at touchdown? How long does the plane take to come to rest? 1 knot = 0·514 m/s.   (16·1 km/h; 47 s)

7.46 The velocity of a point on a mechanism link which is moving in the vertical plane is 1·2 m/s at an angle of 18° to the vertical. What are its horizontal and vertical components?   (0·37 m/s; 1·14 m/s)

7.47 A helicopter is climbing at 60 km/h on a flight path at an angle θ to the vertical. Its rate of climb (vertical) is 500 m/min. What is the angle θ? (60°)

7.16 Freely falling bodies

A body falling freely from rest increases its velocity as it falls. If it falls to earth from a height that is small compared with the radius of the earth, it is found to increase its velocity by an equal amount in each second, i.e. its acceleration downwards is uniform. Similarly if the body is projected upwards its velocity decreases by an equal amount in each second and its deceleration upwards is uniform.

If an experiment were carried out to determine the velocity of a falling body after each second the following table of approximate values would be obtained:

Time t (s)	0	1	2	3	4	5
Velocity v (m/s)	0	9·8	19·6	29·4	39·2	49

The corresponding v–t graph is shown in Fig. 7.18. The graph is a straight line, and hence the acceleration is uniform and is equal to

$$\frac{v - u}{t} = \frac{49 - 0}{5} = 9 \cdot 8 \, \text{m/s}^2$$

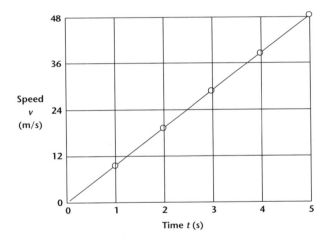

Fig. 7.18

This value is known as the *acceleration due to gravity*, denoted by g. The immediate cause of such an acceleration is the pull due to the gravitational attraction of the earth which exerts a radial inward force on the body. This force is called the *force of gravity* or the *earth-pull*. (*See also* section 8.4.) If acceleration is prevented, or the body is held stationary, the earth-pull is experienced as the *weight* of the body.

The acceleration due to gravity is found to be independent of the weight, size and shape of the body provided that *air resistance is neglected*. It is evident that if a pellet and a feather of the same weight are dropped from the same height, the effect of air resistance on the feather will retard its motion considerably compared with that of the pellet.

That all bodies fall to earth with the same acceleration may be demonstrated as follows. Place an empty matchbox on the edge of the table and on the matchbox place a small piece of chalk. Strike the matchbox (but not the chalk) so as to sweep it *horizontally* away from the table and allow the chalk to fall to the floor. Both box and chalk will strike the floor at the same time. As with all experiments a little practice is necessary.

Now, since the time of flight is the same for both box and chalk, and both have the same initial vertical velocity (zero), the downward acceleration must be the same for both.

It is often stated that Galileo (1564–1642) carried out experiments on falling bodies dropped from the leaning tower at Pisa. There appears to be no evidence that this is in fact true, although similar tests may have been carried out elsewhere. Galileo did, however, perform much more important experiments on bodies moving freely down inclined planes. He was able to show that the maximum speed attained was proportional to the time taken and that the distance travelled from rest was proportional to the square of the time.

The value of g varies slightly over the earth's surface (about 0·5 per cent) but decreases with altitude. For example, at a height of about 70 km above the earth's surface, g falls to 9·6 m/s^2 and at 25 000 km, g is only 0·4 m/s^2. Air resistance, variation in g and the size and shape of a body must be taken into account when carrying out accurate calculations of its motion upwards from the earth's surface. The relationship between the value of g and the *depth* below the surface of the earth is complex because of the differing densities throughout the earth's core and mantle. If the density of the earth is assumed to have a constant value then the relationship between g and depth is linear from zero at the core to the maximum value of 9·8 m/s^2 at the surface.

In Chapter 11 there is further work on gravity forces in connection with space and satellites. It is shown that for a body to 'escape' from the earth's gravity pull it must be given a speed of about 11 km/s from the surface of the earth, ignoring the effect of the atmosphere. This 'escape speed' is equal to the speed that it would reach if it fell to earth from an infinite distance – just as shown in the simple example below that when a body is fired upwards it will have the same speed when it falls back to the ground as it had when leaving the ground.

(7.17) Equations of motion of freely falling bodies

If *air resistance is neglected*, the motion of a freely falling body may be calculated by the formulae for uniformly accelerated motion in a straight line, provided that the acceleration a is replaced by the acceleration due to gravity g.

For motion downwards $a = +g$ (acceleration)

For motion upwards $a = -g$ (deceleration)

The acceleration g may be regarded as a *vector* directed towards the centre of the earth, i.e. vertically downwards.

A common requirement is to find the velocity of a body falling from rest through a height h. Thus $u = 0$, $s = h$, and using the formula $v^2 = u^2 + 2as$, where $a = g$, we have

$$v^2 = 0 + 2gh$$

i.e. $v = \sqrt{(2gh)}$

(**Example**)

A body is fired vertically upwards with initial velocity 100 m/s.

    (a) How far will it rise?
    (b) How long will it take to reach maximum height?
    (c) What is the time taken before it strikes the ground again?
    (d) With what velocity will it strike the ground?

Solution

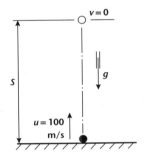

Fig. 7.19

    (a) Motion upward, Fig. 7.19.

Initial velocity $u = 100$ m/s.

Acceleration $a = -g$ (i.e. a retardation of 9·8 m/s²).

When maximum height is reached the velocity is instantaneously zero, thus $v = 0$.

$$v^2 = u^2 + 2as$$

Therefore

$$0 = 100^2 + [2 \times (-9\!\cdot\!8) \times s]$$

i.e. $s = \dfrac{100^2}{2 \times 9\!\cdot\!8} = 510$ m

    (b) Time taken to reach maximum height.

$$v = u + at$$

Therefore

$$0 = 100 - 9\!\cdot\!8t$$

i.e. $t = \mathbf{10\!\cdot\!2\,s}$

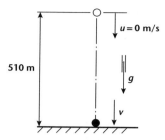

Fig. 7.20

(c) Time to fall 510 m from rest, Fig. 7.20.

Initial velocity $u = 0$; $s = 510\,m$; $a = +g = 9{\cdot}8\,\text{m/s}^2$

Using

$$s = \tfrac{1}{2}at^2$$

$$510 = \tfrac{1}{2} \times 9{\cdot}8 \times t^2$$

i.e. $t = \mathbf{10{\cdot}2\,s}$

Thus the time to fall from maximum height is the same as the time to rise to maximum height. This is a useful fact to remember.

The total time taken to fall back is $10{\cdot}2 + 10{\cdot}2 = \mathbf{20{\cdot}4\,s}$.

(d) Velocity on striking the ground:

$$v^2 = u^2 + 2as$$

Therefore

$$v^2 = 2 \times 9{\cdot}8 \times 510$$

since $u = 0$ at maximum height, i.e.

$$v = \mathbf{100\,m/s}$$

This is the *same* velocity as that with which the body was thrown upwards.

Example

A body is thrown vertically upwards with initial velocity 100 m/s. Calculate the time taken to pass a point 120 m above the ground.

Solution

$$s = 120\,\text{m}; \; u = 100\,\text{m/s}; \; a = -9{\cdot}8\,\text{m/s}^2$$

$$s = ut + \tfrac{1}{2}at^2$$

Therefore

$$120 = 100t - \tfrac{1}{2} \times 9{\cdot}8 \times t^2$$

Rearranging,

$$4 \cdot 9t^2 - 100t + 120 = 0$$

Solving,

$$t = 1 \cdot 3\,s \quad \text{or} \quad 19\,s$$

The first answer, 1·3 s, is the time taken to reach a height of 120 m on the way up; the second is the *total* time taken to pass the same point on the way down. The student should check this by making use of the method of the previous example to find the time of fall from maximum height to a point 120 m above the ground.

Problems

7.48 A hammer initially at rest falls freely a height of 2·4 m on to a pile to drive it into the ground. What is its impact velocity?   (6·9 m/s)

7.49 A shell is fired vertically upwards at a velocity u m/s from the ground. What value of u is required for the shell to reach a maximum height of 1000 m, and how long does it take to travel to this point?   (140 m/s; 14 s)

7.50 A shell is fired vertically upwards with a muzzle velocity of 300 m/s. Find the height reached when its speed falls to 120 m/s. What is the total height attained and the total time that elapses before the shell reaches the ground?
(3·9 km; 4·59 km; 61 s)

7.51 A rocket carrying a cargo lifts off vertically from rest on the ground with a constant acceleration of 8 m/s² for 20 s of flight before rocket and cargo separate. The cargo is released to continue its vertical flight freely upwards under gravity only while the rocket is fired off to earth. For the cargo, find (i) the height at which it is released, (ii) the total height it attains and (iii) the total time that elapses from ground to ground.   (1·6 km; 2·9 km; 61 s.)

7.52 A lunar module approaches the moon's surface to land. At a height of 2 m above the ground it has a vertical downwards velocity of 1·4 m/s and completes the landing by free fall. The acceleration due to gravity at the moon's surface is 1·62 m/s². What is the impact velocity?   (2·91 m/s)

7.18   Projectiles

A projectile is a body given an initial velocity and allowed to move freely through the air. It is then acted upon by gravity and air resistance only. For simplicity, air resistance is neglected and thus the only remaining cause of acceleration is the earth-pull. The projectile therefore accelerates downwards *but has no horizontal acceleration*; though it may have a horizontal velocity due to the initial projection.

7.19 Range

Consider a body projected with initial velocity V at angle θ to the horizontal. The path of its motion will be ACB, Fig. 7.21. In order to find its *range* AB over level ground, we carry out the calculation in the following steps:

1. Resolve V into horizontal and vertical components respectively. Thus

    horizontal component **oa** = $V \cos \theta$

    vertical component **ab** = $V \sin \theta$

2. Calculate the *time of flight t*. Since only the vertical component has any acceleration ($a = -g$ downwards) *the time of flight is the same as that for a body projected vertically upward with velocity V* sin θ; i.e. the time for a body to rise to the maximum height DC and return to earth again.

3. Since the projectile has no horizontal acceleration the horizontal component of velocity $V \cos \theta$ is *constant*. The horizontal range is therefore given by this velocity acting for the time of flight t, i.e.

    range = $V \cos \theta \times t$

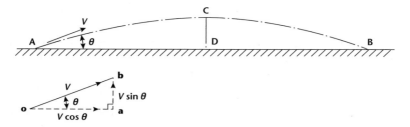

Fig. 7.21

(Example)

A shell is fired at 30° to the horizontal with velocity 20 m/s from the top of a cliff 50 m high. Calculate the distance from the foot of the cliff to the point where the shell strikes the ground.

Solution

The path of the shell is ABC, Fig. 7.22. The range required is DC, and B marks the point of maximum height.

Horizontal component of initial velocity = 20 cos 30°

= 17·32 m/s

If t is the time of flight in seconds,

range = DC = 17·32t m

Vertical component of velocity = 20 sin 30°

= 10 m/s

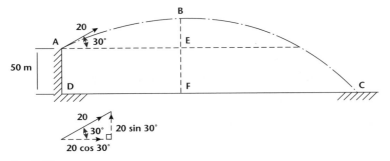

Fig. 7.22

The time of flight required is the time t for a body projected *vertically* upward with velocity 10 m/s to travel from E to B and fall to F.

At B, $v = 0$; the time from E to B is given by

$$v = u + at$$

i.e. $0 = 10 - 9 \cdot 8t$

Therefore

$$t = 1 \cdot 02 \, s$$

To find distance BE $= s$, we use

$$v^2 = u^2 + 2as$$

i.e. $0 = 10^2 + 2(-9 \cdot 8)s$

Therefore

$$BE = s = 5 \cdot 1 \, m$$

Total distance

$$BF = BE + EF$$
$$= 5 \cdot 1 + 50$$
$$= 55 \cdot 1 \, m$$

The time to fall this distance is given by

$$s = \tfrac{1}{2}at^2$$

i.e. $55 \cdot 1 = \tfrac{1}{2} \times 9 \cdot 8 \times t^2$

so that

$$t = 3 \cdot 35 \, s$$

Total time of flight $= 3 \cdot 35 + 1 \cdot 02$

$$= 4 \cdot 37 \, s$$

Therefore

$$\text{range DC} = 17.32 \times 4.37$$

$$= 75.7\,\text{m}$$

Alternatively, the time of flight may be obtained from the formula

$$s = ut + \tfrac{1}{2}at^2$$

where, considering *upwards* motion from E,

$$s = -50\,\text{m for point E}; \quad a = -9.8\,\text{m/s}^2; \quad u = 10\,\text{m/s}$$

Hence

$$-50 = 10t - \tfrac{1}{2}9.8t^2$$

i.e. $4.9t^2 - 10t - 50 = 0$

Therefore

$$t = 4.37\,\text{s, as before}$$

(Example)

Calculate the angle of elevation of a gun to fire 60 km with a muzzle velocity of 1000 m/s.

Solution

Let θ = angle of elevation. Vertical component of muzzle velocity = 1000 sin θ m/s. Time to reach maximum height is given by

$$v = u + at$$

i.e. $0 = 1000 \sin\theta - 9.8t$

Therefore

$$t = 102 \sin\theta \text{ seconds}$$

The total time of flight is twice this value, i.e. 204 sin θ.

Horizontal component of muzzle velocity = 1000 cos θ m/s

$$\text{Range} = 1000 \cos\theta \times \text{time of flight}$$

$$= 1000 \cos\theta \times 204 \sin\theta$$

But also

$$\text{range} = 60\,\text{km} = 60\,000\,\text{m}$$

Therefore

$$1000 \times 204 \sin\theta \cos\theta = 60\,000$$

$$\sin\theta \cos\theta = 0.2941$$

i.e. $2 \sin\theta \cos\theta = 0.5882$

or

$$\sin 2\theta = 0{\cdot}5882$$

$$2\theta = 36° \quad \text{or} \quad 144°$$

i.e. $\theta = 18° \quad \text{or} \quad 72°$

It should be noted that there are two answers for the angle of elevation and that these are complementary, i.e. they add up to 90°. Hence the same range is obtained whether the gun is fired at 18° to the horizontal or at the same angle to the vertical.

Example

A shell is fired from a gun on level ground with a muzzle velocity v. Find the angle of projection θ for the horizontal range to have its greatest value. If $v = 450 \, m/s$ what is the maximum range and the time of flight?

Solution
Using the notation of section 7.19 and referring to Fig. 7.21, the equation for vertical upward flight when the initial velocity is $v \sin \theta$ and the final velocity is zero, is

$$0 = v \sin \theta - gt$$

i.e. $t = \dfrac{v \sin \theta}{g}$

Therefore,

$$\text{\textit{total} time of flight} = 2t = \frac{2v \sin \theta}{g}$$

Horizontal component of velocity (constant) $= v \cos \theta$

Therefore,

range, $s = v \cos \theta \times$ time of flight

$$= v \cos \theta \times \frac{2v \sin \theta}{g}$$

$$= \frac{v^2 \sin 2\theta}{g}$$

since $2\sin \theta \cos \theta = \sin 2\theta$. The range is a maximum when $\sin 2\theta = 1$, i.e. when $\theta = 45°$.

Therefore,

$$\text{maximum range} = \frac{v^2}{g} = \frac{450^2}{9{\cdot}8} = 20{\cdot}7 \times 10^3 \, m$$

$$= 20{\cdot}7 \, km$$

$$\text{time of flight} = \frac{2v \sin \theta}{g} = \frac{2 \times 450 \sin 45°}{9.8}$$

$$= 65 \, \text{s}$$

Problems

7.53 A shell is fired at a velocity of 200 m/s at 30° to the horizontal. Find its range on the horizontal plane. (3·54 km)

7.54 A body is projected horizontally from a cliff 50 m high. It lands a distance 1000 m from the bottom of the cliff face. Calculate the initial velocity of projection. (313 m/s)

7.55 A shell is fired at an angle of 20° to, and above, the horizontal from the top of a cliff 90 m high. The velocity of the shell is 150 m/s. Calculate (a) the distance from the foot of the cliff where it strikes the ground and (b) the highest point above the ground reached by the shell. (1692 m; 225 m)

7.56 A stone is hurled downwards at 30° to, and below, the horizontal with velocity 10 m/s from a roof 15 m high. Calculate the time it is in the air and the horizontal distance it will travel. (1·3 s; 11·4 m)

7.57 Two projectiles are fired at elevations of 15° and 20° respectively to the horizontal. (a) If the initial velocity of projection is 500 m/s, calculate their distance apart when they land. (b) If the angles are 20° and 70°, calculate the new distance apart. (3·7 km; zero)

7.58 A jet of water issues horizontally from an orifice in the side of a water tank. It is seen to pass through a point 15 cm below the centre of the orifice and 45 cm horizontally from the tank. By considering each drop of water to be a projectile, calculate the jet velocity at the orifice. (2·6 m/s)

7.59 A small carton leaves a chute horizontally at 6 m/s. What is the maximum distance a receptacle may be placed at a point 5 m below the chute if it is to catch the carton? (6·1 m)

7.60 Calculate the angle of elevation of a gun to fire 6 km with a muzzle velocity of 600 m/s. (4·7°; 85·3°)

7.61 A gun is fired from level ground at different angles of projection. The greatest range is found to be 30 km. What is the corresponding muzzle velocity, time of flight and maximum height reached? (542 m/s; 78 s; 7·5 km)

Chapter 8

Force and acceleration

In Chapter 1 the quantities mass, weight and force were briefly defined, with their units, in order to deal with problems on rigid bodies at rest or in steady motion. Here, these quantities are again involved when rigid bodies are accelerated or decelerated and the relationships between a force and the change in motion it produces on a body are developed together with inertia, a property of all bodies. This is achieved without reference to Newton's important laws of motion which are deferred to Chapter 9. In our treatment of dynamics at this level we treat all bodies, whether vehicles, satellites or planets, as particles of a certain mass concentrated at the centre of mass of the body. Hoists, trains, vehicles etc, are used as examples of the application of the simple force–mass–acceleration (F = ma) equation and an elementary approach to aircraft and rockets serves to give further examples while introducing the concept of 'drag', the friction force produced by air or fluids.

8.1 Force and acceleration

If a force acting on a body is balanced by an equal and opposite force then the two forces are said to be in equilibrium. Since the forces are in balance, the body will not start to move, or if it is already moving it will not accelerate. On the other hand, an unbalanced force will cause the body to increase its velocity, that is, to accelerate. To calculate the acceleration of a body we use the following experimental facts:

1. If the unbalanced force is constant in magnitude and direction, the acceleration will be uniform.
2. The pull of the earth on a body is its weight W. If this pull is not balanced by an equal and opposite force it causes the body to fall with an acceleration $g = 9.8 \, \text{m/s}^2$ approximately (near the earth's surface).
3. If the acceleration of the *same* body due to a force F is a, then the acceleration a is to the acceleration g as the force F is to the weight W. Mathematically we have the simple proportion (Fig. 8.1)

$$\frac{a}{g} = \frac{F}{W}$$

Solving for the force,

$$F = \frac{W}{g} \cdot a$$

175

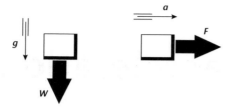

Fig. 8.1

Solving for the acceleration,

$$a = \frac{F}{W} \cdot g$$

When more than one force acts on the body, F is the *resultant* unbalanced force. The direction of the acceleration is always in the direction of the resultant unbalanced force. Both force and acceleration have magnitude, sense and direction and are therefore vector quantities.

8.2 Mass

Consider the ratio $F/a = W/g$. This ratio is found experimentally to be a constant for a given body. This constant will be denoted by m.

Since a force is required to accelerate a body, the quantity m is a measure of its 'reluctance to accelerate'. This reluctance is more properly called the *inertia* of the body and it is measured by the quantity m, which we call its *mass*.

Inertia is a characteristic property of all material bodies. Mass is a magnitude (a quantity of inertia) characteristic of a particular body. Mass is also sometimes defined as a 'quantity of matter'. In dynamics, however, a more satisfactory definition is expressed by the ratio already found:

$$\frac{F}{a} = \frac{W}{g} = m$$

Thus

$$F = ma$$

or

Force = mass × acceleration

Thus the greater the mass of a body the greater the force required to accelerate it (*see also* Ch. 9).

8.3 Units of mass and force

The SI standard of mass is the *kilogram* (kg) and this is a *fundamental* or *base* unit. The kilogram is defined as equal to the mass of the international prototype of the kilogram which is in the custody of the International Bureau of Weights and Measures at Sèvres, near Paris. Other units of mass are the *gram* (g) = 10^{-3} kg, and the *megagram* (Mg) or *tonne* (t) = 10^3 kg. Note that 1 Mg is 1000 units of mass, i.e. 1000 kg, although the prefix M indicates 10^6. This is because the base unit has the prefix *kilo* as the first part of its name.

The SI standard of force is the *newton* (N), and this is a *derived* unit. The newton is defined as *the force required to give a mass of 1 kg an acceleration of 1 m/s²*.

Thus from $F = ma$,

1 (newton) = 1 (kg) × 1 (metre per second per second)

i.e. **1 (N) = 1 (kg) × 1 (m/s²)**

Other units of force are the *kilonewton* (**kN**) = 10^3 N, the *meganewton* (**MN**) = 10^6 N, and the *giganewton* (**GN**) = 10^9 N.

(8.4) Mass and weight

Mass is the amount of matter in a body. Weight is the earth's gravitational force on a body. These are different physical quantities and in the SI system the units are different. However, although the units indicate clearly whether mass or weight is intended, care should be taken in the use of the words 'mass' and 'weight' since colloquially 'weight' is used as a synonym for 'mass'. For example, it is wrong to say that a body 'weighs 1 tonne'; either the body 'has a mass of 1 tonne' or 'weighs 9·8 kN'. The term *load* however may be used in connection with a mass or a force (weight). Statements such as 'a load of 100 kg' or 'a load of 1 MN' should not cause difficulty since the units indicate whether a mass or force is intended.

A beam balance compares the weights of two bodies and hence their masses since the weight of a body is proportional to its mass ($W/m = g$ = constant). A spring balance measures force and when used to compare the weights of bodies and hence their masses, it is being used as a substitute for a beam balance. A spring balance may therefore be calibrated in newtons for measurement of force (weight) or kilograms for measurement of mass. The reading in kilograms is only true for the particular value of g, the acceleration due to gravity, used in calibrating the spring. The error involved is very small since the value of g varies only by about 0·5 per cent over the earth's surface. For practical engineering purposes the value of 9·8 m/s² is satisfactory and the slightly more accurate value 9·81 m/s² may be used when required. For certain purposes a standard international value is required and this is 9·806 65 m/s².

The relation between the weight W of a body and its mass m (*see* page 3 and Fig. 1.2) is

$$W = mg$$

Thus

$$W \text{ (N)} = m \text{ (kg)} \times 9 \cdot 8 \text{ (m/s²)}$$

and

$$W \text{ (N)} = m \text{ (tonne)} \times 10^3 \times 9 \cdot 8 \text{ (m/s}^2)$$

or

$$W \text{ (kN)} = m \text{ (tonne)} \times 9 \cdot 8 \text{ (m/s}^2)$$

8.5 Accelerated motion of a body against a resistance

Let

m = mass of the body

E = applied force or *effort*

a = acceleration produced in the body

R = resistance to motion, assumed constant

The resultant unbalanced or accelerating force F acting on the body is the difference between the applied force E and the resisting force R (Fig. 8.2), i.e.

$$F = E - R$$

and

$$F = ma$$

Therefore

$$E - R = ma$$

or

$$E = ma + R$$

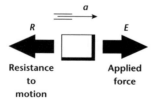

Fig. 8.2

The force E required to be applied to the body is therefore made up of two parts:

1. A force R required to overcome the resisting force; this force is required just to maintain the motion *without acceleration*.
2. A force ma required to accelerate the body, i.e. to overcome its inertia.

   It follows that, if a body is moving at constant velocity (a = 0), the applied force is equal and opposite to the external resisting force.

8.6 Tractive effort of a vehicle

A vehicle moving on the level at constant speed requires only a force to *maintain* the motion of the vehicle against the tractive resistance (*see* section 5.4) since there is no force required to accelerate the vehicle. The tractive effort at constant speed is therefore equal to the tractive resistance. When the vehicle is accelerated on the level, the tractive effort must overcome the tractive resistance and provide the accelerating force. On an incline, the tractive effort must also overcome the effect of the component of the weight down the slope.

Example

A train of total mass 100 t is travelling at 90 km/h on level track. The tractive resistance is 100 N/t. Calculate the braking effort required to stop the train in one minute.

Solution

The braking effort together with the tractive resistance provides the decelerating force.

90 km/h = 25 m/s

The deceleration is 25 m/s in 60 s, i.e. 25/60, or 0·417 m/s². The force required to produce this deceleration is

$ma = 100 \times 10^3 \times 0\cdot417 = 41\,700\,\text{N}$

Total tractive resistance = 100×100

$= 10\,000\,\text{N}$

Therefore braking effort required = $41\,700 - 10\,000 = 31\,700\,\text{N}$

$= \mathbf{32\,kN}$

Example

A train of mass 650 t is drawn by a locomotive of mass 50 t and exerts a tractive effort of 80 kN. If the tractive resistance is 65 N/t, find the time to reach a steady speed of 90 km/h from rest.

Solution

Accelerating force F = tractive effort − tractive resistance

$= (80 \times 10^3) - 65(650 + 50)$

$= 34\,500\,\text{N}$

Since

$$F = ma$$

$$34\,500 = 700 \times 10^3 a$$

Hence

$$a = 0{\cdot}0493 \text{ m/s}^2$$

From $v = at$, since $v = 90 \text{ km/h} = 25 \text{ m/s}$, we have

$$25 = 0{\cdot}0493 \times t$$

Therefore

$$t = 507 \text{ s}, \quad \text{or} \quad \textbf{8{\cdot}45 min}$$

Example

A crane lifts a load of 15 t at a steady speed of 0·1 m/s. Find (a) the power required at the driving motor and (b) the additional power required to give the load an acceleration of 1·5 m/s² when it is moving at 0·1 m/s.

Solution
(a) Since the load moves at *constant speed* the lifting force is equal to the weight of the load, i.e. $mg = 15 \times 9{\cdot}8 = 147 \text{ kN}$.
Therefore

$$\text{power} = \text{force (kN)} \times \text{speed (m/s) kW}$$

$$= 147 \times 0{\cdot}1$$

$$= \textbf{14{\cdot}7 kW}$$

(b) Accelerating force $= ma$

$$= 15 \times 1{\cdot}5$$

$$= 22{\cdot}5 \text{ kN}$$

This force is moving at 0·1 m/s; therefore

$$\text{additional power} = 22{\cdot}5 \times 0{\cdot}1 \text{ kW}$$

$$= \textbf{2{\cdot}25 kW}$$

Example

The cage of a hoist has a mass of 6000 kg when loaded. It is lifted vertically from rest by a wire rope and travels 20 m in 4 s. Assuming the acceleration to be uniform, and neglecting the weight of the rope, find the tension in the rope during the acceleration period. If the cage is braked to rest with a deceleration equal to the acceleration, find the tension in the rope during the braking period.

Solution
The tension in the rope is made up of two parts:

1. The tension required to balance the weight of the load.
2. The tension required to accelerate the load.

To find the acceleration, we have

$$s = \tfrac{1}{2}at^2$$

i.e. $20 = \tfrac{1}{2} \times a \times 4^2$

Hence

$$a = 2\cdot5\,\text{m/s}^2$$

Tension to accelerate cage $= ma = 6000 \times 2\cdot5$

$$= 15\,000\,\text{N}$$

Total tension in rope $= 15\,000 + (6000 \times 9\cdot8)$

$$= 73\,800\,\text{N} = \mathbf{73\cdot8\,kN}$$

Alternatively we can say that the net force in the direction of the acceleration (upwards) is the tension T, minus the dead weight of the cage W, and this is equal to the product *mass* × *acceleration*. Thus

$$T - W = ma$$

Therefore

$$T - (6000 \times 9\cdot8) = 6000 \times 2\cdot5$$

i.e. $T = \mathbf{74\,kN}$

When the cage is braked to produce a deceleration the net force acting in the direction of the acceleration (negative, downwards) is now $W - T$; thus

$$(6000 \times 9\cdot8) - T = 6000 \times 2\cdot5$$

$$T = 43\,800\,\text{N} = \mathbf{44\,kN}$$

Problems

8.1 Find the braking force to bring a car of mass 1 tonne to rest from 90 km/h on the level in 60 m with uniform deceleration.   (5·2 kN)

8.2 Find the average force required to accelerate a planing machine table from rest to a speed of 1·2 m/s in 2 s if the moving parts have a mass of 500 kg.   (300 N)

8.3 A small shaping machine tool is brought to rest in 80 mm from a velocity of 2 m/s. If the moving parts can be considered as a mass of 300 kg moving horizontally, calculate the average force required.   (7·5 kN)

8.4 A vehicle is accelerated from rest to 45 km/h in 15 s. If its mass is 4 t and the tractive resistance is 450 N, calculate the tractive effort and the maximum power exerted.   (3·8 kN; 47·3 kW)

8.5   A hoist lowers a load of 2 t at a uniform speed of 3 m/s by means of a vertical wire rope. Calculate the tension in the rope required to bring the load to rest in 4 s.   (21 kN)

8.6   A train has a speed of 60 km/h. Power is shut off and it comes to rest in a distance of 2 km. Calculate the average tractive resistance to motion in newtons per tonne mass.   (69·5 N/tonne)

8.7   The moving table of a machine tool of mass 300 kg slides on horizontal guides. The coefficient of friction between the table and guides is 0·12. Find the retarding force on the table if it is moving freely, and hence the time to bring it to rest from a speed of 0·8 m/s.   (353 N; 0·7 s)

8.8   A motor-driven winch raises a load of 1 tonne a vertical distance of 6 m from rest in 2 s with constant acceleration. Calculate (a) the maximum power exerted and (b) the average power exerted.   (76·8 kW; 38·4 kW)

8.9   A car exerts a constant tractive effort of 2000 N for 60 s. The mass of the car is 1 tonne and the rolling resistance is 120 N. Calculate the distance travelled from rest in this time.   (3·4 km)

8.10 A planing operation on a casting needs a cutting speed of 0·3 m/s. The machine table's mass is 600 kg and it must reach the required speed from rest in 0·6 s. Find the average acceleration required, the force required to produce this acceleration and the maximum power developed.

(0·5 m/s^2; 300 N; 90 W)

8.11 A lift cage has a mass of 800 kg. Find the tension in the lifting cable when the cage is (i) moving at constant speed, (ii) moving upwards and accelerating at 1·5 m/s^2, (iii) moving upwards and decelerating at 1·5 m/s^2 and (iv) moving downwards and accelerating at 1·5 m/s^2.

(7·84 kN; 9·04 kN; 6·64 kN; 6·64 kN)

8.12 A ship travels at an average speed of 10 knots using 85 per cent of the 13 MW power available. Calculate the average water resistance. 1 knot = 0·514 m/s.

(2·15 MN)

8.13 A loaded mine skip has a mass of 15 t including an allowance for the lifting cables. During winding, the skip has a constant acceleration of 1·2 m/s^2 at the beginning and the same retardation at the end, with a constant velocity of 15 m/s between these periods. Find the tension in the cables during the acceleration, constant velocity and deceleration periods. What is the power required during the period of constant velocity?

(165 kN; 147 kN; 129 kN; 2·21 MW)

(8.7) Body accelerating down an inclined plane

Figure 8.3 shows a body of mass m in motion down an inclined plane. The acceleration of the body is a. The forces acting on the body are the resistance R,

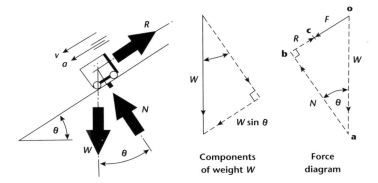

Fig. 8.3

the weight W and the normal reaction N of the plane on the body. Resolving forces parallel to the slope, the resultant unbalanced force is

$$F = W \sin \theta - R$$

or, from the force diagram,

$$F = \mathbf{oc}$$

$$= \mathbf{ob} - \mathbf{bc}$$

$$= W \sin \theta - R$$

The accelerating force is therefore the difference between the component of the weight down the slope and the resistance.

Now

$$F = ma$$

Therefore

$$ma = W \sin \theta - R$$

If there is no resistance to motion, the resultant force F causing the acceleration a is the component $W \sin \theta$ of the weight down the slope, i.e.

$$F = W \sin \theta$$

Therefore

$$ma = W \sin \theta = mg \sin \theta$$

or

$$a = g \sin \theta$$

8.8 Body accelerating up an inclined plane

Figure 8.4 shows a body in motion up an inclined plane under the action of a force E (tractive effort) applied parallel to the plane. The resistance to motion is R. The resultant force F $(= ma)$ causing the body to accelerate up the plane is found by resolving forces parallel to the slope, or alternatively, from the force diagram shown. Thus

$$F = \mathbf{od}$$
$$= \mathbf{oa} - \mathbf{ac} - \mathbf{cd}$$

Therefore

$$ma = E - W \sin \theta - R$$

or

$$E = W \sin \theta + R + ma$$

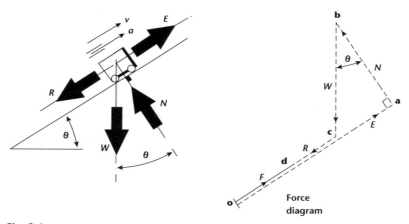

Fig. 8.4

The tractive effort E is seen to be made up of three parts:

1. The force $W \sin \theta$ required to support the body on the plane.
2. The force R required to overcome the external resistance.
3. The force ma required to accelerate the body.

Example

A car for a scenic railway when occupied has a mass of 600 kg. During its run it ascends an incline 25 m long which makes an angle of 15° with the horizontal. If the speeds of the car at the start and finish of the incline are 15 m/s and 18 m/s respectively, find the rail friction resistance in newtons per tonne.

Solution

$s = 25\,\text{m};\ u = 15\,\text{m/s};\ v = 8\,\text{m/s}$

To find the deceleration a,

$v^2 = u^2 - 2as$

$8^2 = 15^2 - (2 \times a \times 25)$

Therefore

$a = 3{\cdot}22\,\text{m/s}^2$

The resultant decelerating force F is made up of two parts, the resistance R and the resolved part of the weight down the slope.
Therefore

$F = R + 600 \times 9{\cdot}8 \sin 15°$

and

$F = ma$

Hence

$R + (600 \times 9{\cdot}8 \times 0{\cdot}2588) = 600 \times 3{\cdot}22$

or

$R + 1522 = 1932$

so that

$R = 410\,\text{N}$

Therefore,

$$\text{resistance per tonne of car} = \frac{410}{0{\cdot}6} = 680\,\text{N}$$

(**Example**) _____

A car of total mass 1100 kg ascends an incline of 1 in 4·4 against a constant frictional and wind resistance of 200 N, and increases its speed uniformly from 54 km/h to 72 km/h in a distance of 175 m. Find the tractive effort employed and the maximum power developed.

Solution

$s = 175\,\text{m};\ u = 15\,\text{m/s};\ v = 20\,\text{m/s}$

To find the acceleration a,

$v^2 = u^2 + 2as$

i.e. $20^2 = 15^2 + (2a \times 175)$

Therefore

$a = 0{\cdot}5\,\text{m/s}^2$

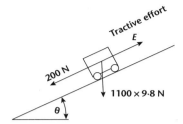

Fig. 8.5

Resolving along the plane, Fig. 8.5, the resultant accelerating force is

$$F = E - (1100 \times 9 \cdot 8) \sin \theta - 200$$

Also

$$F = ma = 1100 \times 0 \cdot 5 = 550 \, \text{N}$$

Hence

$$E - \left(1100 \times 9 \cdot 8 \times \frac{1}{4 \cdot 4}\right) - 200 = 550$$

so that $E = \textbf{3200 N}$.

$$\text{Maximum power} = \frac{\text{force (N)} \times \text{max. speed (m/s)}}{1000} \, \text{kW}$$

$$= \frac{3200 \times 20}{1000} = \textbf{64 kW}$$

Example

A train of mass 500 t ascends a slope of 1 in 150 with a uniform acceleration of $0 \cdot 1 \, \text{m/s}^2$. When the speed is 36 km/h the power is 1200 kW. Calculate the total tractive resistance and express it in newtons per tonne. If the power is suddenly cut off at this speed, calculate the distance the train travels up the incline before coming to rest.

Solution

$$v = 36 \, \text{km/h} = 10 \, \text{m/s}$$

If the tractive effort $= E$ (newtons), then

$$\text{power} = \frac{E \times v}{1000} \, \text{kW}$$

i.e. $1200 = \dfrac{E \times 10}{1000}$

so that

$$E = 120\,000 \, \text{N}$$

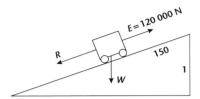

Fig. 8.6

From Fig. 8.6, tractive effort = component of weight down slope + resistance + force required for acceleration.

$$120\,000 = \left(500 \times 10^3 \times 9\cdot8 \times \frac{1}{150}\right) + R + (500 \times 10^3 \times 0\cdot1)$$

$$= 32\,700 + R + 50\,000$$

Therefore

$$R = 37\,300\,\text{N} = \mathbf{37\cdot3\,kN}$$

Hence

$$\text{resistance per tonne} = \frac{37\,300}{500} = \mathbf{74\cdot6\,N}$$

When power is cut off the tractive effort is zero and the train slows down due to the resistance and the component of the weight down the incline.
Therefore

$$\text{decelerating force} = 32\,700 + 37\,300 = 70\,000\,\text{N}$$

From

$$F = ma$$

$$70\,000 = 500 \times 10^3 \times a$$

Hence

$$a = 0\cdot14\,\text{m/s}^2$$

To find distance s travelled after power is cut off,

$$v = 0;\ u = 10\,\text{m/s}$$

$$v^2 = u^2 - 2as$$

$$0 = 10^2 - (2 \times 0\cdot14 \times s)$$

Therefore

$$s = \mathbf{357\,m}$$

8.14 Calculate the time taken for a wheeled vehicle to roll freely 300 m from rest down an incline of 1 in 100. Road resistance may be neglected.   (78.2 s)

8.15 Find the tractive effort required parallel to the slope to pull a wagon of mass 3·5 t a distance 300 m up an incline of 1 in 100 in 78·2 s. Neglect resistance.
(686 N)

8.16 A vehicle of mass 3·5 t is driven 300 m from rest down an incline of 1 in 100. If the road resistance amounts to 100 N/t, calculate the tractive effort required if the time taken is 78·2 s. If it is driven 300 m from rest up the incline in the same time, determine the new effort required.
(350 N; 1·03 kN)

8.17 A load of 3·5 t is dragged up an incline of 1 in 100 a distance of 300 m from rest in 78·2 s. Calculate the tractive effort required parallel to the slope if the coefficient of friction is 0·1. What is the work done in *accelerating* the load?
(4 kN; 103 kJ)

8.18 A load of 1 t is pulled along level ground by a force of 5 kN at an angle of 30° to, and above, the ground. The coefficient of friction between the load and the ground is 0·1. Calculate (a) the friction force acting and (b) the horizontal acceleration of the body.   (730 N; 3·6 m/s^2)

8.19 A haulage rope is to pull a load of 10 t up an incline of 1 in 50 with an acceleration of 1·2 m/s^2. Find the pull in the rope parallel to the incline (a) neglecting resistance and (b) if the frictional resistance to motion is 250 N/t.
(14 kN; 16·5 kN)

8.20 Calculate the braking effort required to stop a train of mass 400 t in 400 m when travelling at 80 km/h down an incline of 1 in 120. The tractive resistance is 100 N/t.   (240 kN)

8.21 A horizontal piston of diameter 40 mm is subjected to a pressure of 7 MN/m^2. The force in the connecting rod is 7 kN when inclined at 15° to the horizontal. If the piston's mass is 2 kg calculate its acceleration at this instant. Neglect friction.   (1020 m/s^2)

8.22 A steel component of mass 5 kg is held on an inclined machine slide by a ratchet. If the inclination of the slide to the horizontal is 25° and the coefficient of friction between component and slide is 0·08, find the time taken for the component to travel 2 m down the slide when the ratchet is lifted.   (1·1 s)

8.23 A train of mass 500 t ascends a slope of 1 in 75 increasing its speed from 15 to 45 km/h in 100 s. The tractive resistance to motion is 100 N/t. Calculate the power developed at maximum speed.   (1.96 MW)

8.9 Forces on an aircraft in flight

The equation of motion $F = ma$ and the work on equilibrium of forces will now be applied to the basic mechanics of an aeroplane in flight.

Introduction

To be airborne and fly, an aeroplane must be provided by the wings with a lifting force equal to, or greater than, its deadweight. The flow of air over the wings in streamlines generates a force because of a slight difference in pressure at the top and bottom skins. The resultant force due to this air pressure acts through the *centre of pressure*, a point which is usually fairly close to the centre of gravity (*see* section 22.8). The component of this force at right angles to the airflow, or more simply the line of flight, is the *lift*, and the component parallel to the line of flight, opposing the motion of the aeroplane is the *drag*. The drag force has to be overcome by the propelling force or *thrust* produced by the engine. As the speed of an aeroplane increases the resultant force increases and therefore both the lift and drag forces increase. The total drag force is the resultant of several types of drag and acts along a mean drag line. The lift and drag forces depend on a number of factors such as the shape, frontal area and speed of the aeroplane, as well as the density of the air.

Another condition for flight depends on the *angle of attack*, the angle at which the wings meet the oncoming airstream into which the aeroplane points. The angle of attack is relative to the airflow, not to the ground. The pressure distribution of the air on the wings changes when the angle of attack alters, hence the lift and the position of the centre of pressure alter in flight.

For an aeroplane to fly at any altitude, therefore, the airspeed and the angle of attack must be such as to ensure sufficient lifting force. The maximum speed is limited by the thrust available from the engine. The minimum speed is determined by the greatest angle of attack possible before the airflow over the wings becomes turbulent. At this angle (about 15°) the aeroplane stalls aerodynamically when the lift is not large enough to maintain it in its flight condition. In straight, level flight, there is a range of speed possible with the maximum speed several times greater than the minimum and there is a speed where the lift/drag ratio is a maximum. The designer's aim is to achieve as large a lift as possible with minimum drag for a required range of speed, usually at an angle of attack of about 4°. A typical lift/drag ratio for a light aeroplane in level flight is 10:1, but the ratio for some aircraft in certain conditions may be as low as 3:1.

The four forces

Four main forces act on an aeroplane in flight, Fig. 8.7:

1. **Weight, *W***, acting vertically downwards through the centre of gravity and assumed to be the all-up weight at take-off although there is a gradual loss of weight due to burning of fuel.

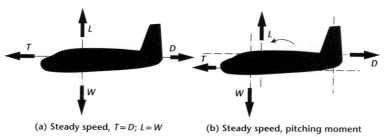

(a) Steady speed, $T = D$; $L = W$ (b) Steady speed, pitching moment

Fig. 8.7

2. **Thrust, T,** due to the jets or propellers, assumed to act through the central longitudinal axis of the aeroplane. The calculation of thrust is dealt with in Chapter 9.
3. **Lift, L,** the component of the resultant force due to the air, acting at right angles to the line of flight through the centre of pressure.
4. **Drag, D,** the component of the resultant force due to the air, acting parallel to the line of flight through the centre of pressure, opposing the motion of the aeroplane.

The four forces seldom act through the same point. The position of the centre of gravity depends on passenger and freight loading and alters as fuel is burned up in flight. As explained above, the centre of pressure also moves its position in flight. Design and loading arrangements try to ensure that these two points only shift within prescribed limits. Figure 8.7(b) shows the usual relative positions of the lines of action of the four forces for straight and level flight at constant speed. The lift acts behind the weight and the thrust below the drag. The effect, since the lift and weight forces are larger than the other forces, is a nose-down pitching moment on the aeroplane. This necessitates a balancing moment provided by a vertical downwards force at the tailplane. This force is produced by the airflow past elevators hinged on at the trailing edges of the tailplane. The moment-arm of the correcting tailplane load is relatively large so that only a small force is needed. (See the worked example on equilibrium of forces, page 36.)

The tailplane elevators control fore-and-aft movement and therefore maintain longitudinal stability. To maintain level flight at any speed requires the adjustment of the elevators to give a tailplane load to match the conditions. If the thrust exceeds the drag, the aeroplane accelerates but the increase in speed generates greater lift and the machine starts to climb. The elevators are then lowered producing an upwards tailplane load thereby depressing the nose of the aeroplane to sustain level flight.

For further background information on the mechanics of flight students are referred to A. C. Kermode (1995) *Mechanics of Flight* 10th edn, Harlow: Addison Wesley Longman.

In the problems set here the lift is assumed to act through the centre of gravity and the thrust and drag forces to directly oppose each other.

Level flight

In straight, level flight at constant speed, for balance of forces, Fig. 8.7(a):

Thrust T = drag D

and

Lift L = weight W

When the thrust exceeds the drag the plane accelerates, Fig. 8.8, and the accelerating force is given by

$F = T - D = ma$

where m is the mass of the aeroplane and a the acceleration.

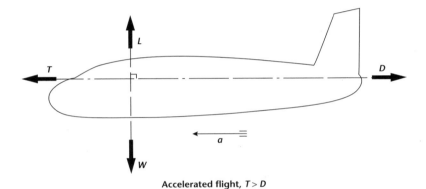

Accelerated flight, $T > D$

Fig. 8.8

Climbing

When climbing at constant speed, the thrust has to overcome the component of the weight parallel to the line of flight, as well as the drag force, and the lift is equal to the component of the weight ($W \cos \theta$) at right angles to the line of flight, Fig. 8.9. Thus

$T = D + W \sin \theta$

and

$L = W \cos \theta$

For an accelerated climb, the thrust must exceed the drag and component of weight combined, i.e.

$F = T - D - W \sin \theta = ma$

and as before

$L = W \cos \theta$

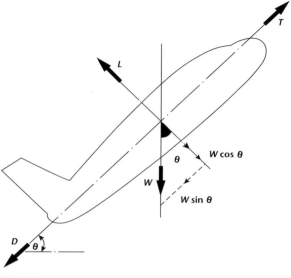

Fig. 8.9

Thrust and power

There is no output shaft in a jet engine so that measurement of power cannot be carried out as in the case of a reciprocating engine. Thrust, therefore, is used instead of power to assess the performance of a jet engine. For a propeller-driven aeroplane, the power output at the engine shaft can be measured and also the power delivered to the propeller shaft. The efficiency with which a propeller converts the work done by the turning-moment on the shaft to work done by the forward thrust can be estimated and hence the thrust determined. For any aeroplane the power output is given by the product *thrust × forward speed*.

Example

A jet aircraft of mass 6·43 t travels in straight, level flight at constant speed, Fig. 8.7(a). The lift:drag ratio is 13. Find the lifting force and the thrust. If the thrust is increased by 50 per cent and the aircraft is maintained in level flight what is its acceleration? Assume the drag remains constant.

Solution

At constant speed the lift must equal the weight and the thrust equal the drag. Thus the lift is

$$L = mg = 6·43 \times 9·8 = 63 \text{ kN}$$

Since the lift:drag ratio is 13, therefore

$$\text{drag} = \frac{63}{13} = 4·85 \text{ kN}$$

Therefore

  thrust = **4·85 kN**.

When the thrust is increased by 50 per cent, the accelerating force along the line of flight is given by

  F = thrust − drag

   = 1·5 × 4·85 − 4·85

   = 2·425 kN

and

    $F = ma$

i.e.   2·425 = 6·43 × a

therefore

    $a = \mathbf{0{\cdot}4\,m/s^2}$

(It is convenient in equation $F = ma$ to take the force F in kN and the mass m in tonnes.)

(**Example**)

A jet-plane of mass 9 t climbs in a straight line at an attitude of 40° to the horizontal, Fig. 8.9(b). At the start of the climb the speed is 60 m/s, the thrust 70 kN and the drag force 8 kN. Find the lifting force, acceleration and vertical distance attained in 30 s, assuming the thrust, drag and angle of climb remain constant.

Solution
Resolving forces normal to the line of flight:

  $L = mg\cos\theta = 9 \times 9{\cdot}8\cos40° = \mathbf{67{\cdot}6\,kN}$

Consider forces acting along the line of flight. The accelerating force is given by

  F = thrust − drag − weight component ($mg\sin\theta$)

   = 70 − 8 − (9 × 9·8 × sin 40°)

   = 5·3 kN

and

    $F = ma$

hence

    5·3 = 9 × a

therefore

    $a = \mathbf{0{\cdot}59\,m/s^2}$

The distance travelled along the line of flight in 30 s is

$$s = ut + \tfrac{1}{2}at^2$$

$$= 60 \times 30 + \tfrac{1}{2} \times 0\cdot59 \times 30^2$$

$$= 2065 \, \text{m} \quad \text{or} \quad 2\cdot07 \, \text{km}$$

Hence the vertical distance attained is $2\cdot07 \sin 40° = \mathbf{1\cdot33\,km}$.

(**Example**)
_____

A propeller-driven aircraft of all-up weight 24 kN flies at a steady 180 km/h in straight, level flight. The power delivered by the engine to the propeller-shaft is 140 kW and the propeller efficiency is 80 per cent. Find the thrust and lift : drag ratio. If the wing area is 36 m², estimate the wing-loading, i.e. the normal pressure upwards on the wings.

Solution

Power output = power at propeller-shaft × propeller efficiency

$$= 140 \times 0\cdot8 \, \text{kW}$$

$$= 112 \, \text{kW}$$

The power output is given by the product *thrust × forward speed*, hence

Thrust × forward speed $= 112 \times 10^3 \, \text{W}$

i.e.   thrust $\times \dfrac{180}{3\cdot6} = 112 \times 10^3$

Therefore

$$\text{thrust} = 2\cdot24 \times 10^3 \, \text{N} = \mathbf{2\cdot24\,kN}$$

In straight, level flight at constant speed, the lift is equal to the deadweight and the drag is equal to the thrust, therefore

$$L = 24 \, \text{kN}$$

$$D = 2\cdot24 \, \text{kN}$$

and

$$\frac{L}{D} = \frac{24}{2\cdot24} = \mathbf{10\cdot7}$$

$$\text{Wing-loading} = \frac{\text{lift}}{\text{wing area}} = \frac{24}{36} = \mathbf{0\cdot67\,kN/m^2}$$

8.24 A four-engined airliner of mass 90 t cruises in straight, level flight. Each engine develops a thrust of 22 kN. Find the lift : drag ratio. (10)

8.25 Find the acceleration of a 5 t jet-plane in straight, level flight at a point when the thrust is 12 kN and the lift : drag ratio is 14. $(1.7 \, \text{m/s}^2)$

8.26 An aircraft of mass 5 t climbs steadily at 20° to the horizontal, the average drag force is 5 kN. Find the thrust and lifting force. (21·8 kN; 46 kN)

8.27 The average resisting force during the take-off run of 1·7 km for a 190 t transporter is 40 kN and the average thrust 350 kN. Find the take-off speed. (268 km/h)

8.28 A 10 t jet-plane touches down in still air at 30 m/s on a level runway. The forward thrust on approach is 11 kN. From the moment of landing the thrust is deflected to produce a reverse thrust giving a braking force of 40 per cent of the forward thrust. Find the minimum length of runway required and the time taken for the plane to come to rest neglecting rolling resistance. (1·02 km; 68 s)

8.29 An aircraft of weight 50 kN climbs steadily at 15° to the horizontal. The thrust is 19 kN. What is the lifting force normal to the line of flight and the drag force along the line of flight in these conditions? Find the thrust needed to produce an acceleration of 1·5 m/s² assuming the drag force and the angle of climb to be constant. (48·3 kN; 6·1 kN; 26·7 kN)

8.30 A 20 t jet-plane climbs at a constant speed of 120 km/h against a drag force of 15 kN when propelled by a thrust of 100 kN. Find the angle of climb. If the thrust is increased by 10 per cent without change in the drag force or angle of climb, find the acceleration and the distance travelled along the line of flight in one minute from the moment the plane accelerates. (26° to horizontal; 0·5 m/s²; 2·9 km)

8.31 An aircraft of mass 68·5 Mg touches down at 210 km/h on a level runway with negligible vertical component of velocity. The average braking force is 85 kN. What is the minimum runway required and time taken to come to rest? (1·37 km; 47 s)

8.32 A 10 t aeroplane is diving at 50° to the vertical at 100 m/s when the engine cuts out. The aerodynamic drag and braking force amount to 70 kN. Find the speed of the plane when it loses 1 km in height. (88·4 m/s)

8.33 An aeroplane of mass 5·5 t lands on a level runway at 75 knots horizontal speed and takes 42 s to come to rest. Find the average braking force. 1 knot = 0·514 m/s. (5·05 kN)

8.34 A light aeroplane cruises in straight, level flight. The lift : drag ratio is 9. The total wing area is 50 m² and the wing-loading, i.e. the normal pressure upwards on the wings, is 0·85 kN/m². Find the thrust. (4·72 kN)

8.35 A propeller-driven aircraft travels in straight, level flight at a steady speed of 220 km/h. The engine delivers 900 kW to the propeller-shaft and the propeller efficiency is 80 per cent. What is the thrust exerted? If the plane weighs 25 kN and climbs steadily at 15° to the horizontal, what is then the thrust required, assuming the drag force remains unchanged? (11·8 kN; 18·3 kN)

8.36 A light aircraft has a groundspeed of 240 km/h against a headwind of 48 km/h. What is the plane's airspeed? If the thrust is 10.3 kN and the efficiency of the propellers is 90 per cent, what is the power supplied to the propeller shaft? (288 km/h; 916 kW)

8.37 A single-engined propeller-driven aeroplane flies at 600 km/h in still air in straight, level flight. The forward thrust of the propeller is 4 kN and the efficiency of the propeller in converting the power supplied to the propeller-shaft to useful work done in the forward direction is 85 per cent. Find the power supplied to the propeller-shaft. (784 kW)

8.38 A helicopter of mass 8 t is travelling at a steady forward speed in level flight. What is the lifting force of the air? If the total reaction of the air on the machine is $R = 90$ kN acting at angle θ to the vertical, Fig. 8.10, find the value of θ and the resisting force (drag) of the air along the line of flight.
(78·4 kN; 29·4°; 44·2 kN)

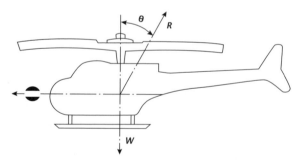

Fig. 8.10

8.10 **Forces on a rocket**

A rocket motor generates a propulsive force or *thrust* by burning fuel and oxygen (called the *propellants*), and ejecting the burnt gases as a high-speed jet. The calculation of the thrust and further work on rockets will be found in Chapter 9. Enormous thrusts can be generated to launch several thousand tonnes of space vehicle and, at the other end of the scale, small 'thrusters' are used to control the motion of satellites and space probes. To increase the performance of a rocket system so as to enable a space craft to leave the earth's gravitational field, *staging* is necessary. A stage consists of one or more rockets which burn out and the casings, equipment and debris are then jettisoned. For example, in the case of the Space Shuttle two rocket boosters fire during the first stage of ascent for about 130 s and then the empty casings together with cameras and other such equipment, totalling about 150 t, are jettisoned. Some minutes later the main external tank of mass 30 t is discarded. The propellants make up the bulk of any rocket system and as they burn up there is a continuous reduction in mass. The thrust for a stage remains fairly constant and thus the acceleration of the reducing mass greatly increases as the flight proceeds. The acceleration due to gravity g decreases with altitude so for this further reason the weight decreases, the effect of the thrust therefore increases and greater acceleration results. The air resistance, however, decreases with altitude but increases with speed in the earth's atmosphere. For these reasons, the calculation of the acceleration of a rocket system is complex. However, providing the forces and mass involved are known exactly at any point, the equation of motion $F = ma$ can be applied directly as in previous problems in this chapter. Two such points are lift-off and, less clearly, the end of a stage where one set of rockets burn out and the vehicle then coasts momentarily before another set of rockets fire. The exact calculation of velocity achieved requires knowledge of the acceleration over the entire flight, having taken account of air resistance, changes in mass and acceleration due to gravity and other complex factors. In a simple approach an average or constant acceleration is assumed.

Calculation of acceleration at lift-off and burn-out

When a rocket of mass m is fired from its pad, two forces act on it immediately – the thrust T and its self-weight $W = mg$, Fig. 8.11. The rocket lifts off with acceleration a when the thrust exceeds the deadweight (usually in practice by about 20 per cent). The accelerating force at lift-off is

$$F = T - W = ma$$

The same equations may be applied to the conditions at the instant of burn-out. The thrust may be taken as constant from lift-off although it reduces shortly after clearing the pad to avoid over-stressing of members. The mass of rocket remaining at burn-out is the all-up mass minus the mass of propellants consumed together with the equipment and debris jettisoned. The acceleration due to gravity g at burn-out after a first stage is assumed to be the same as that at sea-level, 9.8 m/s^2. The value of g decreases with altitude but the effect is only significant at very great altitudes. For example, a height of over 160 km has to be reached before the value of g drops by 5 per cent of its value at sea-level. Air resistance does, however, vary greatly with altitude. The resistance of air to a body moving through it depends firstly on air being present and its density as well as on the body's shape, frontal area and speed. The density of air at sea-level at normal temperature and pressure is 1.2 kg/m^3 and it drops to about 0.3 kg/m^3 at 12 km above sea-level. At an altitude of 70 km, the density is less than $4 \times 10^{-7} \text{ kg/m}^3$, yet it is at this altitude that aerodynamic reheating of a returning spacecraft has to be considered. Needless to say, in the problems set here air resistance and variation in g are neglected.

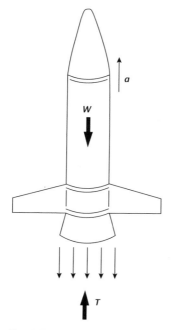

Fig. 8.11 Lift-off

Similar calculations apply to a rocket being fired from the surface of the moon except that g on the moon's surface is only about $1.62\,\text{m/s}^2$. There is no atmosphere on the moon and therefore no resistance to motion.

(**Example**)
_____

A space vehicle of mass 60 t is launched vertically from its pad using two booster rockets each exerting 360 kN thrust, Fig. 8.11. If the rockets burn out in 30 s, estimate the velocity and height reached, assuming constant acceleration from lift-off.

Solution
At lift-off the accelerating force is

$$F = \text{thrust} - \text{deadweight}$$

$$= (2 \times 360) - (60 \times 9.8)$$

$$= 132\,\text{kN}$$

and

$$F = ma$$

i.e. $132 \times 10^3 = 60 \times 10^3 \times a$

$$\therefore \quad a = 2.2\,\text{m/s}^2$$

The velocity after 30 s is given by

$v = at$

$= 2.2 \times 30$

$= 66\,\text{m/s}$ or   **238 km/h**

Height attained,

$s = \tfrac{1}{2}at^2$

$= \tfrac{1}{2} \times 2.2 \times 30^2$

$= 990\,\text{m}$

Example

A space vehicle of mass 200 t is launched at 25° to the vertical, Fig. 8.12. After 15 s of flight, the thrust is 2.2 MN along the line of flight, 15 t of propellants have been burned up, and 20 t of debris jettisoned. What is the acceleration at this point? If the average acceleration up to this point is 2.6 m/s², estimate the velocity and height reached.

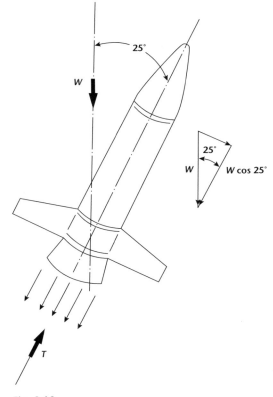

Fig. 8.12

Solution

After 15 s of flight,

$$\text{mass of vehicle, } m = 200 - 15 - 20 = 165\,\text{t}$$

$$\text{weight of vehicle, } W = 165 \times 9{\cdot}8 = 1617\,\text{kN}$$

$$\text{thrust, } T = 2{\cdot}2\,\text{MN} = 2{\cdot}2 \times 10^3\,\text{kN}$$

$$\text{accelerating force, } F = T - W \cos 25^\circ$$

$$= 2{\cdot}2 \times 10^3 - 1617 \cos 25^\circ$$

$$= 735\,\text{kN}$$

and

$$F = ma$$

i.e. $735 \times 10^3 = 165 \times 10^3 \times a$

$$a = 4{\cdot}45\,\text{m/s}^2$$

After 15 s, the velocity reached with average acceleration $2{\cdot}6\,\text{m/s}^2$ is

$$v = at$$

$$= 2{\cdot}6 \times 15$$

$$= 39\,\text{m/s} \quad \text{or} \quad 140\,\text{km/h}$$

The distance travelled along the line of flight is

$$s = \tfrac{1}{2}at^2$$

$$= \tfrac{1}{2} \times 2{\cdot}6 \times 15^2$$

$$= 292{\cdot}5\,\text{m}$$

Therefore the height reached is $292{\cdot}5 \cos 25^\circ = 265\,\text{m}$.

Problems

8.39 A space vehicle of mass $3 \times 10^6\,\text{kg}$ is launched vertically from its pad. Find the thrust (i) when it is lifted just free of its supports and (ii) when the lift-off acceleration is $2\,\text{m/s}^2$.   (29·4 MN; 35·4 MN)

8.40 An astronaut of mass 120 kg hovers freely in space tethered to a spacecraft and is initially at rest relative to the craft before firing a hand-held rocket gun. If the gun gives a thrust of 60 N for 3 s, how far does the astronaut travel?   (2·25 m)

8.41 A rocket of mass 1000 kg is fired vertically from rest. If the average acceleration is $11\,\text{m/s}^2$ for the first 50 s of flight before the fuel is exhausted, find the average accelerating force and the height achieved. If the rocket continues after the burn-out point as a free body moving upwards, find the maximum height reached and the time taken from rest to maximum height.

(11 kN; 13·8 km; 29·2 km; 106 s)

8.42 A rocket of mass 210 t is launched vertically from rest with a thrust of 2·4 MN. Find the initial acceleration and estimate the velocity at a height of 500 m, assuming constant acceleration from rest. What is the time taken to reach this height? (1·63 m/s^2; 145 km/h; 25 s)

8.43 A rocket of initial mass 30 t travelling vertically reaches an altitude where g = 9·2 m/s^2. At this point the thrust is 120 kN, 8 t of equipment debris has been jettisoned and 18 t of propellants consumed. Find the acceleration at this instant. What is the acceleration if g had been taken as 9·8 m/s^2? (20·8 m/s^2; 20·2 m/s^2)

8.44 A rocket has an all-up weight of 196 kN on the launch pad. Vertical lift-off takes place when the thrust exceeds the deadweight by 20 per cent. Assuming constant acceleration from lift-off, estimate the velocity and height achieved in the first 30 s of flight. (212 km/h; 882 m)

8.45 A rocket of weight 200 kN lifts off vertically when the thrust exceeds the deadweight by 20 per cent. The thrust remains constant until burn-out and the propellants account for 40 per cent of the initial mass. Before burn-out 3 t of rocket parts are discarded. Find the accelerating forces and accelerations at lift-off and burn-out. (40 kN; 149 kN; 1·96 m/s^2; 16·2 m/s^2)

8.46 A space vehicle of mass 8 t is manoeuvring prior to docking with a probe which it is overtaking. Its speed relative to the probe is 4 m/s. Four small thrusters, each exerting a force of 300 N, fire in the line of flight to slow the vehicle down to a relative speed of 0·5 m/s ready for docking. Find the firing time. (23 s)

8.47 A missile of mass 950 kg is fired at 40° to the horizontal. Its rockets fire for 23 s giving a constant thrust of 18 kN along the line of flight. Find the acceleration and height reached at the end of firing. (12·7 m/s^2; 2·15 km)

8.48 A space vehicle of mass 180 t is launched at 20° to the vertical with a thrust of 2120 kN along the line of flight. Assuming constant acceleration, estimate the height reached in 40 s. (1·93 km)

8.49 What is the minimum thrust needed to lift a module of mass 400 kg vertically off the surface of the moon where the acceleration due to gravity is 1·62 m/s^2? If an additional thrust of 300 N is provided at lift-off, what height will the module attain in one minute? (648 N; 1·35 km)

8.50 A lunar module of mass 300 kg is launched at 50° to the horizontal at the surface of the moon. The thrust is constant at 1200 N and acts for 20 s along the line of flight. Assuming constant acceleration from lift-off, find the height reached. Take g = 1·62 m/s^2. (423 m)

Chapter 9

Momentum and energy

....

Newton's laws of motion are stated, together with the principles of conservation of momentum and energy to which they lead; these are the fundamental laws forming the basis of engineering dynamics and they relate to the effects produced when forces change the motion of bodies. We have already used the simple F = ma equation but the momentum principle is particularly useful when collisions, explosions or reaction propulsion are involved. We are concerned here only with **linear** *momentum for which the relationship between impulse and change of momentum is derived. Examples are given of impulsive forces; jet aircraft and rockets are used to illustrate reaction propulsion. Energy, the capacity to do work, the various forms of energy and the application of the principle of conservation of energy follow on from the work on momentum. Many problems in dynamics may be solved by employing the two important laws of conservation but care must be taken with the energy principle to ensure that all forms of energy are accounted for, including heat energy.*

9.1 Momentum

It is found experimentally that the average force required to bring to rest a small, quickly moving mass may be as great as that to bring to rest a large, slowly moving mass, the time being the same in each case.

The 'quantity of motion' possessed by either body is called its *momentum* and is measured by the product of its mass m and its velocity v. Thus

Momentum = mv

The unit of mass is the kilogram (kg) and the unit of velocity is the metre per second (m/s). Hence, the unit of momentum is the *kilogram metre per second* (**kg m/s**) or *newton second* (**N s**). However, numerically it may be more convenient to work in tonnes and kilometres per hour.

Momentum has magnitude, and its direction and sense correspond to that of the velocity. Momentum is therefore a vector quantity. A change of momentum may be produced in a body:

- by a change in mass, as in a rocket ejecting burnt fuel;
- by a change in the *direction* of the velocity, as in a body rotating in a circular path;
- by a change in the *magnitude* of the velocity, as in an accelerated body;
- by a change in both mass and velocity.

9.2 Laws of motion

First law of motion

Consider a body projected with an initial velocity along a horizontal path. Experience suggests that it is slowed down and eventually comes to rest due to friction forces opposing the motion. Galileo (1564–1642) showed, however, that as the resistance to motion is reduced, so the distance travelled by the projected body before stopping increases. It is possible to imagine a perfectly smooth path offering no resistance to motion; once started, motion would continue indefinitely. This is an instance of Newton's *first law of motion*, which states that

Every body continues in a state of rest, or of uniform motion in a straight line, except when compelled by impressed forces to change that state.

A rocket ship far out in empty space, with its motor shut off and remote from the gravitational pull of the earth or sun, will continue in its motion at constant velocity. Evidently there is negligible force available to change the motion it already has.

Prior to Galileo, it was thought that, in order for a body to continue to move with a uniform velocity, a force was required. It can now be seen that this idea is correct if the force is applied to overcome only the resistance to motion. There is then no resultant out-of-balance force and no acceleration.

Second law of motion

A resultant out-of-balance force F causes an acceleration a according to the relation

$$F = ma$$

If the acceleration a is uniform, and the mass m is *constant*, then

$$a = \frac{v - u}{t}$$

where u and v are the initial and final velocities respectively and t is the time taken for the change. Hence

$$F = \frac{m(v - u)}{t}$$

$$= \frac{mv - mu}{t}$$

$$= \frac{\text{change of momentum}}{\text{time taken}}$$

i.e. **the applied force is equal to the rate of change of momentum.**

This is a special case of Newton's *second law of motion* which states that

The rate of change of momentum is proportional to the resultant out-of-balance force and takes place in the direction of that force

if no force acts on the body, $F = 0$, and there is, therefore, no change of momentum. This agrees with the first law of motion, since if there is no change of momentum the body continues with uniform motion in a straight line.

When the acceleration a is not uniform, equation (1) gives the *average* force F acting for time t, i.e.

$$\text{Average force} = \frac{\text{change of momentum}}{\text{time taken}}$$

Example

A planing machine has moving parts of mass 1·5 t and is brought to rest from a speed of 0·12 m/s in 0·8 s. Calculate the average retarding force.

Solution

$$\text{Change in momentum} = m(v - u)$$

$$= 1·5 \times 10^3 \times (0·12 - 0)$$

$$= 180 \text{ kg m/s}$$

$$\text{Average force exerted} = \frac{\text{change in momentum}}{\text{time taken}}$$

$$= \frac{180}{0·8}$$

$$= 225 \text{ N}$$

Example

A 400 t train attains 120 km/h from rest on a gradient of 1 in 100. The constant tractive effort is 240 kN and the average track resistance is 110 N/t. Find the momentum of the train at its maximum speed and the time taken to reach it.

Solution

$$\text{Momentum of train at 120 km/h} = mv$$

$$= 400 \times 10^3 \times \frac{120}{3·6}$$

$$= 13·3 \times 10^6 \text{ kg m/s}$$

The student should sketch the train ascending the gradient showing the forces acting on it, i.e. the tractive effort, track resistance and the weight component down the slope (*see* Fig. 8.4).

Average accelerating force on train up the slope is given by

$$F = \text{tractive effort} - \text{track resistance} - \text{weight component}$$

$$= 240 \times 10^3 - 110 \times 400 - 400 \times 10^3 \times 9·8 \times \frac{1}{100}$$

$$= 157 \times 10^3 \text{ N}$$

Average force × time taken = change in momentum

$$157 \times 10^3 \times t = 13 \cdot 3 \times 10^6 - 0$$

$$\therefore \quad t = 85\,\mathbf{s}$$

This problem has been solved using 'momentum'. The student should rework the solution using the equation of motion $F = ma$.

Problems

9.1 A train of total mass 100 t is travelling at 90 km/h. Find the total resisting force required to stop the train in one minute. If the rolling resistance to motion is 100 N/t, what is the braking effort required?
(41·7 kN; 31·7 kN; compare example on p. 179)

9.2 A train of total mass 700 t reaches a steady speed of 90 km/h from rest when the net tractive force exerted by the locomotive is 34·5 kN. Find the time taken. (8·45 min; compare example on p. 179)

9.3 A machine tool table is driven by a constant force of 250 N. The table, of mass 2070 kg, reaches a speed of 0·1 m/s from rest. What is the time taken?
(0·83 s)

9.4 A shunting locomotive jolts a stationary 12 t wagon which moves off freely at 4 m/s. After 18 s the wagon slows down to 2·9 m/s. Find the change in momentum of the wagon and the average track resistance per tonne of wagon. (13·2 × 10³ kg m/s; 61·1 N/t)

9.5 A 20 t truck reaches a speed of 54 km/h in 94 s when climbing a gradient from rest. Find its momentum at this speed and hence the average net force on the truck. If the gradient is 1 in 20 and the tractive effort is 18 kN, what is the average track and air resistance per tonne of vehicle?
(3 × 10⁵ kg m/s; 3·19 kN; 250 N/t)

9.6 An aircraft of mass 68·5 Mg touches down at 210 km/h on a level runway with a negligible vertical component of velocity and takes 47 s to come to rest. Find the momentum of the aircraft at touch-down and the average retarding force (due to braking, rolling friction and air drag). What is the minimum length of runway required?
(4 × 10⁶ kg m/s; 85·1 kN; 1·37 km; compare problem 8·31)

Third law of motion

It is known from statics that a single force cannot exist alone. For equilibrium, forces come in pairs. This is an oversimplification of Newton's *third law of motion*, which states that

action and reaction are equal and opposite.

For example, the pull of gravity on a body (its weight) is the *action* of the earth's pull. By the third law of motion, there is an equal and opposite pull exerted by the body on the earth. This is termed a *reaction*. Examples of 'action and reaction' are rocket and jet aircraft propulsion, the recoil of a gun when fired and the recoil of a fire-hose.

An *active* force is one that can cause an acceleration. For example, the weight of a load hung from a crane hook is an active force exerted by the earth on the body, for if the load were removed from the hook, it would accelerate downwards.

A *reactive* force is one that cannot *of itself* cause acceleration. The force exerted by a crane hook which supports a load is a reactive force. (The force exerted by the crane hook produces an upward acceleration ($-g$) to cancel the downward acceleration ($+g$) due to the gravitational pull.)

Note that active and reactive forces are two equal and opposite forces on the *same* body. Action and reaction are two equal and opposite forces exerted between two *separate* bodies.

We are now in a position to define *force* as *that which produces or tends to produce an acceleration in a body*.

The first two laws of motion appear to have been known to Galileo, but the first formal statement of the three laws as forming the basis of mechanics was made by Newton in 1687.

9.3 Reaction propulsion

An example of 'action and reaction' is rocket propulsion. (*See* Chapter 8 for work on the forces acting on aircraft and rockets.) A rocket carries its own oxygen and fuel (the *propellants*) which burn in a chamber open at one end. A high-speed jet of exhaust gases is ejected from the opening via an exit convergent–divergent nozzle, Fig. 9.1(a), and the rocket is accelerated by the action of the gases. To accelerate the gases and eject them requires that a force be exerted by the rocket on the gases. By Newton's third law, there is an equal and opposite force exerted by the gases on the rocket and it is this reaction or *thrust* which propels the rocket forward. This is the principle of *reaction propulsion*. The same principle applies to the working of propeller and jet-driven aircraft and ships, helicopters and rocket thrusters for satellite control. Apart from the mechanical aspects and body structure, each type of vehicle differs greatly in regard to the mass of fluid dealt with and the speed of the jet formed.

The engine of a turbo-jet aircraft is distinguished from a rocket motor by requiring air to burn its fuel. The forward flight of the aircraft forces air into the engine where it mixes with fuel, followed by combustion and expulsion of a high-speed jet through a nozzle, Fig. 9.1(b). The force on the jet is equal to the rate of change of momentum of the air (and fuel) passing through the engine. The equal and opposite force, i.e. the force propelling the plane forward, is the pressure of the combustion gases on the inside surfaces of the combustion chamber.

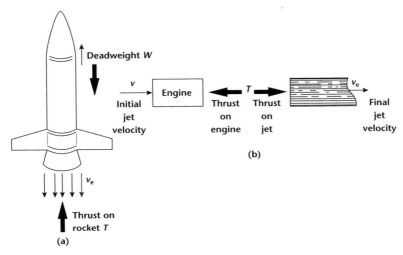

Fig. 9.1

Thrust of a jet

Let v, v_e be the fluid velocities relative to the engine, at entry and exit respectively, Fig. 9.1(b). Then, *relative to the engine* the initial momentum of m kg of fluid is mv, the final momentum mv_e and the change in momentum $m(v_e - v)$. If $\dot{m}$ is the mass of fluid passing through the engine per second, the force on the jet is equal to its change in momentum per second, i.e.

Force $= \dot{m}(v_e - v)$

The thrust T on the engine is equal and opposite to the force on the jet, hence the propulsive force on the vehicle is

engine thrust = (mass of fluid/second)
× (change in fluid velocity in passing through the engine)

i.e. $T = \dot{m}(v_e - v)$

If the mass flow rate $\dot{m}$ is in kg/s, v and v_e in m/s, then T is in newtons. This formula applies to any vehicle which uses a jet of fluid for propulsion. Note that the exhaust jets of aircraft and rockets may have speeds well above the speed of sound (340 m/s at sea-level).

Rockets

The propellants are initially at rest relative to the rocket since they are carried with the rocket, i.e. $v = 0$, and the thrust is

$T = \dot{m}v_e$

The mass flow rate $\dot{m}$ is the mass of propellants consumed divided by the time taken, assuming a steady rate of burning. The thrust for any propulsion stage is fairly constant.

A rocket does not require surrounding air in order to operate and works more efficiently where there is no atmosphere since any air present would resist and slow down the exhaust jet and the motion of the rocket. This makes them the ideal vehicles for high-altitude and space travel. Enormous thrusts can be generated by giving large quantities of propellants a very high speed relative to the rocket in the shortest possible time. The initial thrust of a Saturn V rocket system, for example, was 34 MN for a burning time of 160 s, and the total rated thrust of a Shuttle's three main engines, 5 MN, boosted at lift-off by two rockets, each giving 3 MN of thrust for two minutes of flight. Again, on a different scale, the rockets used for the precise adjustment of the motion of the Orbiter craft give a thrust as low as 100 N.

Jet aircraft

Section 21.10 gives a brief description of a gas turbine engine for aircraft propulsion. The engine takes in a large quantity of air and utilizes a high air-to-fuel ratio. The velocity of the air changes from v at intake to v_e in the jet at exit, both velocities being *relative to the engine*. The fuel (already in the engine) changes in speed from zero to v_e, relative to the engine. Since the air:fuel ratio is very large (a typical figure for a turbo-jet is 70:1), the thrust due to the fuel in the jet is small in comparison with that due to the airflow and may be neglected with little error. The value of $\dot{m}$ may therefore be taken as the mass flow rate of air at intake or gas in the jet.

When an aeroplane flies through *still air*, the velocity of the intake air relative to the engine, v, is also the forward speed of the plane relative to the ground (otherwise the effect of wind velocity has to be taken into account). Thus the thrust for a jet aircraft in still air is

$$T = \dot{m}(v_e - v)$$

where v is the *forward speed* of the plane.

(Example)
_____

A rocket in level flight ejects 100 kg of burnt fuel in 8 s at a velocity of 500 m/s relative to the rocket. Calculate (a) the change in momentum of the burnt fuel and (b) the average thrust on the rocket due to the ejection.

Solution

(a) The burnt fuel is originally moving with the rocket at some unknown speed and therefore has some unknown momentum. Its change in velocity is 500 m/s, therefore the change in momentum

$= 100 \times 500$

$= 50\,000\,\text{kg m/s}$

(b) The average force required to produce this change in momentum

$$= \frac{\text{change in momentum}}{\text{time taken}}$$

$$= \frac{50\ 000}{8}$$

$$= 6250\,\text{N} = \textbf{6·25 kN}$$

The average thrust on the rocket is equal and opposite to this.
   Alternatively, using the formula devised above, then

$$T = \dot{m}\,v_e$$

$$= \frac{100}{8} \times 500$$

$$= 6250\,\text{N}$$

(**Example**)
_____

A rocket has a weight of 27 MN at lift-off from its launch pad. The propellants are burned steadily at the rate of 12 t/s and the initial acceleration is to be 2·4 m/s². Find the thrust and the velocity of the exhaust jet relative to the rocket.

Solution
Refer to Fig. 9.1(a). At lift-off the thrust must overcome the deadweight and accelerate the vehicle. The accelerating force is

$$F = ma$$

i.e. $\qquad T - W = \dfrac{W}{g}a$

i.e. $\quad T - 27 \times 10^6 = \dfrac{27 \times 10^6}{9\cdot8} \times 2\cdot4$

Therefore

$$T = 33\cdot6 \times 10^6\,\text{N}$$

$$= \textbf{33·6 MN}$$

and

$$T = \dot{m}\,v_e$$

i.e. $\quad 33\cdot6 \times 10^6 = 12 \times 10^3 \times v_e$

Therefore

$$v_e = 2\cdot8 \times 10^3\,\text{m/s} \quad \text{or} \quad \textbf{2·8 km/s}$$

(**Example**)

A turbo-jet aircraft cruises at 270 m/s in still air when the jet issues from the engine at 650 m/s relative to the engine, with a mass flow rate of 99 kg/s. Find the thrust.

Solution

$v = 270$ m/s; $v_e = 650$ m/s; $\dot{m} = 99$ kg/s

Thrust $T = \dot{m}(v_e - v)$

$$= 99(650 - 270) \times \frac{1}{1000} \text{ kN}$$

$$= 37 \cdot 6 \text{ kN}$$

(**Example**)

A 5 t jet-plane approaches a runway at 30 m/s horizontal speed, Fig. 9.2. Fuel consumption is 0·4 kg/s and the exhaust jet issues at 400 m/s relative to the engine. If the thrust is 8 kN find the air/fuel ratio.

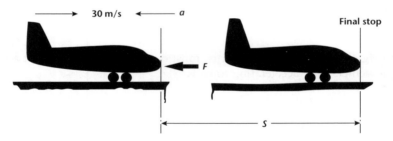

Fig. 9.2

The jet is deflected immediately on touch-down to produce a constant braking thrust of 40 per cent of the forward thrust on approach. Find the minimum runway required, assuming no other resisting forces to be acting.

Solution

$T = 8000$ N; $v = 30$ m/s; $v_e = 400$ m/s

$$T = \dot{m}(v_e - v)$$

i.e. $8000 = \dot{m}(400 - 30)$

i.e. $\dot{m} = 21 \cdot 6$ kg/s

$$\frac{\text{air}}{\text{fuel}} = \frac{21 \cdot 6}{0 \cdot 4} = 54$$

The braking force is 40 per cent of forward thrust, thus

$$F = 0.4 \times 8 \times 10^3 = 3.2 \times 10^3 \, \text{N}$$

and

$$F = ma$$

i.e. $3.2 \times 10^3 = 5 \times 10^3 \times a$

Therefore

$$a = 0.64 \, \text{m/s}^2$$

Using the equation $v^2 = u^2 + 2as$ where s is the length of runway required, v the final velocity is zero, and *a the deceleration* is $0.64 \, \text{m/s}^2$. Thus

$$0 = 30^2 - 2 \times 0.64 \times s$$

Therefore

$$s = 703 \, \text{m}$$

Problems

9.7 A rocket ejects 2000 kg of burnt gases at a velocity of 2000 m/s relative to the rocket. Calculate the average force on the rocket if the time taken is 100 s.

(40 kN)

9.8 A rocket of mass 10 t is mounted vertically on its test bed and ejects exhaust gases at a uniform rate of 0.1 t/s with a velocity of 300 m/s. If the rocket does not move during the test, calculate (a) the force exerted on the rocket by the exhaust gases and (b) the load on the test bed foundation at the start of the test (assume that the exhaust gases do not impinge on the foundation).

(30 kN; 68 kN)

9.9 An Ariane rocket burns propellants at the rate of 1300 kg/s and lifts off from rest with an all-up mass of 210 t. The speed of the exhaust jet relative to the rocket is 2 km/s. Find the thrust and acceleration at lift-off.

(2.6 MN; 2.6 m/s²)

9.10 A spacecraft has an all-up mass of 3000 t on the launch pad. The propellants in the rockets which fire at lift-off represent 62 per cent of the total mass and they burn out in 155 s. If the lift-off acceleration is to be 2.2 m/s², what is the thrust required and the minimum velocity of the exhaust jet relative to the craft?

(36 MN; 3 km/s)

9.11 A communications satellite has a mass of 180 kg and orbits the earth. Its four adjuster rockets fire in the line of flight for 5 s to slow it down. Each rocket discharges a jet at the rate of 0.25 kg/s with a velocity of 360 m/s relative to the satellite. What is the reduction in speed achieved?

(10 m/s)

9.12 A space vehicle leaves its launching pad with an all-up mass of 2800 t. At a point in its vertical flight the burn-rate is 12 t/s and the mass of the vehicle is reduced by 1885 t from burning of propellants, jettisoning of rocket casings and other debris. The jet issues at a speed of 2·5 km/s relative to the vehicle. Find the thrust and acceleration at this instant.   (30 MN; 23 m/s^2)

9.13 A four-engined jet airliner is cruising in straight, level flight at 300 m/s. Each engine takes in air at the rate of 45 kg/s and its jet issues at 900 m/s relative to the engine. What is the resisting force due to the air, i.e. the drag force?
(108 kN)

9.14 A turbo-jet engine is tested on a stationary rig. Air is supplied at the rate of 27 kg/s and velocity 80 m/s to the combustion chamber and the speed of the exhaust jet is 665 m/s. What is the thrust on the supports?   (15·8 kN)

9.15 A twin-engined aircraft of mass 8·5 Mg is cruising in straight, level flight at 185 m/s in still air and draws air into each engine at the rate of 40 kg/s. The jets leave the engines at 500 m/s relative to the aircraft. Find the thrust. What is the additional thrust needed (i) to give the aircraft an acceleration of 1·5 m/s^2 while maintaining straight, level flight and (ii) to enable the craft to climb at a steady speed of 185 m/s at 20° to the horizontal? Assume the drag force to remain constant.   (25·2 kN; 12·8 kN; 28·5 kN)

9.16 A jet-plane of mass 8 t flies in straight level flight in still air at 972 km/h and the exhaust jet issues at the rate of 30 kg/s with a velocity of 950 m/s relative to the plane. The drag force is 5 kN. Find the thrust and acceleration.
(20·4 kN; 1·93 m/s^2)

9.17 A jet-plane travels at 864 km/h in still air with the exhaust jet issuing at 900 m/s relative to the engine at the rate of 80 kg/s. Find the thrust.
If the mass of the plane is 30 t and the acceleration is 1·5 m/s^2 whilst maintaining straight level flight with this thrust, estimate the drag force.
(52·8 kN; 7·8 kN)

9.18 A ramjet missile takes in a large amount of air at its front end, burns it with a small amount of fuel and ejects a high-speed jet. If the air is drawn in at the rate of 70 kg/s when the flight speed is 250 m/s, what must be the exit velocity of the jet for a thrust of 24 kN?   (593 m/s relative to missile)

9.19 A four-engined jet-plane of mass 155 t has to reach a take-off speed of 252 km/h. On the run-up, each engine takes in air at the rate of 74 kg/s and the jetspeed at the exit nozzle is 700 m/s relative to the plane. Find the total thrust.
Assuming the thrust to be constant from the beginning of the run-up and a track and wind resistance of 40 kN, find the minimum runway required.
(187 kN; 2.6 km)

9.4 Conservation of momentum

In principle, all problems in mechanics can be solved using the three laws of motion. In practice the solution may be very difficult. For example, in the case of an impact between two bodies, the forces involved are unknown, may be very large and may act for a very short time. A further principle, based on the first law of motion, which simplifies the solution of such problems is the *principle of conservation of momentum*.

Momentum can be destroyed only by a force and can be created only by the action of a force. If no *external* force acts on a body or system of bodies then the momentum remains constant, i.e. constant in both magnitude and direction. This is termed the *principle of conservation of momentum*.

9.5 Application of conservation of momentum to collision of two bodies

A body of mass m_1 moving with velocity u_1 collides with a second body of mass m_2 and velocity u_2, as shown in Fig. 9.3. Both bodies are moving in the same straight line.

During the collision, the only forces acting are the equal and opposite forces between the two bodies. There is no *external* force acting on the *two-body system* immediately before, during or after impact. Hence *the total momentum of the system remains constant*, i.e.

momentum before impact = momentum after impact

If the bodies move off after impact with velocities v_1, v_2 respectively, then

total initial momentum $= m_1u_1 + m_2u_2$

total final momentum $= m_1v_1 + m_2v_2$

Therefore

$$m_1u_1 + m_2u_2 = m_1v_1 + m_2v_2$$

or

$$-m_1(v_1 - u_1) = m_2(v_2 - u_2)$$

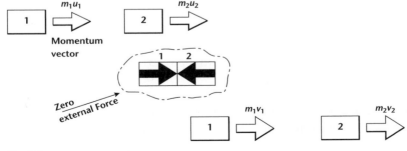

Fig. 9.3

That is, the magnitude of the change in momentum of each body is the same, or the gain in momentum of one body is equal to the loss in momentum of the other body.

If the bodies lock together after impact and move off with a common velocity V, then

$$v_1 = v_2 = V$$

and

$$m_1u_1 + m_2u_2 = (m_1 + m_2)V$$

If the bodies are moving initially *towards* each other then, since the total initial momentum is the vector sum of the separate momenta, one of the quantities must be considered negative. The final direction of motion then depends on which body has the greater momentum.

It should be noted that the assumption that the bodies move after impact with a common velocity is a special case, as for example in the coupling together of railway wagons. Otherwise the two bodies will in general have different final velocities. Further information is then necessary since the momentum equation alone cannot give both these velocities. In the questions set here, this information will be given.

Example

A railway wagon of mass 35 t travels along a level track at 15 km/h and collides with another wagon of mass 15 t travelling in the opposite direction at 20 km/h. After impact, the first is seen to travel in the same direction as before with a speed of 3 km/h. Find the speed of the second wagon.

Solution

It will be convenient to use the tonne (1000 kg) as the unit of mass and the kilometre per hour as the unit of velocity.

Initial momentum of first wagon = 35×15

Initial momentum of second wagon = -15×20

Total initial momentum of system = $(35 \times 15) - (15 \times 20)$

$$= 225$$

Final momentum of first wagon = 35×3

Final momentum of second wagon = $15 \times V$

where V is the velocity of the second wagon after impact. The second wagon must move in the same direction as the first in this case.

Final total momentum = $225 = (35 \times 3) + (15 \times V)$

Hence

$$15V = 225 - 105$$

so that

$$V = 8\,\text{km/h}$$

Example

A 10 t wagon is set moving at 6 m/s and travels against a track resistance of 90 N/t over a distance of 100 m. It then strikes a 5 t wagon moving at 3 m/s in the same direction. What is the velocity of the smaller wagon immediately after impact if the 10 t wagon moves on immediately at 2 m/s?

Solution
For the 10 t wagon the initial velocity $u = 6$ m/s, the distance travelled $s = 100$ m, and the decelerating force F is $90 \times 10 = 900$ N, Fig. 9.4. Then

$$F = ma$$

$$900 = 10 \times 10^3 \times a$$

$$a = 0{\cdot}09\,\text{m/s}^2$$

Using $v^2 = u^2 + 2as$, where $a = -0{\cdot}09$ m/s^2,

$$v^2 = 6^2 - 2 \times 0{\cdot}09 \times 100$$

$$v = 4{\cdot}24\,\text{m/s}$$

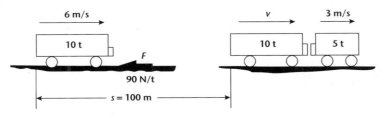

Fig. 9.4

The 10 t wagon therefore has a velocity of 4·24 m/s on impact. Let V be the velocity of the 5 t wagon immediately after impact. Equating the momentum before and after impact, working in tonne, metre, second units:

$$(10 \times 4{\cdot}24) + (5 \times 3) = (10 \times 2) + 5V$$

$$V = 7{\cdot}5\,\text{m/s}$$

9.20 A railway wagon of mass 12 t and travelling at 15 km/h collides with a second similar wagon at rest. If both wagons travel on coupled together, calculate their common velocity. If the second wagon were moving at 15 km/h in the opposite direction to the first, what would then be their common velocity?   (7·5 km/h; 0)

9.21 A truck of mass 8 t travelling due E with velocity 10 km/h collides with a second truck of mass 12 t travelling due W at 1 km/h, both trucks moving on the same straight line. (a) Find the magnitude and direction of their common velocity if they are locked together after impact. (b) If the resistance to motion is 90 N/t, find how far the trucks will move before coming to rest.   (3·4 km/h due E; 5 m)

9.22 A 10 t wagon moving at 4 m/s strikes a 20 t wagon moving at 2·5 m/s in the same direction on the same track. After the collision the 10 t wagon travels backwards at 3·6 m/s. Find the velocity of the 20 t wagon immediately after impact.   (6·3 m/s)

9.23 A rail wagon of mass 7 t is checked at rest on a slope inclined at 2° to level ground. It starts to roll freely down the slope against a track resistance of 50 N/t. After 5 s it strikes a stationary 12 t wagon held in check until the moment of impact when it immediately moves off down the slope at 0·5 m/s. Find the velocity of the 7 t wagon immediately after impact.   (0·6 m/s)

9.24 A shunting locomotive strikes a wagon of mass 10 t and sets it in motion with an initial speed of 4 m/s. The wagon travels freely along a level track against a track resistance of 60 N/t. After 10 s it collides with a second wagon of mass 12 t moving in the opposite direction at 0·5 m/s. If the wagons move off immediately coupled together, find their common speed and the time taken for the coupled wagons to come to rest.   (1·3 m/s; 21 s)

9.6 Explosions: recoil

When a body explodes freely and instantaneously, as, for example, a shell from a gun, a payload ejecting from a rocket, or simply a body fragmenting, then the initial momentum is redistributed among the parts. The momentum principle may be applied since there are no external forces acting except for gravity which can be neglected, as an explosion is over before it can have any effect.

The total momentum, although redistributed, remains unchanged in any given direction before and immediately after an explosion.

When a shell is fired from a gun, the gun kicks back due to the force of recoil. Equal and opposite forces are exerted on the shell and gun during the time the shell takes to traverse the barrel, hence both shell and gun have the same change of momentum, i.e. the *forward* momentum of the shell equals the *backwards* momentum of the gun. The initial momentum of shell and gun is zero, therefore the final momentum immediately after firing is zero.

A rocket in flight may be separated from its payload, such as a communications satellite, by a small explosion or spring-release mechanism. If the rocket and satellite continue along the line of flight then the sum of their momenta along the line of flight after the explosion must equal the initial momentum of rocket and satellite. The average force and time of separation, i.e. time during which the force acts, is the same for both rocket and satellite. For each body, therefore

average force × time of separation = change of momentum

Note that when the time of action of a very large force is extremely short, as in an explosion, the force is called an *impulsive force* and the change in momentum, i.e. *force × time*, is called an *impulse*, the units being N s, the same as momentum.

See also sections 9.9 and 9.10 but note here that an explosion adds energy to the system.

(Example)

A shell is fired from a gun at 600 m/s at an angle of 30° to the horizontal. If the shell's mass is 10 kg and the velocity of recoil of the gun is to be limited to 1·2 m/s, calculate the minimum mass of gun required.

Solution

Before firing, the total momentum is zero.

$$\text{After firing, the momentum of shell} = 10 \times 600$$

$$= 6000 \, \text{kg m/s}$$

$$\text{Horizontal component of this momentum} = 6000 \times \cos 30°$$

$$= 5196 \, \text{kg m/s}$$

Let m = mass of gun in kg.

Horizontal momentum of gun at recoil = $-m \times 1·2$ kg m/s, Fig. 9.5.

For gun and shell together,

$$\text{initial (horizontal) momentum} = \text{final (horizontal) momentum}$$

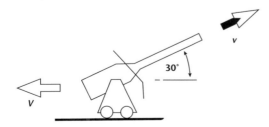

Fig. 9.5

$$0 = 5196 - (m \times 1.2)$$

i.e. $m = 4330\,\text{kg}$

$$= 4.33\,\text{Mg}$$

Example

A satellite is released to the rear of its host rocket in the line of flight by triggering off a small explosion when the rocket is coasting in orbit. The separating force is 480 N acting for 1·2 s. The mass of satellite is 960 kg and the mass of rocket after separation is 240 kg. Find the speed of recession.

Solution

Let x_1 (m/s) be the *reduction* in speed of the satellite and x_2 (m/s) the *gain* in speed of the rocket, Fig. 9.6. The speed of recession is the relative velocity of the two parts, i.e. the velocity of the satellite assuming the rocket to be at rest.

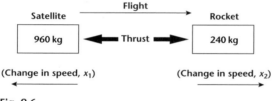

Fig. 9.6

For the satellite,

average force × time of separation = change of momentum

$$= \text{mass} \times \text{change in speed}$$

i.e. $480 \times 1.2 = 960 \times x_1$

therefore

$$x_1 = 0.6\,\text{m/s (reduction)}$$

For the rocket,

$$480 \times 1.2 = 240 \times x_2$$

$$x_2 = 2.4\,\text{m/s (gain)}$$

The speed of recession is $x_2 + x_1$, i.e.

$$(2.4 + 0.6) = 3\,\text{m/s}$$

Note that the initial speed of the rocket with satellite is not required since we are dealing here with *changes* in momentum.

Example

A satellite-carrying rocket of total mass 1280 kg coasts in space at 8 km/s. An explosion ejects the satellite (320 kg) rearwards in the line of flight and its speed falls by 18 m/s immediately after separation. Find the final speed of the rocket and the speed of recession.

If the duration of the explosion is 1·5 s, what is the force of the explosion?

Solution

The mass of the rocket remaining after separation is (1280 − 320) kg, i.e. 960 kg. Let v be the velocity of the rocket immediately after the explosion. Equating the initial and final momenta (noting that rocket and satellite continue in the *same* direction, the final speed of the satellite is 7982 m/s and momentum is *mass × speed*):

$$1280 \times 8000 = (320 \times 7982) + 960v$$

therefore

$$v = 8006 \text{ m/s}$$

The rocket gains in speed by 6 m/s, and the speed of recession is therefore 6 + 18 = **24 m/s**.

For the satellite,

average force × time = change of momentum

i.e.   average force × 1·5 = 320 × 18

∴   average force = **3840 N**

Checks

1. Equate the change of momentum of satellite to change of momentum of rocket to confirm that change in speed of rocket is 6 m/s.
2. For the rocket,

average force × 1·5 = 960 × 6

∴   average force = 3840 N

Problems

9.25 A gun fires a shell of mass 5 kg in a horizontal direction. The gun recoils at 0·5 m/s and its mass is 3 t. Calculate the velocity of the shell.   (300 m/s)

9.26 A gun fires a 12 kg shell at 20° to the horizontal with a velocity of 300 m/s. If the gun's mass is 4 t and it recoils in the horizontal direction, find (a) the velocity of recoil and (b) the resistance to motion if the gun comes to rest in 150 mm.   (0·85 m/s; 9·52 kN)

9.27 A gun of mass 10 t fires a 50 kg projectile with a horizontal velocity of 750 m/s. Calculate the initial recoil velocity of the gun and if the shell takes 0·01 s to leave the barrel, find the average force on the gun.

(3·75 m/s; 3·75 MN)

9.28 A bullet, mass 17 g, strikes a wood-block target of mass 16 kg and remains embedded in the wood after impact. The target is suspended by a light string and at the moment of impact it is moving towards the bullet at 0·6 m/s. Target and bullet then move along the line of flight of the bullet at 0·4 m/s. What is the striking speed of the bullet? (942 m/s)

9.29 A rocket travelling in a horizontal straight line at 1 km/s explodes and breaks up into two parts of mass, 800 kg and 200 kg respectively. Both parts travel on in the same direction as before but the lighter section is moving 100 m/s faster than the heavier part. Calculate the velocity of each section.

(980; 1080 m/s)

9.30 A projectile of mass 42 kg is travelling at constant speed 25 m/s when it explodes into two parts. One part of mass 28 kg is immediately at rest. Find the speed of the second part if it continues along the line of flight.

(75 m/s)

9.31 If the projectile of problem 9·30 explodes into three parts such that a 28 kg fragment is immediately at rest, a second fragment of 8 kg continues upwards at 30° to the line of flight with a speed of 40 m/s, find the speed and direction of the third part. (132 m/s at 11·7° to line of flight; downwards)

9.32 A 56 kg projectile is moving at constant speed 8 m/s when it explodes into three pieces, A, B and C. Figure 9·7 shows the masses, speed and direction of A, B and C, immediately after the explosion. Confirm that the total momentum along the line of flight remains at 448 kg m/s and that the momentum at right angles to the line of flight remains zero immediately after the explosion.

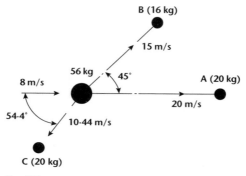

Fig. 9.7

9.33 A rocket is coasting in space when a small explosion discharges a satellite of mass 180 kg in the line of flight. The separating thrust is 1200 N and the satellite's speed increases by 10 m/s after the thrust period of duration t s. Find t.　　　　　　　　　　　　　　　　　　　　　　　　　　　(1·5 s)

9.34 A satellite is launched from a spacecraft by firing a cluster of small rockets to give a total separating thrust of 4·8 kN, acting for 1 s. The mass of the satellite is 1200 kg and the mass of the craft remaining after separation is 800 kg. Find the speed of recession.　　　　　　　　　　　(10 m/s)

9.35 An astronaut of mass 90 kg floats in space attached by a cord to the craft, and carries a piece of rocket debris of mass 8 kg. He is at rest relative to the craft and hurls the debris away from him. If the debris travels at 45 m/s relative to the craft, what is the speed of recession of astronaut and debris?
　　　　　　　　　　　　　　　　　　　　　　　　　　　　　(49 m/s)

9.36 A rocketcraft of mass 1800 kg including a payload is coasting in space when holding bolts are broken by an explosion to release the payload rearwards in the line of flight. The speed of the payload falls by 24 m/s and the craft gains by 6 m/s. Find the mass of the payload.　　　　　　　　　　(360 kg)

9.37 A spacecraft of total mass 1000 kg travelling at 10 km/s releases a cargo of mass 200 kg using a powerful spring-loaded tilt table. The cargo recedes rearwards in the line of flight at a speed of 1 km/s relative to the craft. Find the final speeds of craft and cargo immediately after separation.
　　　　　　　　　　　　　　　　　　　　　(10·2 km/s; 9·2 km/s)

9.7　Principle of conservation of energy

Energy is defined as the capacity to do work. *The principle of conservation of energy states that energy can be neither created nor destroyed.* This principle enables us to obtain directly the answer to certain complex problems where the use of the laws of motion would be tedious. Before discussing in detail the nature of energy, it is convenient to study the connexion between work and energy, using as a starting point the second law of motion.

9.8　Energy and work

Consider the work done in uniformly accelerating a body of mass m from rest to speed v over a distance s along a horizontal straight line, Fig. 9.8. The acceleration required is given by

$$v^2 = 2as$$

or

$$a = \frac{v^2}{2s}$$

Fig. 9.8

By the second law of motion the *average* force required is

$F = ma$

$= m \times \dfrac{v^2}{2s}$

The work done by the force F in distance s is

$F \times s = m \times \dfrac{v^2}{2s} \times s$

$= \tfrac{1}{2}mv^2$

The expression on the right-hand side is known as the *kinetic energy* (KE), or *energy of motion* of the body at velocity v; it is the energy possessed by the body by virtue of its mass and velocity. The units of KE are the same as those of work, i.e. *newton metre* (N m) or *joule* (J). (Nm is normally reserved for 'torque').

9.9 Change of kinetic energy

The work done in accelerating a body from velocity u *to velocity* v *is equal to the change in its kinetic energy.*

Thus the acceleration required is given by:

$v^2 - u^2 = 2as$

or

$a = \dfrac{v^2 - u^2}{2s}$

The uniform (or average) force required is

$F = ma$

$= m \times \dfrac{v^2 - u^2}{2s}$

and the work done over distance s is

$F \times s = m \times \dfrac{v^2 - u^2}{2s} \times s$

$= \tfrac{1}{2}mv^2 - \tfrac{1}{2}mu^2$

which is the change in kinetic energy.

Example

A car of mass 1 tonne is accelerated from a speed of 24 km/h to a speed of 48 km/h in 50 m. Find the average tractive effort required. What would be the average braking force required to bring the car to rest in 50 m from 48 km/h?

Solution

$u = 6.67$ m/s; $v = 13.33$ m/s

Let E be the tractive effort; then the work done = $E \times 50$ J.

Change in kinetic energy = $\frac{1}{2}m(v^2 - u^2)$

$$= \frac{1}{2} \times 1 \times 10^3 \times (13.33^2 - 6.67^2)$$

$$= 66\,600 \text{ J}$$

Since work done = change of kinetic energy

$E \times 50 = 66\,600$

so that

$E = 1333 \text{ N} = \mathbf{1.33\,kN}$

Let F be the braking force, then the work done in bringing the car to rest is $F \times 50$ J.
Therefore

$$F \times 50 = \frac{1}{2} \times 1 \times 10^3 \times (13.3^2 - 0)$$

i.e. $F = 1780 \text{ N} = \mathbf{1.78\,kN}$

Example

An aircraft carrier is cruising at 18 knots while heading into a wind of 10 knots when a 4.5 t aircraft touches down on deck against the wind with an airspeed of 84 knots. It is brought to a halt in a distance of 70 m by the arrestor gear, a reverse engine thrust of 8 kN and wheel braking force of 10 kN. Find (a) the speed of the aircraft relative to the ship, (b) the kinetic energy to be absorbed in halting the aircraft and (c) the average force needed from the arrestor gear. 1 knot = 0.514 m/s.

Solution

(a) *Airspeed* is the speed of a body relative to the speed of the surrounding air:

airspeed at carrier deck = 18 + 10 = 28 knots

airspeed of aircraft = 84 knots

Therefore,

speed of aircraft relative to ship = 84 − 28 = **56 knots**

(b) The kinetic energy to be absorbed is $\frac{1}{2}mv^2$ where v is the speed of the aircraft relative to the ship. Thus

$$\text{kinetic energy} = \frac{1}{2} \times 4.5 \times (56 \times 0.514)^2$$

$$= 1864\,\text{kJ} \quad \text{or} \quad \textbf{1·86 MJ}$$

(c) If the average force exerted by the arrestor gear is P kN then the total retarding force is

$$F = P + 8 + 10 = P + 18\,\text{kN}$$

The work done by the force F in a distance of 70 m is equal to the loss of KE of the aircraft, therefore

$$F \times 70 = 1864$$

i.e. $(P + 18) \times 70 = 1864$

i.e. $P = \textbf{8·6 kN}$

(9.10) **Potential energy**

The work done in lifting a body of weight W through a vertical height h is $W \times h$. This quantity is known as the *potential energy* (PE) of the body referred to its original position. The body possesses this energy by virtue of its weight and position (height). Potential energy is a relative quantity since the original position is chosen arbitrarily for convenience in solving a given problem, i.e. we calculate *changes* in potential energy.

If m is the mass of the body in kg and h is the height in metres then the potential energy is mgh joules.

More correctly, we are dealing here with the gravitational potential energy of a body in the particular gravitational field caused by the earth. Near the surface of the earth its gravity field force is fairly uniform and the variation in g can be ignored.

(**Example**)
_____

A drop hammer is lifted a height 6 m above the ground and then allowed to fall from rest on to a forging at ground level. Calculate the downward velocity of the hammer when it strikes the forging.

Solution
Let the mass of the hammer be m kg; then the work done in raising it h m is mgh J. This is then its potential energy above ground level.

In falling h m the earth-pull does an equal amount of work in accelerating the hammer to a velocity v m/s from rest. In effect the potential energy of the hammer is converted into kinetic energy of motion (principle of conservation of energy).

Thus

kinetic energy at ground level = initial potential energy

i.e. $\frac{1}{2}mv^2 = mgh$

or

$$v = \sqrt{(2gh)} = \sqrt{(2 \times 9\cdot8 \times 6)}$$

Hence

$$v = 10\cdot9\,\text{m/s}$$

Example

In a stamping machine the die has a mass of 35 kg and falls from a height of 2 m on to a metal block. If the depth of indentation is 10 mm find the average stamping force assuming the die does not rebound.

Solution

Loss of potential energy of die = mgh

$$= 35 \times 9\cdot8 \times (2 + 0\cdot01)$$

$$= 689\,\text{J}$$

The average stamping force is equal to the average resisting force of the metal against which the die moves through 10 mm.

Work done against resisting force = loss of potential energy of die

i.e.

average resisting force $\times$ 0·01 = 689

so that

average resisting force = 68 900 N

or

average stamping force = **69 kN**

Problems

9.38 A mass of 18 kg moving at 20 m/s is acted on by a constant force for 5 s as it covers a distance of 125 m. What is the gain in its kinetic energy and the force?   (4·5 kJ; 36 N)

9.39 A railway truck of mass 12 t moving at 2 m/s strikes a second truck of mass 18 t moving in the same direction at 1·2 m/s and they move off, coupled together, with a common velocity of 1·1 m/s. What is the loss of kinetic energy to the system?   (18·8 kJ)

9.40 A ship moves through the water at 4 knots and when its engines are stopped, it travels a distance of 60 m before coming to rest. What is the kinetic energy per unit mass of the ship at 4 knots and the average water resistance in kN/t? 1 knot = 0·514 m/s.   (2·1 kJ/t; 0·035 kN/t)

9.41 Calculate by an energy method the braking force required to stop a car in 15 m when travelling at 100 km/h. The car has a mass of 1 tonne.   (25·7 kN)

9.42 An aircraft of mass 2000 kg is to be catapulted from the deck of a stationary aircraft carrier with a speed of v m/s in a distance of 70 m. The engine propelling force on the plane during take-off is 20 kN; the estimated forces exerted by the catapult and drag resistance are 8 kN and 4 kN respectively. Find the speed v and the kinetic energy of the plane when launched.

(41 m/s; 1·68 MJ)

9.43 An aircraft of mass 3000 kg lands on the deck of a carrier against the wind with an airspeed of 110 knots. The ship is cruising at 26 knots into a head wind of 28 knots. The aircraft is brought to rest on the deck by the combined action of wheel brakes, reverse thrusters and arrestor gear. Find (i) the speed of the aircraft relative to the ship, (ii) the kinetic energy to be absorbed in halting the aircraft and (iii) the total braking force necessary and the time taken to bring the aircraft to rest in a distance of 60 m. 1 knot = 0·514 m/s.

(56 knots; 1·24 MJ; 20·7 kN; 4·2 s)

(9.11) Notes on conservation of energy

We recall that energy is the capacity to do work. In dynamics we deal only with kinetic and (gravitational) potential energy which are the two forms of mechanical energy. Kinetic energy is the energy possessed by a body by virtue of its mass and velocity. Potential energy is the energy possessed by a body by virtue of its weight and its position (height). Another form of 'potential' energy is *strain energy*, when the work done in stretching or compressing a bar or spring is stored and available as energy, i.e. there is no permanent deformation of the bar or spring. All forms of energy are transferable; that is, the work done in accelerating a body reappears as kinetic energy; similarly the potential energy lost by a falling body reappears as kinetic energy as its velocity increases. For all practical purposes energy cannot be destroyed. However, it is sometimes converted into a form which may be of no further practical use. For example, work done against friction is converted into heat, which may be of no further value but which is nevertheless a form of energy. In such a case, the energy may be considered 'lost', for it may not always be recovered to do useful work, as can potential energy or work done against gravity.

Similarly, in a collision, kinetic energy is generally 'lost'. Hence it is not ever possible to write that the kinetic energy after impact is equal to the kinetic energy before impact. The initial kinetic energy of two bodies before a collision occurs is recoverable only if both are perfectly elastic. This is not ever the case, the energy reappearing partly as heat, sound and mechanical vibration, none of which is recoverable.

The principle of conservation of energy has been illustrated by using the second law of motion to obtain a relation between kinetic energy and work. However, the principle is independent of the laws of motion. It stands alone and for bodies of constant mass no exception has ever been found.

(Example)
_____

A train of mass 200 t is moving up an incline of 1 in 100 at 60 km/h. The resistance to motion is 50 N/t. If power is shut off calculate the distance travelled along the incline before the train comes to rest.

Solution
No collision is involved; hence the energy principle may be used provided that allowance is made for work done in overcoming resistance to motion. Let s metres be the distance travelled after power is shut off.

$$\text{Gain in height} = \frac{s}{100}$$

$$\text{Work done against gravity} = 200 \times 10^3 \times 9\cdot8 \times \frac{s}{100}$$

$$= 19\,600s\,\text{J}$$

$$\text{Total resistance to motion} = 50 \times 200\,\text{N}$$

$$= 10\,000\,\text{N}$$

$$\text{Work done against resistance} = 10\,000 \times s\,\text{J}$$

$$\text{Initial velocity} = 60\,\text{km/h} = 16\cdot7\,\text{m/s}$$

$$\text{Initial kinetic energy} = \tfrac{1}{2} \times 200 \times 10^3 \times 16\cdot7^2$$

$$= 2\cdot79 \times 10^7\,\text{J}$$

Initial kinetic energy = work done against gravity + work done in overcoming resistance

i.e. $2\cdot79 \times 10^7 = 19\,600s + 10\,000s$

so that

$$s = \frac{2\cdot79 \times 10^7}{2\cdot96 \times 10^4}$$

$$= 943\,\text{m}$$

(9.12) Momentum and energy

It has been seen that, in addition to the second and third laws of motion, two further principles are available to solve problems in mechanics:

- *The principle of conservation of momentum.*
- *The principle of conservation of energy.*

For their correct application it must be realized that:

- Momentum is not conserved if an external force acts.
- Mechanical energy is not conserved if work is done against a resistance, or energy is put into the system, as in the burning of a fuel. Nevertheless, when the work done against a resistance is known, it may be allowed for and the energy equation used.

In the recoil of a gun total momentum is conserved, since no external forces act upon the system of gun and shell during the explosion. The propelling force on the shell is equal and opposite to the recoil force of the gun. These equal and opposite forces are internal forces. They act for the same *time*; hence the momentum given to the shell is equal and oppositely directed to the momentum given to the gun. The mechanical energy is not constant, since firing of the charge converts chemical energy into heat, sound, and mechanical vibration together with useful kinetic energy of motion.

The velocity of the shell is greater than that of the gun since it is of smaller mass. The time for which the propelling force acts is the same for both; hence the distance travelled by the shell is greater than the distance travelled by the gun in the same time. Since the forces on shell and gun are of equal magnitude, the work done on the shell is greater than the work done on the gun. Hence the kinetic energy given to the shell is greater than that given to the gun.

Similarly a rocket which loses the greater part of its mass in burnt fuel will receive as kinetic energy a greater proportion of the chemical energy liberated than will the ejected burnt gases.

Note: for *energy* and *power* see section 19.3.

(Example)

A hammer (or 'monkey') of mass 100 kg falls 4 m on to a pile of mass 300 kg and drives it 80 mm into the ground, Fig. 9.9. Calculate (a) the loss of energy on impact, (b) the work done by the resistance of the ground and (c) the average resistance to penetration. Assume the hammer does not rebound on impact.

Solution

(a) Initial potential energy of hammer = mgh. Let v be velocity of hammer just before impact; then kinetic energy of hammer

$$= \tfrac{1}{2}mv^2 = \tfrac{1}{2} \times 100v^2 = 50v^2 \, \text{J}$$

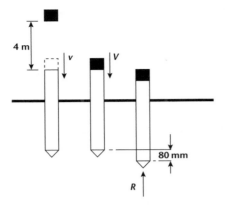

Fig. 9.9

Equating kinetic energy gained to potential energy lost by the hammer,

$$50v^2 = mgh = 100 \times 9\cdot8 \times 4 = 3920\,\text{J}$$

so that

$$v = 8\cdot86\,\text{m/s}$$

In collisions between perfectly elastic bodies, both momentum and energy are conserved. This gives sufficient information for such a problem to be solved. In general, however, for bodies in collision energy is *always* lost on impact, as in this problem, so that energy is never conserved. It is necessary in this case therefore to use the momentum principle to find the common velocity after impact.

Let V be the common velocity of hammer and pile after impact. The total mass moving at velocity V is therefore $(100 + 300)\,\text{kg}$, i.e. $400\,\text{kg}$.

Momentum before impact = momentum after impact

i.e.

$$100 \times 8\cdot86 = 400 \times V$$

so that

$$V = 2\cdot22\,\text{m/s}$$

$$\text{Kinetic energy after impact} = \tfrac{1}{2} \times 400 \times 2\cdot22^2$$

$$= 986\,\text{J}$$

Kinetic energy before impact = initial potential energy

$$= 3920\,\text{J}$$

Loss of kinetic energy on impact

$$= 3920 - 986 = 2934\,\text{J}$$

$$= 2\cdot93\,\text{kJ}$$

(b) Work done by resistance of ground in bringing pile and hammer to rest

= total energy of pile and hammer *after impact*

= kinetic energy of pile and hammer + potential energy of pile *and* hammer

= 986 + (400 × 9·8 × 0·08) = 1300 J

$$= 1·3 \, kJ$$

(c) Let average resistance to penetration be *R* newtons; then work done = *R* × 0·08 newtons. Therefore

$$R × 0·08 = 1300$$

and

$$R = 16·3 \, kN$$

Problems

9.44 A pendulum swings through an angle of 30° to either side of the vertical. Calculate the maximum velocity of the bob if its mass may be considered concentrated at a point 4 m from the point of suspension.   (3·24 m/s)

9.45 Calculate by an energy method the braking force required to stop a car in 15 m when travelling at 100 km/h. The car has a mass of 1 tonne.

(25·7 kN)

9.46 A car of mass 500 kg is uniformly accelerated from rest for a distance of 100 m up an incline of 1 in 30 (sine). If its final velocity is 80 km/h, calculate:

(a) the work done against gravity;
(b) the change in kinetic energy;
(c) the total work done by the car;
(d) the tractive effort required.

(16·4 kJ; 124 kJ; 140 kJ; 1·4 kN)

9.47 The head of a drop hammer is raised 2 m and then falls freely to strike the die. If the net mass of the hammer is 300 kg and the penetration of the hot metal is 15 mm, what is the average force?   (392 kN)

9.48 A gun mounted on a wheeled carriage has a mass of 2 t. It fires a projectile at an angle of 60° to the horizontal at a velocity of 1000 m/s. If the shell's mass is 12 kg, find the velocity with which the gun starts to recoil. If the gun carriage comes to rest in 150 mm, find the average resistance to motion.

(3 m/s; 60 kN)

9.49 The hammer of an impact testing machine has a mass of 8 kg and is concentrated at 1 m radius from a frictionless pivot. The hammer is held in a horizontal position and allowed to swing freely in a vertical circle. Calculate the speed of the hammer and its kinetic energy at its lowest position.

If at this point it strikes and breaks a metal specimen which absorbs 20 J of energy, calculate the angle the arm will make with the vertical when the hammer comes instantaneously to rest.   (4·43 m/s; 78·4 J; 75°)

9.50 A pile driver of mass 200 kg falls with velocity 3 m/s on to a pile of mass 100 kg. After impact both travel on together. Calculate the loss of kinetic energy.   (300 J)

9.51 A mass of 2000 kg falling 9 cm is used to drive a pile of mass 1000 kg vertically into the ground. Assuming no rebound, find the common velocity of driver and pile at the end of the blow and the loss of kinetic energy.

If the resistance of the ground is constant, find its value if the pile is driven 2 cm.   (0·9 m/s; 590 J; 88·1 kN)

9.52 A mass of 100 kg is dropped on to a pile of mass 1000 kg from a height of 1·5 m and drives it 6 cm into the ground. Calculate (a) the initial velocity of pile and driving weight immediately after impact and (b) the average resisting force exerted by the ground.   (0·49 m/s; 13 kN)

9.53 A rifle of mass 20 kg fires a 14 g bullet with a muzzle velocity of 900 m/s. What is the velocity of recoil of the rifle and the energy of the explosion?
(0·63 m/s; 5·7 kJ)

9.54 A 10 g bullet travelling at 800 m/s embeds itself in a stationary wood target of mass 2 kg. Find the speed with which the target moves on and the loss of energy at impact.   (4 m/s; 3.2 kJ)

9.55 A motor vehicle of mass 1800 kg climbs a slope of 1 in 10 at a steady speed of 10 m/s. Find (i) its kinetic energy and (ii) the change in its potential energy when it has travelled 200 m up the slope.   (90 kJ; 353 kJ)

9.56 A truck of mass 10 t travels freely along a track at 2 m/s and strikes a 12 t truck moving at 0·5 m/s in the same direction on the track. The trucks move on coupled together. Find their common speed and the loss of energy on impact.   (1·2 m/s; 6·13 kJ)

9.57 A small road vehicle of mass 500 kg is stationary at the top of a slope of length 300 m. The inclination of the slope to the horizontal is $\sin^{-1} 0·2$. What is the potential energy of the vehicle measured relative to the foot of the slope? If the vehicle rolls freely down the slope against a constant track resistance of 250 N, find (i) its acceleration, (ii) its kinetic energy at the foot of the slope and (iii) the work done against the track resistance.
(294 kJ; 1·46 m/s²; 219 kJ; 75 kJ)

9.58 A space shuttle of initial mass 2000 t is fired vertically from rest and may be assumed to have a constant acceleration of 5 m/s² until the propellants in the first stage burn out after 130 s. At burn-out its mass has been reduced by 885 t due to consumption of propellants and debris jettisoned. Find its kinetic and potential energies at burn-out.   (236 GJ; 462 GJ)

Chapter 10

Uniform motion in a circular path

....

Equations for linear motion have already been derived and now angular motion and the relationship between the vector quantities angular displacement and uniform angular velocity have to be considered; angular acceleration is not dealt with at this level. In particular, the motion of a body moving in a circular path at constant speed is examined to show that there is always a radially inwards acceleration produced by a radially inwards (centripetal) controlling force. The kinetic energy of rotation is defined and examples on flywheels, clutches, ships, fluid couplings, etc. are used to illustrate the approach to problems involving angular motion at constant speed.

10.1 Angular velocity

Consider a line OP, of length *r*, rotating about a fixed point O, Fig. 10.1. When the point P has travelled round the circle a distance AB equal to the radius *r*, the angle swept out by the line OP is defined as *one radian*, i.e.

The radian is the angle subtended by an arc equal to the radius.

In general, the angle swept out when the line moves from position OA to position OP is

$$\theta = \frac{\text{arc AP}}{\text{radius of circle}}$$

$$= \frac{s}{r} \text{ radian}$$

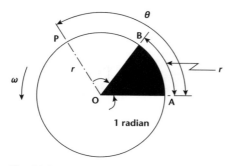

Fig. 10.1

where s is the distance measured round the arc AP.

In one complete revolution the angle is $\frac{2\pi r}{r}$, i.e. 2π radians. Hence

360° = 2π radians

or

1 radian ≃ 57·3°

The *angular velocity* ω of the line OP is defined as the number of radians swept out per second. If ω is constant, the angle θ swept out in t seconds is therefore

$\theta = \omega t$ radians

or

$$\omega = \frac{\theta}{t} \text{ rad/s}$$

Thus the angular velocity of the line is ω rad/s.

If the line OP makes n rev/s, this is equivalent to $2\pi n$ rad/s:

$$\omega = 2\pi n \text{ rad/s}$$

(10.2) Relation between angular and linear velocity

Let point P, Fig. 10.1, travel round the circle with uniform speed v; then, if s is the arc traversed in time t and θ the angle turned through by the line OP, we have

$$\text{speed} = \frac{\text{distance}}{\text{time}}$$

i.e. $v = \frac{s}{t}$

and

$s = r\theta$

Therefore

$$v = \frac{r\theta}{t}$$

and

$$\frac{\theta}{t} = \omega$$

Therefore

$$v = \omega r \quad \text{or} \quad \omega = \frac{v}{r}$$

Note that here the symbol ω will always denote angular velocity in radians per second, and therefore θ must be in radians and t in seconds. The velocity v is numerically equal to the speed but is always directed along the tangent to the circle.

Example

A flywheel of 1·2 m diameter rotates at 300 rev/min. Calculate the velocity of a point on its rim.

Solution

$$\omega = 2\pi n = \frac{2\pi \times 300}{60} = 31\text{·}42 \text{ rad/s}$$

$$v = \omega r = 31\text{·}42 \times 0\text{·}6 = \mathbf{18\text{·}9 \text{ m/s}}$$

Example

A car travels at 72 km/h. Calculate the angular velocity of the wheels in revolutions per second if they are of 80 cm diameter. How many revolutions are turned through by each wheel in 1 km? Assume that the wheels rotate without slip.

Solution

72 km/h = 20 m/s

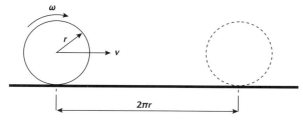

Fig. 10.2

In one revolution of the wheels the car moves forward a distance equal to the circumference of the wheel, Fig. 10.2; therefore

distance travelled per revolution = $2\pi r$

no. of revolutions per second $n = \dfrac{\text{speed of car}}{\text{distance travelled per rev.}} = \dfrac{v}{2\pi r}$

and 1 rev/s = 2π rad/s; therefore

$$\omega = 2\pi n$$

$$= 2\pi \times \frac{v}{2\pi r} = \frac{v}{r}$$

$$= \frac{20}{0 \cdot 4} = 50 \text{ rad/s}$$

$$= \frac{50}{2\pi} = 7 \cdot 95 \text{ rev/s}$$

Time to travel $1 \text{ km} = \dfrac{1000}{20} = 50 \text{ s}$

Revolutions turned through by wheel in $1 \text{ km} = 7 \cdot 95 \times 50 = \textbf{398.}$

(**Example**)

A car travels at 110 km/h. Its wheels are 700 mm in diameter. Calculate the velocity of a point on the wheel rim when the point is:

(a) vertically above the axle;
(b) in contact with the ground.

Solution
If the wheel does not slip, the velocity of any point on the wheel is made up of two parts:

1. Its velocity due to the linear motion of the car axle.
2. Its tangential velocity due to the rotation of the wheel about the axle.

Linear velocity of axle $v = 110 \text{ km/h} = 30 \cdot 6 \text{ m/s}$

Angular velocity of wheel $\omega = \dfrac{v}{r} = \dfrac{30 \cdot 6}{0 \cdot 35} = 87 \cdot 5 \text{ rad/s}$

(a) If the axle is stationary and the wheel revolving, the velocity of a point P vertically above O (Fig. 10.3) is

$$\omega r = 87 \cdot 5 \times 0 \cdot 35 = \textbf{30} \cdot \textbf{6 m/s}$$

This velocity is from right to left and is equal to v, the linear velocity of the axle.

When the axle moves with velocity v,

total velocity of $\text{P} = v + v = 2 \times 30 \cdot 6 = \textbf{61} \cdot \textbf{2 m/s}$

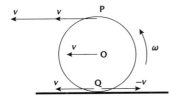

Fig. 10.3

(b) If the axle is stationary, the point Q, vertically below O, moves with velocity $\omega r = v$, from left to right; then

total velocity of $Q = v - v = 0$

The point Q on the rim in contact with the road is therefore at rest.

Example

In Fig. 10.4 the drum A rotates with the drum B, about a common axle O. A third drum C is pressed against the drum B, as shown. The diameters are: A, 40 cm; B, 120 cm; C, 90 cm.

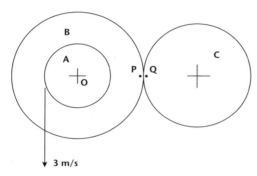

Fig. 10.4

A thin cord is fastened to, and wrapped round, the circumference of drum A and is pulled downwards at 3 m/s so as to rotate the drum. If there is no slip between B and C, or between the cord and drum A, find the speed of C in revolutions per minute.

Solution

Velocity of rim of A = velocity of cord

$$= 3\,\text{m/s}$$

Angular velocity of B = angular velocity of A

$$= \frac{3}{0 \cdot 2}$$

$$= 15\,\text{rad/s}$$

Let angular velocity of C be ω.
Velocity of point P on B = velocity of point Q on C
Therefore

$$15 \times 0 \cdot 6 = \omega \times 0 \cdot 45$$

so that

$$\omega = 15 \times \frac{0 \cdot 6}{0 \cdot 45}$$

$$= 20 \, \text{rad/s}$$

and

$$n = 20 \times \frac{1}{2\pi} = 3 \cdot 18 \, \text{rev/s} \quad \text{or} \quad \textbf{191 rev/min}$$

Problems

10.1 A body moves in a circular path of radius 10 m. In an interval of 0·5 s the radius from the centre of the circle to the body sweeps out an angle of 18°. Calculate the average angular velocity and the average linear speed of the body.   (0·63 rad/s; 6·3 m/s)

10.2 A car has a wheel diameter of 80 cm and a brake drum diameter of 30 cm. Calculate the velocity of a point P on the brake drum circumference (a) when P is vertically above the axle and (b) when P is vertically below the axle. The speed of the car is 16 m/s.   (22 m/s; 10 m/s)

10.3 A clutch plate consists of a flat plate 30 cm in diameter. Calculate the speed of a point at a radius of 7 cm when the plate rotates at 10 rev/s about its centre. What is the greatest linear velocity of any point on the clutch face?   (4·4 m/s; 9·4 m/s)

10.4 A wheel and axle runs with the axle on two parallel tracks. The wheel rotates with the axle and the linear speed of the wheel centre is 0·6 m/s. The wheel diameter is 60 cm and the axle diameter is 3 cm. Calculate the greatest linear velocity of (a) any point on the wheel, and (b) any point on the axle.   (12·6 m/s; 1·2 m/s)

10.5 A heavy rope is wrapped round a pulley of 60 cm diameter. The rope is 3 cm thick. Calculate the angular velocity of the pulley in revolutions per second if the rope is pulled off the pulley without slip at a speed of 2 m/s tangent to the pulley. (Allow for the effect of rope thickness on the effective pulley diameter.)   (1·01 rev/s)

10.6 A grinding wheel rotates at 2800 rev/min. What is the cutting speed in m/s if the wheel diameter is 200 mm?   (29·3 m/s)

10.7 A ship travels at a constant speed of 20 knots round a curve of 300 m radius. Find how long it would take to turn through 180° of arc. (1 knot = 1·85 km/h.)   (92 s)

10.8 A load is propelled at a speed of 1·5 m/s over level ground by running it on rollers of 50 mm diameter. What is the speed of the rollers in revolutions per minute?   (573 rev/min)

10.3 Kinetic energy of rotation

A body rotating about a fixed point possesses kinetic energy by virtue of its speed of rotation. In Fig. 10.5, the body A rotates about the fixed point O with angular velocity ω. The linear speed of the body is therefore $v = \omega r$, where r is the radius of rotation OA. If the body has mass m its kinetic energy is $\frac{1}{2}mv^2$ and since $v = \omega r$

kinetic energy $= \frac{1}{2}m\omega^2 r^2$

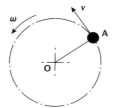

Fig. 10.5

Hence the kinetic energy of rotation is proportional to the square of the radius of rotation, and proportional to the square of the speed.

If the body is in the form of a thin ring then all parts of the mass are at the same radius of rotation. *This only applies when the thickness of the ring is small compared with its radius.*

The units of kinetic energy are the same as those of work, i.e. joules.

10.4 The engine flywheel

The purpose of the engine flywheel is to even out changes in crankshaft speed during each revolution. It consists essentially of a heavy metal rim mounted on the crankshaft. The kinetic energy of rotation of the flywheel serves as a source of energy if the shaft tends to slow down. Similarly, if the shaft tends to speed up, the flywheel stores the excess energy. Changes in speed during each revolution due to the uneven action of the engine mechanism are therefore minimized.

Example

An engine flywheel is formed by a thin rim of steel of mass $\frac{1}{2}$ tonne and diameter 1·2 m. Calculate its kinetic energy of rotation at 600 rev/min.

If a constant braking torque of 500 Nm is applied to the shaft, calculate the number of revolutions turned through before coming to rest.

Solution

$$\omega = \frac{2\pi \times 600}{60}$$

$$= 62 \cdot 83 \text{ rad/s}$$

$37 \cdot 7 \times 6 = 226 \cdot 2$

Kinetic energy of rotation $= \frac{1}{2}m\omega^2 r^2$

$= \frac{1}{2} \times (\frac{1}{2} \times 1000) \times 62\cdot83^2 \times 0\cdot6^2$

$= 355\,000\,\text{J} = 355\,\text{kJ}$ *172686.74 J = 17 kJ*

If θ is the angle in radians turned through by the flywheel before coming to rest, then

work done by braking torque T = decrease in kinetic energy of rotation

or

$$T\theta = 355\,000\,\text{J} \quad \textit{172686.74 J}$$

Therefore

$$\theta = \frac{355\,000}{500}$$

$$= 710\,\text{rad}$$

$$= \frac{710}{2\pi}\ \text{revolutions}$$

$$= 113\,\text{revolutions}$$

(handwritten) 172686.76 / 226.2 = 763.4

Problems

10.9 An engine flywheel in the form of a thin rim of 1 m diameter is to be designed to store 200 kJ of kinetic energy at a speed of 20 rev/s. Calculate the mass of metal required in the rim.

If the diameter is increased to 1·2 m, what is then the mass required?

(102 kg; 70·3 kg)

10.10 An engine flywheel is equivalent to a thin rim of 900 mm diameter and of mass 400 kg. Calculate its kinetic energy at 8 rev/s. What constant torque would be required to attain this speed from rest while turning through 50 complete revolutions?

(103 kJ; 326 N m)

10.11 The rim of an engine flywheel has a mass of 7 t and mean radius 2 m. When rotating at 100 rev/min, the energy of the flywheel increases by 900 kJ. What is its new speed?

(126 rev/min)

10.12 A flywheel used in a shearing machine is of mass 100 kg and consists essentially of a thin rim of mean radius 0·5 m. What is the kinetic energy available in the flywheel at 450 rev/min?

If the shearing operation absorbs 20 per cent of the energy of the flywheel, what will be the flywheel speed immediately after the shearing stroke?

(28 kJ; 403 rev/min)

10.5 Centripetal force

If a body moves in a straight line with uniform velocity, the first law of motion states that it will continue to do so unless acted on by an external force. If the body travels in a circular path with uniform speed, the second law of motion states that a force is necessary in order to make the body deviate from a straight line. The magnitude of this force is determined by the rate of change of momentum of the body. This is illustrated by the following example.

Example

A body of mass 10 kg rotates anticlockwise in a horizontal path with uniform speed 30 m/s. It makes one complete revolution per second. Find the magnitude and direction of the average force required to make the body turn through one right angle.

Solution

The body is initially at point A, and after turning through 90°, it is at B, Fig. 10·6(a).

At A, momentum $= mv = 10 \times 30 = 300$ kg m/s

This momentum is in the direction of the velocity at A, i.e. vertically upwards, as shown.

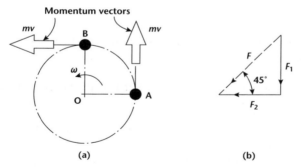

Fig. 10.6

At B, the *magnitude* of the momentum is the same since the speed is the same, but its direction has been turned through 90°. When the body has reached B the whole of the vector momentum possessed by the body at A has been destroyed and this momentum has been re-created in the horizontal direction at B.

The line OA turns through one revolution, or four right angles, in one second; hence the time taken to travel from A to B is $\frac{1}{4}$ s. The average force, over $\frac{1}{4}$ s, required to *destroy* the momentum at A is

$$F_1 = \frac{\text{change of momentum}}{\text{time taken}} = \frac{300}{\frac{1}{4}} = 1200 \text{ N} = 1\cdot2 \text{ kN}$$

In the same time the average force F_2 over $\frac{1}{4}$ s, required to re-create the momentum at B is also 1·2 kN. This force acts in the same direction as the velocity at B, i.e. horizontally to the left.

From the force diagram, Fig. 10.6(b), the resultant average force required is

$$F = \sqrt{(F_1^2 + F_2^2)}$$

$$= \sqrt{(1{\cdot}2^2 + 1{\cdot}2^2)}$$

$$= 1{\cdot}7 \, kN$$

and it is seen that this force acts at 45° to the line OA, since $F_1 = F_2$, i.e. *towards the centre of rotation*.

However, this solution gives no information concerning the magnitude and direction of the force required to maintain circular motion *at any point* between A and B. A method of calculating this force is now given.

(10.6) Centripetal force formula

The force required at a point A to cause a body to move in a circular path of radius r is the average value of the force on the body between point A and a point B *very close to A*.

Let m be the mass of the body, v its uniform speed, and Δt the time taken for the body to travel from point A to point B, Fig. 10.7(a).

At A the momentum of the body is mv tangential to the circle as shown.

At B the momentum has the same magnitude but is changed in direction through the small angle $\Delta\theta$; therefore

momentum destroyed at A = mv

Therefore the average force required at A,

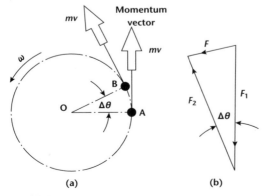

(a)   (b)

Fig. 10.7

$$F_1 = \frac{\text{change of momentum}}{\text{time taken}}$$

$$= \frac{mv}{\Delta t}$$

This force must be oppositely directed to the velocity at A.

In the same time Δt the momentum created at B = mv, and therefore the average force required at B is

$$F_2 = \frac{mv}{\Delta t}$$

This force must act in the *same* direction as the velocity at B. The resultant F of the forces F_1 and F_2 at A and B is shown in the force triangle, Fig. 10.7(b). As $\Delta\theta$ becomes smaller, B moves closer to A and it can be seen that, in the limiting condition when B is almost coincident with A, the direction of F coincides with the line OA, i.e. the resultant force at A is radially inwards towards the centre of rotation.

Assuming that, in the limiting condition, the force triangle approximates to the sector of a circle, then

$$F = F_1\Delta\theta \quad \text{or} \quad F_2\Delta\theta$$

i.e. $$F = \frac{mv}{\Delta t} \times \Delta\theta$$

But

$$\Delta\theta = \omega\Delta t, \text{ and } v = \omega r$$

Therefore

$$F = \frac{m \times \omega r}{\Delta t} \times \omega\Delta t$$

i.e. $$F = m\omega^2 r$$

or, since $\omega = v/r$,

$$F = m\frac{v^2}{r}$$

This force F, required to cause the body to move in a circular path, is called the *centripetal force* and it *acts radially inwards towards the centre of rotation*.

10.7 Centripetal acceleration

Consider the formula

centripetal force, $F = m\omega^2 r$

Comparing this with the formula for the second law of motion,

$$F = ma$$

we see that $a = \omega^2 r$ is the acceleration of the body due to the force F. This acceleration is called the *centripetal acceleration* and is directed *radially inwards* towards the centre of rotation, i.e.

centripetal acceleration $= \omega^2 r = \dfrac{v^2}{r}$

An alternative derivation of the formula for centripetal acceleration is given on page 246.

The centripetal force represents the product *mass × acceleration* and is in the same direction as the acceleration. This force must be provided by some means external to the body. For example, when a body is carried on a rotating arm the centripetal force is the tension in the arm; if a car moves in a curved path the inward force is provided by the sideways friction force at the tyres; train wheels are flanged and, on a curve, the rails provide the side thrust.

(10.8) Centrifugal force

If a body attached to a light rod fixed at O, Fig. 10.8, rotates in a circle of radius r, the centripetal force is the tension T in the arm, acting inwards on the body, i.e.

$T = m\omega^2 r$

At the fixed point O there will be an equal and opposite force T acting radially outwards. This is known as the *centrifugal force*.

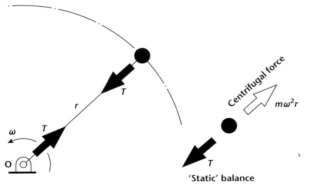

Fig. 10.8

Alternatively, the body may be considered as in 'static' balance under the action of the radially inward force T and a 'force' $m\omega r^2$ in a direction opposite to the centripetal acceleration, i.e. radially outwards. This outward 'force' is also referred to as the *centrifugal force*. Thus,

centrifugal force, $F = m\omega^2 r = m\dfrac{v^2}{r}$

(**Example**)

A car of mass 2 tonne travels at 108 km/h round a circular track of 100 m radius. Calculate the side thrust on the tyres.

Solution

The side thrust on the tyres is the centripetal force required to maintain the car moving in its circular path.

$$v = \frac{108}{3 \cdot 6} = 30 \, \text{m/s}$$

$$\text{Side thrust} = m\frac{v^2}{r}$$

$$= \frac{2 \times 1000 \times 30^2}{100}$$

$$= 18\,000 \, \text{N} = \mathbf{18 \, kN}$$

(**Example**)

A vertical shaft carries a rotor of mass 800 kg rotating at 600 rev/min. If the centre of gravity of the rotor is displaced 2 mm from the axis of the shaft, find the total force on the bearings. If the shaft is horizontal, state the maximum and minimum total forces on the bearings.

Solution

The mass of the rotor may be taken as concentrated at the centre of gravity. The force at the bearings is produced by the out-of-balance centrifugal force due to the mass of 800 kg rotating at 2 mm radius, Fig. 10.9. The reaction of the bearings produces the necessary centripetal force.

$$\omega = \frac{2\pi \times 600}{60} = 20\pi \text{ radians per second}$$

$$\text{Force at bearings} = m\omega^2 r$$

$$= 800 \times (20\pi)^2 \times 0 \cdot 002$$

$$= 6300 \, \text{N} = \mathbf{6 \cdot 3 \, kN}$$

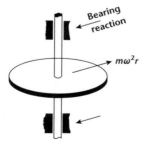

Fig. 10.9

This total force is shared between the bearings, depending on the position of the rotor on the shaft. Since the shaft is vertical, the deadweight reactions do not enter into the problem.

When the shaft is horizontal, the 'deadweight' of the rotor mass is a constant force on the bearings. The maximum total force on the bearings will occur when the centrifugal force acts vertically downwards in line with the deadweight. The minimum total force will occur when the centrifugal force is vertically upwards. Thus,

$$\text{maximum total force} = 6\cdot3 + \frac{800 \times 9\cdot8}{1000}$$

$$= \textbf{14}\cdot\textbf{1}\,\textbf{kN}$$

$$\text{minimum total force} = \frac{800 \times 9\cdot8}{1000} - 6\cdot3$$

$$= \textbf{1}\cdot\textbf{54}\,\textbf{kN (downwards)}$$

Example

A body of mass 30 kg rotates at 10·5 rad/s in a vertical circle at the end of a light rod pivoted 150 mm from the centre of gravity of the body. Calculate the total force on the pivot (a) when the body is vertically above the pivot and (b) when it is vertically below the pivot. Neglect the weight of the rod.

Solution

At the highest point the body has the following forces acting upon it, Fig. 10.10:

- the tension T in the rod, downwards;
- the weight W, downwards;
- the centrifugal 'force' $m\omega^2 r$, upwards.

For 'static' balance

$$T + W = m\omega^2 r$$

Therefore, since $W = mg$,

$$T = m\omega^2 r - mg$$

$$= 30 \times (10\cdot5)^2 \times 0\cdot15 - (30 \times 9\cdot8)$$

$$= 496 - 294$$

$$= \textbf{202}\,\textbf{N}$$

At the lowest point of the circle the weight and centrifugal force together balance the tension and therefore

$$T = m\omega^2 r + mg$$

$$= 496 + 294$$

$$= \textbf{790}\,\textbf{N}$$

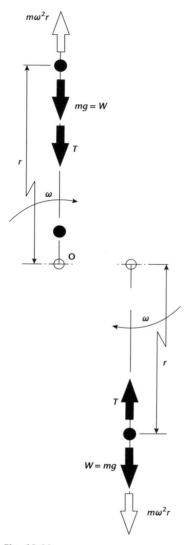

Fig. 10.10

Example

A line OA rotates about a fixed point O with constant angular velocity ω. The point A is at radius OA = r. Prove that the acceleration of A is $a = v^2/r = \omega^2 r$, where v is the linear speed of the point, and that the direction of the acceleration is along the line A to O.

Solution

At A the velocity v of the point is $v = \omega r$, at right angles to OA, Fig. 10.11. This is represented by vector **oa**. Let line OA turn through a very small angle $\Delta\theta$ to

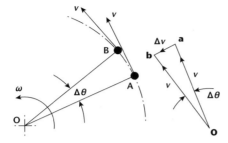

Fig. 10.11

position OB in a short time Δt. The velocity of the point at B is again $v = \omega r$, but the velocity at B is now represented by vector **ob** at right angles to OB.

The vector *change* of velocity Δv in time Δt is therefore represented by the velocity vector **ab**, i.e.

vector change of velocity $\Delta v = \mathbf{ab} = \mathbf{oa} \times \Delta\theta$, since $\Delta\theta$ is small

$$= v\Delta\theta$$

But $v = \omega r$, and $\Delta\theta = \omega\Delta t$; therefore

$$\mathbf{ab} = \omega r \times \omega\Delta t$$

$$= \omega^2 r\Delta t$$

Then

$$\text{acceleration of A, } a = \frac{\text{vector change in velocity}}{\text{time taken}}$$

$$= \frac{\Delta v}{\Delta t}$$

$$= \frac{ab}{\Delta t}$$

$$= \frac{\omega^2 r\Delta t}{\Delta t}$$

$$= \omega^2 r$$

$$= \left(\frac{v}{r}\right)^2 \times r \quad \text{since} \quad \omega = \frac{v}{r}$$

Hence

$$a = \frac{v^2}{r} = \omega^2 r$$

The direction of the acceleration of the point A rotating about O is that of vector **ab**, that is, along the line AO from A to O.

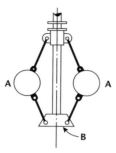

Fig. 10.12

10.9 The centrifugal governor

The purpose of the simple engine governor is to maintain the speed of the engine constant *over a period of time*, regardless of changes of load.

The radius of rotation of a pair of weights A, Fig. 10.12, is determined by the speed of the shaft on which they are mounted. A fall in load tends to allow the engine speed to increase; an increase in speed, however, causes the weights A to move outwards, thus shifting the position of the sleeve B. As B moves so the fuel supply is decreased by a suitable mechanism and the engine tends to slow again.

10.10 The fluid coupling

The combined effects of centrifugal force and kinetic energy of rotation are utilized in the *fluid coupling*, Fig. 10.13. This consists of a casing mounted on the

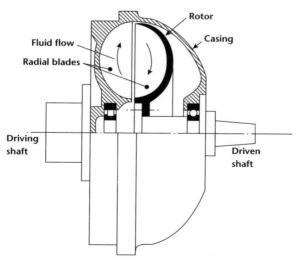

Fig. 10.13

driving shaft and a separate rotor mounted inside the casing on the driven shaft. The casing is almost full of oil. The casing carries a number of straight radial blades which cause the oil to rotate with the casing. Similar radial blades on the rotor are driven by the rotating oil. Otherwise there is no positive connection between the driving and the driven shafts.

The action of the coupling is as follows: rotation of the casing imparts kinetic energy of rotation to the oil between the blades. The centrifugal force of the rotating oil tends to cause it to flow radially outwards and into the spaces between the rotor blades. Since the diameter of the rotor is slightly less than that of the casing, the centrifugal force on the oil in the rotor is less than that on the oil in the casing. Hence the net direction of flow is as shown in the figure. There is therefore a continuous transfer of kinetic energy from the casing to the rotor.

Now, by Newton's third law of motion, the torque exerted by the oil on the rotor blades is equal and opposite to the torque exerted by the oil on the casing blades. Hence the torques at the driving and the driven shafts are equal. On the other hand, there is a net loss of energy due to swirling of the oil between casing and rotor. The rotor speed is therefore slightly less than the speed of the casing. The difference in speed is called the *slip* of the coupling and is about 3 per cent at the design load. At very high torques the loss of energy and the slip increase until when the coupling is overloaded the rotor comes to rest. The rotor is then said to *stall*.

Problems

10.13 A casting is bolted to the faceplate of a lathe. The centre of gravity of the casting is offset 20 mm from the axis of rotation and the faceplate rotates at 200 rev/min. Calculate the *maximum* force on the bearings. The casting's mass is 12 kg and the mass of the faceplate may be neglected.   (223 N)

10.14 A car travels over a humpbacked bridge of 20 m radius. Calculate the speed at which the wheels will leave the ground.   (50·3 km/h)

10.15 A car travelling at 100 km/h passes over a dip in the road which has a radius of 25 m. If the car's mass is 1 tonne, calculate the *total* vertical force on the car springs.   (40·7 kN)

10.16 A vehicle traverses a horizontal circle of radius 70 m at a speed of 70 km/h. What is the minimum coefficient of friction necessary between tyres and road if it is not to skid? The road is unbanked.   (0·55)

10.17 An aircraft 'loops the loop'. At the top of the loop it is travelling at 360 km/h in a vertical circle of radius 300 m. If the pilot's mass is 100 kg, calculate the force pressing him into his seat. At what speed would he fall out of the cockpit if not strapped in?   (2·35 kN; 195 km/h)

10.18 An aircraft pulls out of a dive at 2000 km/h in a vertical circle of radius 1000 m. The pilot's mass is 100 kg. What is the force pressing him into his seat?   (31·9 kN)

10.19 A small carriage travels in a vertical plane around the inside of a circular track of 25 m diameter. What is the least speed it must have if it is not to fall away from the track at the highest point?   (40 km/h)

10.20 A rotor of mass 4000 kg is mounted on a horizontal shaft which is supported by two equally spaced bearings. If the centre of gravity of the rotor is 1 mm out of alignment find the maximum and minimum load on each bearing when the shaft speed is 10 rev/s.   (27·5 kN; 11·7 kN)

10.21 A balance mass of 0·3 kg is bolted to the rim of a wheel at a radius of 200 mm. A test on a similar bolt showed that it would fracture under a tensile load of 10 kN. At what angular speed will the balance weight come away from the wheel?   (3900 rev/min)

10.22 A racing car traverses an unbanked curve of 30 m radius. The coefficient of friction between tyres and road is 0·6. What is the maximum speed if skidding is not to occur?   (48 km/h)

Chapter 11

Gravitation; satellites

....

This chapter is concerned with a special case of motion around a curved track when one body circles another, i.e. a satellite orbiting the earth or a planet. Elliptical orbits are mentioned but the main theme is the relationships among the quantities orbital speed and height, time of orbit, acceleration due to gravity and Newton's universal gravitational constant, for a satellite in circular orbit around the earth. Apart from the aim of solving elementary problems on satellites the treatment is extended to include descriptions of different kinds of orbits and their uses and the satellites launched, in particular by the European Space Agency.

11.1 Newton's law of gravitation: satellites

The pull of gravity obeys *Newton's law of gravitation* – not to be confused with the three laws of motion. This law of gravitation states that the force of gravity is proportional to the mass and inversely proportional to the square of the distance of the mass centre from the earth's centre. Thus a body of mass m at a distance R from the centre of the earth is under a gravitational force F given by

$$F = \frac{GmM}{R^2}$$

where M is the mass of the earth and G a *universal gravitational constant*. In this equation (which, in general, applies to the attractive force between any two objects), it can be seen that F is proportional to m and inversely proportional to R^2, M has a constant value, as has G, which is the same for all masses. The force of gravity is experienced as the *weight* of the body. Thus the weight of the body is inversely proportional to R^2.

If W_0 is the weight of a body at the earth's surface and the radius of the earth is R_0 then the weight W of the body at any other radius R from the centre of the earth is found from the proportion

$$\frac{W}{W_0} = \frac{R_0^2}{R^2}$$

Also, since W/g is constant (i.e. the mass m is constant) and the acceleration due to gravity at radius R is $g = W/m$, then g is proportional to W, or

$$\frac{g}{g_0} = \frac{R_0^2}{R^2}$$

where g_0 is the acceleration due to gravity at the surface of the earth (sea level) ($9.8 \, \text{m/s}^2$ approximately).

Thus

$$g = g_0 \left(\frac{R_0}{R}\right)^2$$

Also, since $F = W = mg$ then

$$mg = \frac{GmM}{R^2}$$

i.e. $g = \dfrac{GM}{R^2}$

This formula gives the acceleration due to gravity at radius R from the earth's centre in terms of the universal gravitational constant and the mass of the earth.

As an example, consider the motion of the first artificial satellite, Sputnik 1 (burned up in the earth's atmosphere after seven months in orbit). Assume that it rotated around the earth's centre at a constant radius R. At this radius

$$g = g_0 \times \frac{R_0^2}{R^2}$$

Now, the earth's surface radius R_0 is about 6370 km, and assuming the mean height of Sputnik 1 was about 600 km then $R = 6970$ km. Thus at this height

$$g = 9 \cdot 8 \times \left(\frac{6370}{6970}\right)^2$$

$$= 8 \cdot 2 \, \text{m/s}^2$$

The centripetal force required for motion in a circle is $m \, v^2/R$, where v^2/R is the radially inwards acceleration of the satellite. This force is, of course, provided by the gravitational pull, i.e. the weight mg of Sputnik 1. Thus

$$m\frac{v^2}{R} = mg$$

Therefore

$$v^2 = gR$$

i.e. $v = \sqrt{(gR)}$

$$v = \sqrt{(8 \cdot 2 \times 6970 \times 1000)}$$

$$= 7570 \, \text{m/s} \quad \text{or} \quad 27\,200 \, \text{km/h}$$

The mass of Sputnik 1 was about 84 kg, hence its weight in orbit, i.e. the gravitational pull of the earth on the satellite, was

$$mg = 84 \times 8 \cdot 2 = 689 \, \text{N}$$

The orbital period is the time taken for a complete circuit of the earth. Thus

$$t = \frac{\text{distance travelled in orbital circuit (m)}}{\text{linear speed in orbit (m/s)}}$$

i.e. $t = \dfrac{2\pi R}{v}$

For Sputnik 1,

$$t = \frac{2\pi \times 6970 \times 10^3}{7570}$$

$$= 5785\,s \quad \text{or} \quad 1\,h\ 36\,min$$

A person travelling in such a satellite would appear to be weightless; the gravitational pull on the person is exactly balanced by the centrifugal force due to rotation. If weighed by a spring balance on the satellite, his or her weight would be zero.

A similar effect can occur when 'looping the loop' in an aircraft. If travelling at the correct velocity at the top of the loop the centrifugal force on the pilot may just balance the pull of the earth; the pilot will simply float, apparently weightless.

Again, if a load is hung from a spring balance in a freely falling lift the balance will always record zero weight: both load and lift are falling freely together and there is no tendency for the load to move relative to the lift or the spring. Hence there is no force between load and lift. The spring will be neither stretched nor compressed and the load is apparently weightless.

11.2 Space

The earth's atmosphere may be considered to end completely at a height of about 350 km but space is usually considered to begin at about 100 km. Once a rocket or spacecraft is at such altitudes, it will continue its motion almost indefinitely since there are no friction forces to slow it down. A body continuing outwards in space will, in due course, reach close to some other large space body and be captured by its gravitational field unless redirected by rocket forces. From a height of about 160 km upwards the effect of the atmosphere is negligible and although the solar cells and other instruments have a limited life, a satellite or spacecraft in orbit will continue to circle the earth for hundreds of years. Many satellites have circular or near circular orbits but a great number have elliptical orbits. It was assumed above that Sputnik 1 was moving in a circular orbit, but its nearest point to earth was 218 km and its furthest point 930 km.

11.3 Kepler's laws

A satellite orbiting the earth obeys *Kepler's laws of planetary motion*, one of which states that the satellite will move in an elliptical orbit determined by the speed of the body. When a satellite is launched into orbit, depending on its speed and direction, it takes up a circular or elliptical orbit or travels out into space in a special trajectory. A circular orbit is a particular limiting case of an ellipse. If a satellite is moving in a circular orbit and receives a forward rocket thrust increasing its speed, it moves into an elliptical orbit; if already in an elliptical orbit it moves into a larger one. The earth's centre is the focal point of the ellipse; its nearest point to earth is its *perigee* and its furthest point, its *apogee*. A typical example is a satellite with a 12 hour period, perigee 500 km and apogee 40 000 km.

Energy of a satellite

A satellite in circular or elliptical orbit around the earth possesses both kinetic and potential energy and in the absence of atmospheric drag its total energy is constant; in the case of the elliptical orbit there is a continuous exchange of its store of KE and PE as its height and speed alter. The KE is proportional to the square of its speed; the PE depends on its position in the earth's gravitational field relative to some arbitrary datum level. To simplify discussion we will take the PE as zero at the surface of the earth.* The PE of the satellite is the work done in raising it to its orbital position, and to calculate its value requires the use of the differential calculus to take account of the variation in g with altitude.

Consider a satellite in circular orbit given an impulse to send it into a lower orbit. In its new position, the PE is reduced since it is nearer to the earth but the force of attraction of the earth is increased so that the outwards directed centrifugal force must increase to maintain equilibrium; this requires an increase in speed if the correct speed for the orbital radius is to be reached. Thus the KE of the satellite increases and the gain of KE can be shown to exceed the loss of PE so that the total energy is greater in the lower orbit.

When a satellite in circular orbit is subject to atmospheric drag, work is done on it, resulting in heat losses and a drop in total energy; unless controlling action is taken, the *circumferential* speed is reduced leading to a reduction in the centrifugal force and the satellite then starts to descend, causing an increase in its *resultant* velocity as it follows a downwards spiral path to earth.

11.4 Orbital plane

The *orbital plane* of a body in orbit is a flat plane which may be inclined to that of the equator. The *angle of inclination* is the angle at which the orbital plane cuts that of the equator. This angle is determined by the purposes for which the satellite is intended. Typically, it could be to make systematic surveillance of some area of the earth's surface. Together with the orbital period, (which determines the speed and height of the satellite) the angle of inclination, the beamwidth of the

---

* Scientists dealing with bodies in space take the datum level for PE as infinity so that for a satellite its *absolute gravitational potential energy* (GPE) is the work done *on* it *by* the force of gravity in bringing it from infinity to its orbital radius position; the GPE is therefore a *negative* quantity because of the choice of datum level. Using differential calculus it can be shown that the work done and hence its GPE is $-mgR$ or $-GmM/R$, since $g = GM/R^2$.

  Note: (i) If the satellite is raised from its orbital position to infinity the work done is mgR and similarly, *from the earth's surface*, it is $mg_0 R_0$. (ii) The KE is $\frac{1}{2}mv^2$ and since $v^2 = gR$ it can be written $\frac{1}{2}mgR$ or $GmM/2R$ hence the total energy is $-\frac{1}{2}mGR$ or $-GmM/2R$. (iii) From the expressions in terms of G, the effect on the KE, PE and total energy of the satellite due to changes in the value of R can be followed through.

sensor used, and other factors govern the surface coverage of the earth by the satellite's instruments, i.e. its radiated signal 'footprint'.

11.5 Control of satellites in orbit

The motion of a satellite in orbit is affected by the following influences:

- the gravitational fields of the sun, moon and stars;
- variations in the earth's gravitational field due to its shape since the equatorial radius is slightly greater (by 21 km) than the polar radius; the earth *precesses*, i.e. wobbles slowly on its axis with a motion like that of a spinning-top, because of the pulls of the sun and moon on its equatorial bulge;
- the earth's orbit round the sun is not truly circular but slightly elliptical;
- the direction of the earth's polar axis is fixed in space but it is tilted at an angle of 23·5° away from the vertical to the plane of its orbit round the sun; the orientation of the polar axis with respect to the sun therefore alters throughout the year and the poles in turn incline towards, or away from, the sun;
- air resistance (drag), for low orbiters particularly;
- intense solar radiation belts.

Ground controllers command tiny rocket thrusters which direct the satellite to its final position or adjust its speed, orbital height and orientation when in flight. Satellites need internal power and this is provided by storage batteries, fuel and solar cells. In addition, besides its payload of tracking beacons, sensors, impact microphones, etc., a satellite carries antennas and large solar panels. Antennas must point towards the receivers or command stations and panels must face the sun as long as possible. The orientation in space of the satellite, antennas and panels requires a variety of control devices including gyroscopes and momentum wheels.

There are three methods of navigation in use for all types of spacecraft, including satellites:

1. *Inertial* – the most common method which utilizes the effects on gyroscopes of changes in direction or acceleration of the craft.
2. *Celestial* – which uses the stars for guidance.
3. *Radio, radar* – limited to earth-orbiters.

11.6 Transmission time of a signal

An *active* satellite receives and processes signals of all kinds in a range of very high frequencies whereas a *passive* one simply emits or reflects signals. Signals are radiated electromagnetic waves, and receivers convert them into voice, data and pictures. The waves travel in straight lines and because of the earth's curvature transmissions must be by 'line-of-sight' and therefore over a limited range, before being boosted onwards by high relay stations on earth or by satellite. A single satellite is able to link stations over continents For low-frequency waves, originating from earth, the range may be extended using the reflecting properties of the earth's ionosphere (*see* page 391).

The *slant range* distance between a ground station (or receiver) and a satellite in orbit depends on their relative positions as regards longitude, latitude and height, but they are seldom at their shortest distance apart. There is a difficulty about satellite systems, in comparison, for example, with fibre optic cables – the *time of transmission*, in relation particularly to telephone communications. For a low orbiter circling the earth at a height, say, of 600 km, the shortest up-and-down link distance is 1200 km between satellite and ground receiver. All electromagnetic waves have the same speed, that of light; the velocity of light in vacuum is 3 × 10^5 km/s and in the atmosphere it is only slightly different. The minimum transmission time, therefore, for the 600 km low orbiter, assuming the station is directly below the satellite, is

$$\frac{1200}{3 \times 10^5} = 0.004\,\text{s} = \textbf{4 milliseconds}$$

For deep space probes the transmission times are measured in minutes or hours; for example, the mean distance from earth of the relatively close planet Mars is 228 × 10^6 km and the up-and-down link time is therefore about 25 min.

(11.7) Escape speed

In order to circle the earth at just above sea-level, assuming no atmosphere or obstacles, a satellite's speed is

$$v = \surd(g_0 R_0)$$

where $g_0 = 9.8\,\text{m/s}^2$ and $R_0 = 6370$ km. Thus

$$v = \surd(9.8 \times 6370 \times 10^3)$$

$$= 7.9 \times 10^3\,\text{m/s} \quad \text{or} \quad \textbf{7.9 km/s}$$

Theoretically then, if a satellite reaches an orbital velocity of about 8 km/s, it circles the earth close to the surface; if it fails to reach this speed, it soon strikes the ground and if it exceeds 8 km/s it goes into elliptical orbit. Should the speed continue to increase, then at a particular value v_E, known as the *escape speed*, the satellite will finally 'escape' the pull of the earth's gravity although the force of gravity will always be present. The term 'speed' is used and not 'velocity' since the direction is irrelevant except that it must be outwards from the earth. The escaping satellite gradually decelerates and never falls back to earth but in due course becomes attracted to the sun or a planet. Becoming free of the earth's gravity field can only be achieved if the kinetic energy given to a body is equal to the total work done in raising the body to infinity. We showed above that the work done in raising a body from the earth's surface to infinity is $mg_0 R_0$. If the kinetic energy required by the body in order to escape is $\frac{1}{2}mv_E^2$ then equating it to $mg_0 R_0$ gives

$$\tfrac{1}{2}mv_E^2 = mg_0R_0$$

i.e. $v_E = \sqrt{(2g_0R_0)}$

Since $\sqrt{(g_0R_0)}$ is the circular velocity required to orbit close to the earth's surface (7.9 km/s) then

v_E = (circular velocity required at earth's surface) × $\sqrt{2}$

$\quad = 7.9 \times \sqrt{2}$

$\quad = 11.2 \text{ km/s}$

This speed, 11.2 km/s, is that at which the body would escape to infinity and also the speed at which it would strike the earth if it fell from infinity. If the body's speed fails to reach 11.2 km/s the body falls back to earth and the maximum height reached depends on the initial velocity and angle of projection.

If a body is launched at a radius R from the earth's centre the escape speed becomes

$v_E = \sqrt{(gR)} \times \sqrt{2}$

$\quad = $ (circular velocity at radius R) × $\sqrt{2}$

For example, in the special orbit dealt with in the next paragraph it is shown that $R = 42\,200$ km, $g = 0.22$ m/s^2 so that its circular velocity is about 3 km/s and the escape speed if launched from this orbit becomes 4.2 km/s.

(11.8) Geostationary orbit

A *geostationary* satellite or orbiter (GEO) is one travelling **eastwards** in a circular equatorial orbit synchronous* with the earth's motion. Thus its angular velocity is equal to that of the earth and the inclination of its orbital plane to that of the equator is zero. Its eastwards motion about the earth's polar axis means that the satellite appears to be stationary above a fixed point on the earth's surface at the equator.

Consider the motion of a GEO. Its period is 24 hours approximately and the equations for its motion are

$$g = g_0\left(\frac{R_0}{R}\right)^2$$

where $g_0 = 9.8$ m/s^2 and $R_0 = 6370 \times 10^3$ m,

$$t = \frac{2\pi R}{v}$$

where $t = 24 \times 3600$ s,

* The geostationary orbit is a special case of an earth-synchronous orbit which has a period of about 24 hours (23 h 56 min 4 s, to be exact). The orbital plane of a synchronous orbit is inclined to that of the equator and although it appears to be stationary in *longitude* its inclination causes it to travel in a figure-of-eight pattern, once every 24 hours. For example, if the inclination is 20° the pattern would have its limits in latitude of 20° N and 20° S.

and

$$v^2 = gR$$

It is left as an exercise in mathematics for the student to substitute the known values of g_0, R_0 and t in these equations to show that for a GEO the results are:

- its radially inwards acceleration, $g = \mathbf{0{\cdot}22\,m/s^2}$
- its orbital radius, $R = \mathbf{42\,200\,km}$;
- its height above the earth's atmosphere $(R\text{-}R_0) = \mathbf{35\,830\,km}$
- its linear velocity, $v = \mathbf{3070\,m/s,\ 3\,km/s\ or\ 11\,050\,km/h}$

Thus a satellite circling the earth in this important orbit at 36 000 km has to have a speed of 3 km/s. The shortest up-and-down link distance for a radiated signal between a GEO and a ground station directly beneath it on the equator is 72 000 km resulting in a long transmission time of 240 ms. Nevertheless, this orbital path has a number of advantages; a geostationary orbit provides

- a stable body for antennas and instruments;
- uninterrupted 'line-of-sight between satellite and ground station (signals from GEOs beam into a ground station in this country at around 40° above the southern horizon);
- entire coverage (theoretically) of the globe[†] by using only three satellites, 120° apart, Fig. 11.1;
- a position suitable for the use of satellites as signal relay 'towers'.

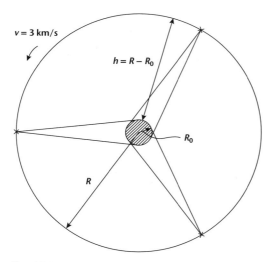

Fig. 11.1 Geostationary orbit. Theoretical three-satellite global coverage; 120° apart, $R \approx 42\,370\,km$, $R_0 = 6370\,km$, $h \approx 36\,000\,km$

[†] Arthur C. Clarke, engineer and scientific writer, foresaw this in 1945 in connection with television and communications.

However, with radiant energy the beams being sent out from ground station transmitters to satellites diverge with distance and those reaching the GEOs are very wide indeed, resulting in signal strength 'spreading loss'. Deep space probes present more difficulties requiring the provision of high-powered radiation beams and very large diameter, accurately formed antennae.

Since the launch of Intelstat's first geostationary satellite, Early Bird, in 1965, its satellite communication system now has a fleet of 23 to 31. In total, in circulation around the globe, there are now in position in geostationary orbit over 150 satellites primarily concerned with communications, weather and earth-surface monitoring. International agencies allocate 'slots' in arcs of 1° or less of the orbital plane; the GEOs are more numerous over the land masses and are accompanied in their path by defunct satellites and debris.

Rocket launch

A geostationary satellite may have a mass on the launch pad of over 4 t and to propel it 36 000 km high requires a multi-stage rocket together with thruster rockets on the satellite for final positioning. It is usually fired **eastwards** from a site near the equator or possibly from a rig in equatorial waters, to take advantage of the earth's maximum surface speed of nearly 500 m/s or 1450 km/h. The European Space Agency (ESA) launches the Ariane satellite-carrying rockets from Kourou, French Guiana, on the north-east coast of S. America. Several transfer elliptical orbits are used when the objective is a geostationary orbit, the first being as close as 200 km to earth then each following orbit has its apogee and perigee increased before the satellite reaches its required speed of 3 km/s in circular orbit.

11.9 Use of the Universal Gravitational Constant

The velocity and periodic time of a satellite in circular orbit around the earth may be found using the basic statement of Newton's law of gravitation given at the beginning of this chapter. The law states the pull of gravity F in terms of the mass of a body, m, the universal gravitational constant G, the mass of the earth, M, and the distance R of the body from the centre of the earth. Thus, given that $G = 6\cdot67 \times 10^{-11}\,\mathrm{N\,m^2/kg^2}$ and that the mass of the earth $M = 6 \times 10^{24}\,\mathrm{kg}$, and working throughout in metres, seconds and kilograms, we have, for a body orbiting at linear speed v,

$$F = \frac{mv^2}{R} = \frac{GmM}{R^2}$$

i.e.
$$v = \sqrt{\left(\frac{GM}{R}\right)}$$

$$= \sqrt{\left(\frac{6\cdot67 \times 10^{-11} \times 6 \times 10^{24}}{R}\right)}$$

$$= \frac{2 \times 10^7}{\sqrt{R}}\ \mathrm{m/s\ approx.}$$

where R is in metres.

The period t is given by

$$t = \frac{2\pi R}{v}$$

$$= 2\pi\sqrt{\left(\frac{R^3}{GM}\right)}$$

$$= 3\cdot14 \times 10^{-7}\sqrt{R^3}\,\text{s approx.}$$

Since the force F is the weight of the body, mg, we may write for a body at the surface of the earth where $g = g_0$ and $R = R_0$,

$$mg_0 = \frac{GmM}{R_0^2}$$

i.e. $$M = \frac{g_0 R_0^2}{G}$$

which gives the mass of the earth in terms of g_0, R_0 and g. The values of g_0 and G can be determined experimentally and the value of R_0 may be measured. Thus taking $g_0 = 9\cdot8\,\text{m/s}^2$, the values of R_0 and G as above, we may obtain a value for the mass of the earth, i.e.

$$M = \frac{9\cdot8 \times (6370 \times 10^3)^2}{6\cdot67 \times 10^{-11}} = 5\cdot96 \times 10^{24}\,\text{kg}$$

Determination of G

The first accurate figure for G was arrived at from the results of an experiment carried out around 1800 using a *torsion pendulum* in the 'Cavendish experiment', which was based on an apparatus designed by an earlier scientist; this experiment, very much refined, gives the current internationally accepted value for G, apparently agreed to at least three significant figures with an uncertainty of 128 parts per million (ppm). A variety of experiments to find the mean density of the earth were performed prior to the pendulum method.

Henry Cavendish also set out to estimate the mass of the earth. He used a light barbell carrying a small lead sphere at each end, oscillating in the horizontal plane, when suspended from a very long, thin wire. Two similar, much larger spheres were arranged, one on either side of the rod, so that they could exert gravitational forces on the small spheres and affect the oscillations. From the experiment he determined the gravitional attraction force between the small and large spheres. By comparing the gravitational force of the earth on the small sphere (its weight) with the force exerted on the small sphere by the larger one, he did obtain a fairly accurate value for the mass of the earth and hence its mean density. Using the experimental data the values corresponding to F, m, M and R in Newton's law above were obtained and hence a fairly accurate value for G calculated; F is the gravitational force between the small and large spheres, m and M their respective masses and R the distance between their mass centres when force F is acting. Inherent difficulties in the experiment arise from the gravitational attraction between the spheres being extremely weak, the need to shield the apparatus from disturbing influences and ascertaining the torsional properties of the material of the wire. The refinements to the experiment include the use of a vacuum chamber, improved temperature control, tungsten spheres of uniform density, a modern quartz fibre for the suspension and enhanced methods of measurement.

The figure for G is not yet to the accuracy that scientists seek in a fundamental constant and the experiment has been repeated many times with improvements in techniques. Other varieties of experiment have been performed; one attempt repeated an alternative 'beam-balance' method from the past and different arrangements, size and shape of test masses have been tried without yet improving on the pendulum method. In obtaining the value of G by an experiment of any kind, reasons for the result being at variance with the accepted value can usually be produced. However, in a most recent determination no weakness could be ascertained to explain a critical difference. The latest proposal to resolve the problem requires a new type of experiment to be carried out in a drag-free satellite orbiting the earth.

Note:

- Care must be taken in numerical problems to work throughout in metres and seconds.
- The formulae obtained above apply to any body orbiting another, for example, a body orbiting a planet, the moon orbiting the earth or the earth orbiting the sun. Thus for a lunar module in orbit the values of $g_0 = 1.62\,\text{m/s}^2$ and $R_0 = 1740\,\text{km}$ for the moon must be used.
- The worked examples and the problems below assume circular orbits.

Example

A navigation satellite of mass $240\,\text{kg}$ orbits the earth with a linear speed of $18\,000\,\text{km/h}$. Assuming circular orbit, find the height of its path, orbital period and the radially inward force controlling its motion. Radius of earth's surface $= 6370\,\text{km}$.

Solution

$$v = 18\,000\,\text{km/h} = 5000\,\text{m/s};\ R_0 = 6370 \times 10^3\,\text{m};\ g_0 = 9.8\,\text{m/s}^2$$

The acceleration due to gravity at the satellite is

$$g = g_0\left(\frac{R_0}{R}\right)^2$$

$$= 9.8\left(\frac{6370 \times 10^3}{R}\right)^2$$

$$= \frac{397.7 \times 10^{12}}{R^2}\,\text{m/s}^2$$

where R the orbital radius is in metres; and

$$v^2 = gR$$

i.e.

$$5000^2 = \frac{397 \cdot 7 \times 10^{12} \times R}{R^2}$$

therefore

$$R = 15\,900 \times 10^3\,\text{m} = 15\,900\,\text{km}$$

Height of orbit $= R - R_0$

$$= 15\,900 - 6370\,\text{m}$$

$$= 9530\,\text{km}$$

Orbital period $t = \dfrac{2\pi R}{v}$

$$= \frac{2\pi \times 15\,900 \times 10^3}{5000}$$

$$= 20 \times 10^3\,\text{s} \quad \text{or} \quad \textbf{5 h 33 min}$$

Acceleration due to gravity at the satellite is

$$g = 9 \cdot 8 \left(\frac{6370 \times 10^3}{15\,900 \times 10^3} \right)^2$$

$$= 1 \cdot 57\,\text{m/s}^2$$

Radially inward force $= mg$

$$= 240 \times 1 \cdot 57$$

$$= \textbf{377 N}$$

(**Example**)
_____

A sun-synchronous satellite has to have an exact orbital period of 101 min in circular orbit around the earth. What must be its orbital height? If the mass of the satellite is 120 kg, what is its weight in orbit? Take $R_0 = 6370$ km.

Solution

$R_0 = 6370 \times 10^3\,\text{m}; \; g_0 = 9 \cdot 8\,\text{m/s}^2$

At radius R m from the centre of the earth, the acceleration due to gravity is

$$g = g_0 \left(\frac{R_0}{R} \right)^2 = 9 \cdot 8 \left(\frac{6370 \times 10^3}{R} \right)^2 \; \text{m/s}^2$$

Orbital speed, $v = \sqrt{(gR)}$

$$= \sqrt{\left[9 \cdot 8 \left(\frac{6370 \times 10^3}{R} \right)^2 \times R \right]}$$

$$= \frac{19 \cdot 94 \times 10^6}{\sqrt{R}}$$

Orbital period, $t = \dfrac{2\pi R}{v}$

Therefore

$$101 \times 60 = \frac{2\pi R \times \sqrt{R}}{19 \cdot 94 \times 10^6}$$

i.e. $R^{3/2} = 19.23 \times 10^9 \, \text{m}$

i.e. $R = 7178 \times 10^3 \, \text{m}$ or $7178 \, \text{km}$

Height of orbit, $h = 7178 - 6370 = \textbf{808 km}$

In orbit, $g = 9\cdot8 \times \left(\dfrac{6370 \times 10^3}{7178 \times 10^3}\right)^2 = 7\cdot72 \, \text{m/s}^2$

Therefore

its weight in orbit $= mg$

$$= 120 \times 7\cdot72 \, \text{N}$$

$$= \textbf{926 N}$$

Example

The GPS satellites orbit at a height of 10 900 nautical miles. Given that the mass of the earth, $M = 5\cdot97 \times 10^{24} \, \text{kg}$, its mean radius $R_0 = 6370 \, \text{km}$ and the universal gravitational constant $G = 6\cdot67 \times 10^{-11} \, \text{N m}^2/\text{kg}^2$, find the orbital velocity and periodic time of the satellites. 1 n.m. $= 1852 \, \text{m}$.

Solution

The orbital height is $10\,900 \times 1\cdot852 = 20\,187 \, \text{km}$, hence the orbital radius, $R = 20\,187 + 6370 = 26\,557 \, \text{km}$.

Gravitational force, $F = mg = \dfrac{GmM}{R^2}$

hence the acceleration due to gravity at the satellite is

$$g = \frac{GM}{R^2}$$

$$= \frac{6\cdot67 \times 10^{-11} \times 5\cdot97 \times 10^{24}}{(26\,557 \times 10^3)^2}$$

$$= 0\cdot565 \, \text{m/s}^2$$

Orbital speed, $v = \sqrt{(gR)}$

$$= \sqrt{(0.565 \times 26\,557 \times 10^3)}$$

$$= \textbf{3870 m/s}$$

$$\text{Periodic time} = \frac{2\pi R}{v}$$

$$= \frac{2 \times \pi \times 26\,557 \times 10^3}{3870}$$

$$= 43\,100\,\text{s}$$

$$= 12\,\text{h}$$

Example

A moon-landing probe of mass 110 kg circles the moon at a height of 120 km. Find the gravitational pull of the moon on the probe, and the probe's orbital period and linear speed. For the moon's surface, take the acceleration due to gravity as 1·62 m/s² and the radius of its surface as 1740 km.

Solution

Orbital radius $R = (1740 + 120) \times 10^3\,\text{m} = 1860 \times 10^3\,\text{m}$

Acceleration due to gravity of the moon at the probe is

$$g = g_0 \left(\frac{R_0}{R}\right)^2$$

$$= 1\cdot62 \left(\frac{1740 \times 10^3}{1860 \times 10^3}\right)^2$$

$$= 1\cdot418\,\text{m/s}^2$$

Gravitational pull of the moon $= mg = 110 \times 1\cdot418 = \textbf{156 N}$

Equating centripetal force to pull of gravity gives

$$v^2 = gR$$

$$= 1\cdot418 \times 1860 \times 10^3$$

$$v = 1624\,\text{m/s} \quad \text{or} \quad \textbf{5847 km/h}$$

Orbital period $t = \dfrac{2\pi R}{v}$

$$= \frac{2\pi \times 1860 \times 10^3}{1624}$$

$$= 7196\,\text{s} \quad \text{or} \quad \textbf{2 h}$$

Example

The mass of the planet Mars is $6\cdot39 \times 10^{23}\,\text{kg}$ and its mean diameter is 6770 km. Find the acceleration due to gravity at its surface and the speed of escape needed from its surface. If a probe of mass 112 kg circles the planet at a height of 1100 km, find its kinetic energy. $G = 6\cdot67 \times 10^{-11}\,\text{N m}^2/\text{kg}^2$.

Solution

$R_0 = 3385 \times 10^3$ m; mass of planet, $M = 6\cdot39 \times 10^{23}$ kg

The force of gravity, F, on a body of mass m at the surface of the planet is:

$$F = mg_0 = \frac{GmM}{R_0^2}$$

i.e. $g_0 = \dfrac{GM}{R_0^2}$

$$= \frac{6\cdot67 \times 10^{-11} \times 6\cdot39 \times 10^{23}}{(3385 \times 10^3)^2}$$

$$= 3\cdot72 \text{ m/s}^2$$

The orbital velocity to circle the planet close to its surface is

$\sqrt{(g_0R_0)} = \sqrt{(3\cdot72 \times 3385 \times 10^3)} = 3549$ m/s

Escape speed, $v_{\text{E}} = 3549 \times \sqrt{2}$

$$= 5020 \text{ m/s} \quad \text{or} \quad \textbf{5 km/s}$$

At 1100 km, $R = 3385 + 1100 = 4485$ km.
Acceleration due to gravity at radius R is

$$g = g_0\left(\frac{R_0}{R}\right)^2 = 3\cdot72 \times \left(\frac{3385}{4485}\right)^2 = 2\cdot12 \text{ m/s}^2$$

Orbital velocity, $v = \sqrt{(gR)}$

$$= \sqrt{(2\cdot12 \times 4485 \times 10^3)}$$

$$= 3084 \text{ m/s}$$

Therefore,

kinetic energy of probe $= \frac{1}{2}mv^2$

$$= \frac{1}{2} \times 112 \times 3084^2$$

$$= 533 \times 10^6 \text{ J} \quad \text{or} \quad \textbf{533 kJ}$$

Problems

(For earth's surface, $R_0 = 6370$ km; $g_0 = 9\cdot8$ m/s^2. For moon's surface, $R_0 = 1740$ km; $g_0 = 1\cdot62$ m/s^2. All orbits are assumed circular.)

11.1 A global weather satellite circles the earth in a path 7370 km from the centre of the earth. Find its acceleration towards the earth, linear speed and period of revolution.   (7·32 m/s^2; 26 440 km/h; 105 min)

11.2 A low-orbit (short-life) photo-reconnaissance satellite orbits in a path where the acceleration due to gravity is 9·2 m/s^2. Find the height of its path, linear speed and orbital period.   (204 km; 28 000 km/h; 88·5 min)

11.3 A detection satellite travels in a circular path 96 750 km from the centre of the earth. What is its linear speed and the shortest transmission time of the up-and-down signal? (7298 km/h; 603 ms)

11.4 Find the gravitational pull on a 600 kg weather satellite orbiting at 914 km above the earth's surface. (4·5 kN)

11.5 Treating earth as a satellite in orbit around the sun, show that its orbital speed is about 29·7 km/s and its centripetal acceleration towards the sun is about 0·006 m/s². The radius of the earth's orbit around the sun is $1·49 \times 10^8$ km and its orbital period is 365·3 days.

11.6 A Navstar satellite has a 12 hour orbit. If its acceleration due to gravity (radially inwards) is 0·563 m/s², find the radius of its path around the earth and its linear speed. (26 580 km; 13 930 km/h)

11.7 A meteorological satellite intended to be geostationary is wrongly flighted and its distance from the centre of earth is 39 830 km. Find its orbital period. (22 h)

11.8 An Orbiter craft of total mass 82 t carries a payload when it orbits the earth at a speed of 7·72 km/s. Show that its radius of orbit is about 6670 km, orbital period 90·5 mins and weight in orbit, 733 kN.

11.9 What is the theoretical escape speed for a body launched 1100 km from the earth's surface? (10·3 km/s)

11.10 Express the escape speed from the earth's surface in terms of G, M and R_0.

$$\left(\sqrt{\left(\frac{2GM}{R_0} \right)} \right)$$

11.11 A lunar module of mass 360 kg circles the moon at a height of 100 km above its surface. Find the linear speed of the module and its period of revolution. What is the gravitational pull of the moon on the module? (5880 km/h; 118 min; 522 N)

11.12 A space probe orbiting the moon has an orbital period of 111 minutes. Find its linear speed and the height of its path above the surface. (6000 km/h; 26·3 km)

11.13 A space probe being prepared for launching weighs 118 N on the ground. What is its mass? What is its weight, i.e. the pull of gravity (i) when orbiting the earth where the acceleration due to gravity is 0·49 m/s², (ii) when on the moon's surface and (iii) when orbiting the moon at a height of 100 km? (12 kg; 5·9 N; 19·5 N; 17·4 N)

11.14 The moon's surface has no atmosphere so that there is no retarding force on a low-orbiting probe to slow it down. Also an orbital path of 10 km average height would clear the lunar mountains. Assuming such a circular orbit at 10 km height is possible, at what linear speed would the probe travel and how long would it take to make one complete circuit of the moon? (1·67 km/s; 110 mins)

11.15 Show that the escape speed from the moon's surface is 2·4 km/s. At what height from the surface is the escape speed 2 km/s?   (710 km)

11.16 A piece of equipment weighs 950 N on the launch pad on the earth's surface. What is the weight of the equipment (i) when circling the earth at 800 km, (ii) on the surface of Mars and (iii) 150 km above the surface of Mars? For Mars $R_0 = 3385$ km, $g_0 = 3.72$ m/s^2.   (750 N; 361 N; 331 N)

11.17 What is the acceleration due to gravity at a height of 350 km above the surface of Mars? Find the velocity and period of revolution of a probe circling the planet at a height of 350 km and the speed of escape at this height. For Mars, $R_0 = 3385$ km, $g_0 = 3.72$ m/s^2.
                    (3·06 m/s^2; 3·4 km/s; 116 min; 4·8 km/s)

11.18 What is the periodic time for a probe satellite circling close to the surface of Jupiter? For Jupiter, $R_0 = 7 \times 10^4$ km, $g_0 = 25.8$ m/s^2.   (173 min)

11.19 Taking the mass of Jupiter as 1.9×10^{27} kg, its mean radius as 70 070 km and the universal gravitational constant as 6.67×10^{-11} N m^2/kg^2 show that the acceleration due to gravity at its surface is 25·8 m/s^2 and find the speed with which a body will escape the influence of its gravity field when launched from its surface.   (60·1 km/s)

(11.10) Notes on other orbits of interest

Polar and sun-synchronous orbits

Polar orbits serve to cover the high latitudes with instruments on the satellites able to scan a fresh swathe of the earth's surface as the globe rotates beneath them. A polar elliptical orbit allows a satellite to swoop to one of the poles for close study of the region before travelling outwards a great distance. *Near-polar* orbits are launched from a position on the equator as for geostationary orbiters but directed towards the north.

A special case of a low, near-circular, near-polar orbit with advantages for earth-observation tasks is the so-called *sun-synchronous* orbit where the orbital plane of the satellite apparently remains stationary when 'viewed' from the sun. The orbit's parameters are carefully chosen, approximately at right angles to the earth–sun plane, about 96°–99°, with an orbital period of around 100 min in which time the earth has rotated roughly 25° of longitude.

An example of a sun-synchronous satellite is the first of the European Remote Sensing Satellites, ERS-1, still circling the earth although dormant. It has a mass of 2.4 t, length 12 m, mean orbital height 780 km, inclination 98·5°, period 101 min, speed 7·4 km/s and it makes about 29 passes of the equator each day, crossing the equatorial line at 10.30 a.m. local time and again at 10.30 p.m. local time. Its tasks included observing the polar ice sheets, the extent of deforestation and monitoring the ocean currents and coastal changes in regard to the El Niño–Southern

Oscillation. Its earth-pointing instruments have recorded images of the same spot on earth at the same local time each day under similar sun-illumination conditions. The satellite was computer-controlled to alter its altitude slightly and thus to repeat its pattern of tracks after certain intervals using seven ground stations.

ERS-2 is now orbiting 30 min later on a parallel track, monitoring a fresh swathe of the earth's surface at each orbit and continuing ERS-1's observations of the Antarctic 'ozone-hole', the levels of ozone in the northern hemisphere and continuous, high precision, sea-surface temperature measurements.

Low circular and elliptical orbits

The first simple passive 'balloon' type satellites (Echo) used for communications circled the earth at 1500 km and these were followed by the Sputnik 1 and Explorer 1 types using low elliptical paths coming in to earth about 200 km high. There are many low earth orbiters (LEOs) in both circular and elliptical orbits below 2000 km with some concentration of numbers in the 500–600 km range. The 2000–9000 km range is little used mainly because of the intense inner radiation belt. The existence of the Van Allen radiation belts was confirmed by the work on the early Explorer satellites. The belts consist of concentrations of charged particles (protons and electrons) trapped by the earth's magnetic field. There are two zones, roughly 1000–5000 km and 15 000–25 000 km above the equator; from the equator the particles 'migrate' with a spiral motion to and from the poles.

The radiation, together with the effects of layers of the atmosphere such as the ionosphere and troposphere, affect the solar power, instrumentation, etc. of the various satellite systems in different ways. Astronomical satellites are often fairly low orbiters; for example, the Hubble Space Telescope circles at 600 km, a suitable height for the Shuttle to carry out repairs when necessary and high enough to be clear of the atmosphere since the density of the air at this height is only 4×10^{-13} kg/m^3. There is still a drag effect even in this thin atmosphere, resulting in a very slow gradual loss of height requiring periodical correction to the satellite's position. A current example of a handheld telephone LEO system is Motorola's Iridium 40-satellite constellation orbiting at 780 km, just below the radiation belt.

For LEOs with an orbital period of 90–120 min the advantages are:

- short transmission time for radiated signals and reduction in signal losses;
- smaller size satellites can be used;
- lower power costs for launching; clusters of satellites can be fired up by a single-stage satellite.

The disadvantages are:

- a small 'footprint' for radiated signals;
- many satellites are required for uninterrupted coverage for as one satellite disappears over the horizon another must appear;
- drag and tidal forces are relatively strong;
- storage batteries are essential because for a large part of orbital time the satellite is in shadow and the solar cells are not receiving the sun's energy;

- the antennas need to be large and to be continually adjusted;
- shorter satellite lifetime.

There are also other very technical and important advantages and disadvantages in regard to various orbital heights. These involve, for example, propagation of signals, antenna noise, *Doppler* effects and the problems with certain layers of the atmosphere such as the ionosphere and troposphere.

Geostationary satellites are expensive to launch and maintain, have a fairly long transmission time for signals and require higher satellite powers and larger antenna for some puposes. LEOs also have their difficulties so that cheaper and perhaps more efficient alternatives to existing satellite systems are continually being sought; for example, huge stabilized, ground-controlled airships 'floating' in the stratosphere, well above the operational cruising height of airliners. Another project revives and modernizes an old concept by planning to employ large, light-weight, robot aircraft under ground control, which will rely on solar power or the latest design of batteries to orbit the earth high in the upper stratosphere.

Intermediate circular orbits (ICOs)

Intermediate orbits are roughly classified as those in the 5000–15 000 km range. An example is the 'ICO Global Communications' fleet of 10 satellites orbiting clear of the radiation belts at just over 10 000 km with a period of about 6 h and an up-and-down transmission time for signals between 68 and 104 ms.

Global positioning system (GPS)

This important system has a constellation of three-axis stabilized Navstar satellites at an altitude of around 20 000 km or 10 900 nautical miles. It has 24 satellites, including three spares circling the earth every 12 h in six planes, staggered at 60°, inclination 55°, providing highly accurate position fixing for receivers on the ground, at sea or in the air. Each satellite emits signals giving its identity, position and *exact* time of broadcast. A receiver notes the frequency and compares the transmission time with reception from four satellites and then determines its position as regards longitude, latitude and height. The position can be fixed to within less than 100 m and with refinements and high-quality navigation signals, this can be reduced to within a few metres. The system relies on precise measurement of time to within 180 nanoseconds based on atomic clocks and this is the most accurate of all the quantities measured.

Sun-observation satellites (SOHO, ACE)

The European Space Agency (ESA) has for some time had a solar science spacecraft, Ulysses, observing the sun as its line of flight took it past the sun's poles. Now, jointly with NASA another spacecraft SOHO – Solar and Heliospheric Observatory – has been placed in a unique position to provide continuous observation of the sun. This three-axis stabilized observatory was launched by the Space Shuttle and positioned with the aid of Jupiter's gravity field. The uniqueness is brought about by the effects of the gravity fields of the three massive bodies, earth, moon and sun, together with the centrifugal forces caused by the masses spinning about their own axes and rotating in space. There are points in the vicinity of any two massive bodies such as the earth (the *secondary*) and the sun (the *primary*) where the gravity and centrifugal forces produce points of balance. These are called Lagrangian gravity-balance points after the 19th century mathematician Joseph Lagrange who made the calculations concerning the *three-body problem* of gravity fields. For the earth and sun there are five points in all, three of which lie along the centreline joining the mass centres of the earth and sun. The point where SOHO is said to 'sit', but actually moves in a 'halo' orbit, is at the L1 Lagrangian point, 1.5×10^6 km from earth, the only one of the three points on the centreline between earth and sun. In this position it is well clear of two difficult regions, the terrestrial atmosphere and the earth's magnetic field.

SOHO was in its operational position, almost free of gravity, in February 1996 with an expected useful life of at least two years. Some of its instruments continually point at the sun's core and corona and others monitor the solar winds, the stream of electrically charged particles issuing from the sun. In 1998 it was 'lost' for a short period and when found had to be repowered before gaining a further lease of life.

SOHO has now been followed by the 1997 launch of a further spacecraft, the Advanced Composition Explorer (ACE), moving in an elliptical halo orbit (apogee 3×10^5 km, perigee 1.5×10^5 km) around the same gravity-balance point. It was launched by a Delta rocket from Cape Canaveral. A primary objective of ACE is to study the charged particles of the solar wind and of the surrounding matter in deep space so as to provide a one-hour warning of sunstorms to satellite operators and power generating companies. It has an expected mission life of 2–5 years.

For further information on the work of the European Space Agency and on the satellites and spacecraft mentioned above reference should be made to the British National Space Centre, 151 Buckingham Palace Road, London.

Chapter 12

Stress and strain

· ·

So far we have seen that a bar forming part of a structure or a machine may be subject to load. To calculate the load on a bar, we have assumed the bar to be 'rigid' and we have ignored the effects of the load on the size, shape and strength of the bar. We must now answer the question: what is the load that may be allowed on any part of a structure?

12.1 Load

One end of a bar may be subjected to a push or a pull. Since the forces acting on the bar must be equal and opposite a pull on one end must result in an equal and opposite pull on the other end, and the bar is in *tension*. Similarly, a push on one end is accompanied by a push on the other end, and the bar is in *compression*.

12.2 Deformation

When a body is loaded in any manner it deforms. Its length, cross-sectional area and volume are changed because of the load. In particular, if a bar is pulled, it stretches, and the increase in length is called the *extension*. Similarly, when compressed, the bar contracts in length. Extension is accompanied by a decrease in cross-sectional area, whereas compression is accompanied by an increase.

12.3 Relation between load and deformation

Evidently the greater the load on a bar, the greater the extension or compression. If, in an experiment, the load is gradually increased and the extension measured for each load we may plot the load against the extension and obtain a *load–extension graph*.

Figure 12.1 shows a typical load–extension curve for a bar loaded in tension. The shape and scale of the curve depends on the material. Figure 12.1 is sketched for aluminum. The extension increases with the load until the material breaks.

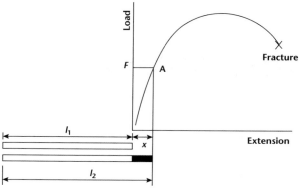

Fig. 12.1

12.4 Elasticity

A material is said to be *elastic* if, after being extended by a load, it returns to its original length when unloaded.

Figure 12.1 shows a bar of elastic material of length l_1 when unloaded and the extension x when loaded by a force F. When the load is removed the bar returns to its original length l_1 and the extension is zero. Many materials, such as iron, brass and copper, are elastic provided they are not overloaded, i.e. for small extensions.

12.5 Hooke's law

For most metals it is found experimentally that the extension is proportional to the load, provided that the load is not too great. The load–extension curve is then a straight line. The metal is said to obey *Hooke's law*. This law states that the extension produced in a *(linearly)* elastic material is directly proportional to the load which produces it.

Figure 12.2 shows part of the load–extension graph for a steel bar. The portion OA of the graph is straight and the steel obeys Hooke's law up to the point A. Beyond A the graph is curved and since the load is no longer proportional to the extension, point A is called the *proportional limit* or *limit of proportionality*. Hooke's law applies only up to point A, and for any point between O and A the relation between load F and extension x is

$$F = (\text{constant}) \times x$$

The value of the constant depends on the material of the bar and on its dimensions.

If the bar is loaded beyond the proportional limit A the metal may still be elastic up to some point B, which is known as the *elastic limit* or *limit of elasticity*. If unloaded it will still return to its original length. The range of load, or extension, up to point B is known as the *elastic range*.

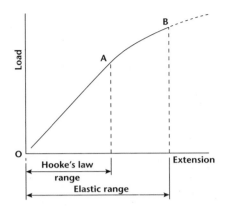

Fig. 12.2

12.6 The plastic range

If, as in Fig. 12.3, the extension x is continued beyond the elastic limit B, the material will not return to its original length when unloaded. If unloaded from a point on the curve beyond B there will be a permanent extension left in the bar at no load. This is called a *permanent set* and is shown by the length s.

Beyond the elastic limit B the material is said to be *plastic*. The part of the load–extension curve beyond B is said to lie in the *plastic range*.

The term plastic should not be confused with the popular idea of a 'plastic' material. A so-called 'plastic' may behave as an elastic material at low loads.

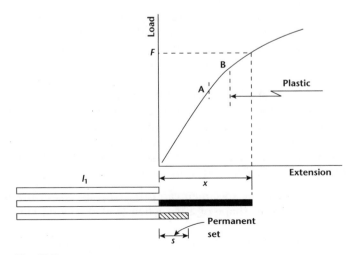

Fig. 12.3

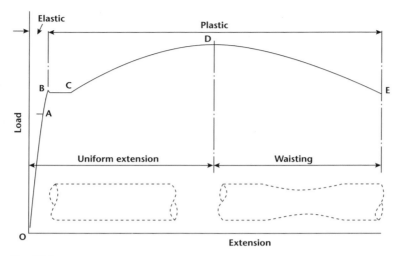

Fig. 12.4

12.7 The tensile test

A complete tensile test on a metal involves loading a specimen bar of the metal and increasing the extension at each value of the load until the bar fractures. Figure 12.4 is a typical load–extension curve obtained for a mild-steel bar but with the initial tiny elastic extension between O and B very much exaggerated for clarity.

The portion ABC of the curve has already been described. A is the proportional limit, B the limit of elasticity. Just beyond B mild steel will extend without further increase in load up to some point C. The metal is said to *yield* and the point at which it occurs is called the *yield point*.

At C the load must be increased to provide any further increase in extension. The load–extension curve continues to rise up to the maximum load at point D. Between no load at O and maximum load at D the specimen, usually a uniform bar, extends uniformly along its length. Its shape does not alter even though its length changes.

At point D the specimen starts to thin rapidly in one region along its length. A 'waist' or 'neck' forms at this region. The area of metal carrying load at the waist decreases rapidly and the load required to maintain the same rate of extension also decreases. The load–extension graph therefore begins to curve downwards along DE. The bar eventually fractures at point E at a load less than the maximum.

Two important facts may be deduced from the load–extension curve:

1. The maximum load which the specimen can carry corresponds to the point D on the curve. The change in shape of a bar when loaded beyond D is such as to make the bar useless.
2. The extension of the specimen in the elastic range is small up to the point B. The extension beyond B is large compared with the tiny extension up to B.

A structural member, however, should not be loaded above the yield point corresponding to point B, otherwise permanent distortion will occur.

12.8 Intensity of loading: stress

It is evident that the load that may be carried by a bar 20 mm in diameter is less than the load that may be carried by a 40 mm diameter bar of the same material. It is found that the load that can be carried is proportional to the cross-sectional area normal to the direction of the load. The greater the area, the greater the load that can be carried. Hence we are interested in the *intensity* of loading rather than the actual load.

The *load intensity* or *stress* (σ) is the load carried by unit area, Fig. 12.5, thus

$$\text{stress} = \frac{\text{load}}{\text{area}}$$

i.e. $\sigma = \dfrac{F}{A}$

where A is the area of section *normal* to the load.

The stress is tensile if the load is tensile, compressive if the load is compressive. The derived SI unit of stress (and of pressure) is the *newton per square metre* ($\mathbf{N/m^2}$). This unit is called the *pascal* (**Pa**) (*see* section 16.1), but in this text we shall restrict ourselves as far as stress values are concerned to N/m^2 and the following forms: kN/m^2, MN/m^2, GN/m^2 and (particularly in the construction industry) N/mm^2, kN/mm^2. Note that

$1\,N/mm^2 = 1\,MN/m^2$

$1\,kN/mm^2 = 1\,GN/m^2$

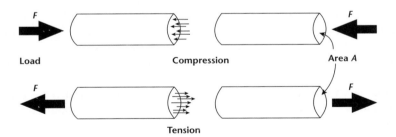

Fig. 12.5

12.9 Strain

The extension of a uniform bar specimen for a given load is found to be proportional to the length of the specimen, provided that 'waisting' does not occur.

The extension ε of unit length is called the *strain*. Thus

$$\text{strain} = \frac{\text{extension}}{\text{original length}}$$

or

$$\varepsilon = \frac{x}{l}$$

For example, if a 100 mm bar extends 1 mm the strain is 0·01 mm/mm or simply 0·01.

(12.10) Relation between stress and strain

For a material that obeys Hooke's law, the extension is directly proportional to the load, up to the limit of proportionality, i.e.

$$\frac{\text{load}}{\text{extension}} = \text{constant}$$

Therefore

$$\frac{\text{stress} \times \text{area}}{\text{strain} \times \text{original length}} = \text{constant}$$

Since the area of section and the original length are constant for a given bar, then

$$\frac{\text{stress}}{\text{strain}} = \text{constant} = E$$

i.e. $\quad \dfrac{\sigma}{\varepsilon} = E$

where E is called the *modulus of elasticity*, or *Young's modulus*.

E is a constant for any one material but takes different values for different materials. The units of E are the same as those for stress but are usually stated in the larger multiples GN/m². In some fields of work the form kN/mm² is used. Thus for mild steel, $E \simeq 200\,\text{GN/m}^2 = 200\,\text{kN/mm}^2 = 200 \times 10^9\,\text{N/m}^2$. In calculations the basic unit N/m² should normally be used.

Approximate values of E, obtained by experiment, are given in Table 12.1. Since there are many types of steels, brasses, cast-irons and alloys of aluminium, these values may vary from one type to another.

Table 12.1 Moduli of elasticity

Material	E (GN/m² or kN/mm²)	E (N/m²)
Mild steel	200	200×10^9
Cast-iron	120	120×10^9
Brass	84	84×10^9
Aluminium	70	70×10^9

An aluminium tie-rod is 48 cm long and 15 mm in diameter (Fig. 12.6). If it is turned down to 7·5 mm in diameter for a length of 6 cm at each end, calculate the total extension due to a tensile load of 40 000 N. $E = 70 \, \text{GN/m}^2$.

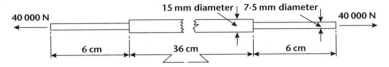

Fig. 12.6

Solution

The total extension of the tie is made up of the sum of the extensions of the two parts.

The extensions must be calculated separately since, although the same load acts on both parts of the rod, the stresses are different because of the change in cross-sectional area.

For the 15 mm diameter portion, 36 cm long,

$$\text{stress} = \frac{\text{load}}{\text{area}}$$

i.e. $\sigma = \dfrac{F}{A}$

$$= \frac{40\,000}{\frac{\pi}{4} 0 \cdot 015^2}$$

$$= 226 \times 10^6 \, \text{N/m}^2$$

Extension = strain × length

$$= \frac{\text{stress}}{E} \times \text{length}$$

$$= \frac{226 \times 10^6}{70 \times 10^9} \times 0 \cdot 36$$

$$= 0 \cdot 001\,17 \, \text{m} = 1 \cdot 17 \, \text{mm}$$

For the 7·5 mm diameter portion, 12 cm long,

$$\sigma = \frac{40\,000}{\frac{\pi}{4} 0 \cdot 0075^2}$$

$$= 904 \times 10^6 \, \text{N/m}^2$$

$$\text{Extension} = \frac{904 \times 10^6}{70 \times 10^9} \times 0{\cdot}12$$

$$= 0{\cdot}001\,55\,\text{m} = 1{\cdot}55\,\text{mm}$$

Therefore

$$\text{total extension} = 1{\cdot}17 + 1{\cdot}55 = \mathbf{2{\cdot}72\,mm}$$

(12.11) Calculation of *E* from the load–extension diagram

Figure 12.7 shows a typical load–extension diagram for a length of mild-steel wire 0·4 mm in diameter. The points plotted represent the load in newtons required to produce an extension over a total length of 4·5 m.

The best straight line is drawn through the plotted points so that they lie equally spaced on either side of the line as shown. It is seen that the straight line does not go through the origin; the reason for this is usually that a small load is necessary to pull the wire out straight before the test is started.

Since most of the points are likely to be slightly in error it would be incorrect to calculate *E* from any one of them. The most accurate method is to calculate *E* from the *slope* of the graph. This eliminates errors due to scatter in the results. Also, it no longer matters what load was put on the wire in order to straighten it out initially. Now

$$E = \frac{\text{stress}}{\text{strain}} = \frac{\sigma}{\varepsilon}$$

$$\text{stress } \sigma = \frac{\text{load}}{\text{area}} = \frac{F}{A}$$

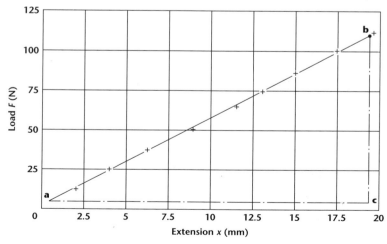

Fig. 12.7

$$\text{strain } \varepsilon = \frac{\text{extension}}{\text{original length}} = \frac{x}{l}$$

Hence

$$E = \frac{\sigma}{\varepsilon}$$

$$= \frac{F/A}{x/l}$$

$$= \frac{l}{A} \times \frac{F}{x}$$

Now l/A is a constant for this wire and F/x is given by the slope of the load–extension graph. Therefore.

$$E = \frac{l}{A} \times \text{slope of the graph}$$

For the wire,

$$\frac{l}{A} = \frac{4 \cdot 5}{\frac{\pi}{4} \times (0 \cdot 0004)^2} = 35 \cdot 8 \times 10^6 \text{ per metre}$$

The slope F/x of the graph is found as follows.

Pick two points, **a**, **b** *on the straight line*, as far apart as possible. Draw a horizontal line through **a**, parallel with the axis of extension, and draw a vertical line through **b**, parallel to the load axis. These two lines cut at **c** as shown. Measure **ac** and **bc**. Then, from the graph,

$$\text{slope of graph} = \frac{F}{x} = \frac{bc}{ac}$$

$$= \frac{105(N)}{0 \cdot 019(m)}$$

$$= 5530 \, N/m$$

Hence

$$E = \frac{l}{A} \times \text{slope}$$

$$= 35 \cdot 8 \times 10^6 \times 5530$$

$$= 198 \times 10^9 \, N/m^2$$

$$= \mathbf{198 \, GN/m^2}$$

Note that this method depends on the best straight line being drawn very carefully through the plotted points. Also the slope should be found by taking two points on the line as far apart as possible.

12.12 Ultimate tensile and compressive stresses

The *ultimate tensile stress* (UTS) is the stress in a test at the maximum load (section 12.7) and the area on which the load acts is taken to be the original area of section of the specimen, i.e.

$$\text{UTS} = \frac{\text{maximum load}}{\text{original area}}$$

Typical average values of UTS are: ordinary cast-iron 150, mild steel 450, brass 120, annealed copper 200, cast aluminium 60, tool steel 1000 MN/m².

For some materials it is the *ultimate compression* or *crushing strength* that is important. Typical values for materials used in compression are: ordinary cast-iron 750, brass 75, cast copper 300, brick 15–50 MN/m².

12.13 Factor of safety

In practice, although the ultimate tensile stress is used as a basis for design the *maximum allowable stress* or *working stress* is always less than this value. The ratio of the UTS to the working stress is called the *factor of safety*. Thus

$$\text{factor of safety} = \frac{\text{UTS}}{\text{working stress}}$$

The factor of safety is a number, always greater than unity and chosen by experience.

The choice of a factor of safety depends on many factors, some of which are as follows:

- possible overloads;
- defects of workmanship;
- possible defects in materials;
- deterioration due to wear, corrosion, etc.;
- the amount of damage resulting if the material should fail;
- the possibility of the loads being applied suddenly or repeatedly, i.e. *dynamic* or *live loading*.

If failure might result in possible loss of life or damage expensive to repair then a high factor of safety should be used. If the properties of the material are known exactly and the nature of the loading is carefully predicted a low factor of safety may be used. If the load is steady (a *dead* load) a lower factor of safety may be used than if the load is cyclic, fluctuating or of an impact nature.

It should be remembered that, if a factor of safety is used that is greater than is necessary, the weight of material used is increased, thus increasing the cost and producing additional stresses due to the weight of the structure. In moving machinery, power is used in driving the moving parts. If too great a factor of safety is used too great a proportion of the power is taken up in driving the moving parts.

The allowable working stress should always be below the elastic limit. Typical factors of safety (for dead loads) used in practice might be as follows:

steel   3
cast-iron 4 (not used in tension)
timber   7

(**Example**)

The maximum load in a tensile test on a mild steel specimen is 90 kN. If the test-piece is 12 mm in diameter calculate the UTS. What is the working stress and greatest allowable load on a bar 25 mm in diameter? Use a factor of safety of 3.

Solution

$$\text{UTS} = \frac{\text{load}}{\text{area}}$$

$$= \frac{90 \times 1000}{\frac{\pi}{4} \times (0.012)^2}$$

$$= 795 \times 10^6 \,\text{N/m}^2 = \textbf{795 MN/m}^2$$

$$\text{Working stress} = \frac{\text{UTS}}{\text{factor of safety}}$$

$$= \frac{795}{3}$$

$$= \textbf{265 MN/m}^2$$

Greatest allowable load on a 25 mm diameter bar

$$= \text{working stress} \times \text{area} = 265 \times \frac{\pi}{4} \times 0.025^2$$

$$= 0.13 \,\text{MN} = \textbf{130 kN approx.}$$

(**Example**)

A copper wire 1.6 mm in diameter, 4 m long, extends 1.7 mm when carrying a mass of 10 kg. Find (a) the stress and strain in the wire at this load, (b) the modulus of elasticity of the copper and (c) the factor of safety if the ultimate tensile strength of the copper is 220 N/mm².

Solution
The force extending the wire is the weight of the load, $10 \times 9.8 = 98 \,\text{N}$.

(a) $\text{Stress} = \dfrac{\text{load}}{\text{area}} = \dfrac{98}{\frac{\pi}{4} \times (0.0016)^2} = 48.8 \times 10^6 \,\text{N/m}^2$

$$= \textbf{49 MN/m}^2$$

$$\text{Strain} = \frac{\text{extension}}{\text{original length}} = \frac{0 \cdot 0017}{4} = \mathbf{0 \cdot 000\,425}$$

(b) Modulus of elasticity $= \dfrac{\text{stress}}{\text{strain}}$

$$= \frac{49 \times 10^6}{0 \cdot 000\,425}$$

$$= 115 \times 10^9\,\text{N/m}^2$$

$$= \mathbf{115\,GN/m^2}$$

(c) UTS $= 220\,\text{N/mm}^2 = 220\,\text{MN/m}^2$

$$\text{Factor of safety} = \frac{\text{UTS}}{\text{working stress}} = \frac{220}{49} = \mathbf{4 \cdot 5}$$

Problems

12.1 A tie-bar in a roof truss is 20 mm in diameter and 4 m long. It carries a tensile load of 80 kN. If $E = 200\,\text{GN/m}^2$, calculate the elongation under this load.
(5·1 mm)

12.2 A cast-iron column is 15 cm in outside diameter and 12 cm in inside diameter. Calculate the maximum load it may carry if the compressive stress is not to exceed $80\,\text{MN/m}^2$.
(510 kN)

12.3 A locomotive exerts a pull of 200 kN on its train through a simple link coupling. If each link is forged from steel 30 mm in diameter, calculate the tensile stress in the link.
($142\,\text{MN/m}^2$)

12.4 The maximum load in a tensile test of a mild steel specimen 15 mm in diameter is 100 kN. What is the greatest allowable tensile load on a bar 25 mm in diameter of the same material if a factor of safety of 6 is to be used?
(46·3 kN)

12.5 In a test on a wrought-iron specimen, the stress at the proportional limit is $150\,\text{N/mm}^2$. What is the greatest tensile load that can be applied to a 10 mm diameter bar of the same material if Hooke's law is to apply?
(11·8 kN)

12.6 The stress at the limit of proportionality of a specimen of mild steel is $300\,\text{N/mm}^2$. If $E = 206\,\text{kN/mm}^2$, calculate the maximum elastic strain.
  If the elongation of a 50 mm length of the same steel is 13 mm at the maximum load, calculate the strain.
  What is the ratio of the strain at the maximum load to the strain at the elastic limit?
(0·001 46; 0·26; 178 to 1)

12.7 A steel specimen 15 mm in diameter reaches its limit of proportionality at a load of 50 kN. If it is 20 cm long, what is the corresponding elastic extension? Take $E = 206\,\text{GN/m}^2$.
(0·28 mm)

12.8 The elastic limit load of a specimen of mild steel 15 mm in diameter is 55 kN. What is the greatest allowable load on a tie-bar of the same material 50 mm in diameter if the elastic range is not to be exceeded? What is the stress at this load in newtons per square millimetre? (611 kN; 311 N/mm^2)

12.9 A specimen of an aluminum alloy, 15 mm in diameter, gave a straight-line load–extension curve under test up to a load of 40 kN. The measured extension on a length of 20 cm was 0·7 mm. Calculate the stress at the proportional limit and the modulus of elasticity for the alloy.
(226 MN/m^2; 64·6 GN/m^2)

12.10 Part of an aircraft undercarriage consists of a hollow alloy tube 80 mm in outside diameter. The axial load on the tube is 80 kN and the allowable working stress is 30 MN/m^2. If the tube is to be as light as possible, determine the minimum thickness of the tube wall. (13 mm)

12.11 A steel rail is effectively clamped along its length, and welded at the ends, so that it is unable to extend. If it were free, a rise in temperature of 1 °C would produce an extension of 0·000 011 m per metre length.
Calculate the compressive stress in the rail due to the clamping if all extension is prevented and the temperature rise is 30 °C. $E = 200$ GN/m^2.
(66 MN/m^2)

12.12 A 50 mm diameter tie-rod, 30 cm long, is turned down to 40 mm diameter over half its length. The rod takes a pull of 100 kN. If $E = 200$ kN/mm^2, calculate the extension of each part and the total extension of the rod.
(0·04; 0·06; 0·1 mm)

12.13 A structure is to be underpinned by a number of wooden struts. The total load to be carried is 1·2 MN and each strut is 100 mm square. If the maximum allowable stress in the timber is 7 N/mm^2, calculate the minimum number of struts required. (18)

12.14 A steel column is 90 cm long and 150 mm in diameter. If a hole 100 mm in diameter and 60 cm long is bored out, calculate the total compression of the column when carrying a load of 1 MN. $E = 206$ GN/m^2. (0·4 mm)

12.15 A cylindrical rubber pad 100 mm in diameter, 40 mm thick, compresses 6 mm when supporting a mass of 200 kg. Find the modulus of elasticity of the rubber in newtons per square millimetre assuming Hooke's law to apply. (1·67 N/mm^2)

12.16 A cylinder of mass 320 kg is supported at two points 500 mm apart on its longitudinal axis by two vertical wires. Each wire is 4 mm in diameter and 3 m long before attachment to the cylinder. One wire is of steel, the other of brass. If the wires are fixed at the upper end to a horizontal joist, what is the inclination of the cylinder axis to the horizontal? Take E for steel as 210 GN/m^2 and for brass 85 GN/m^2. (18′)

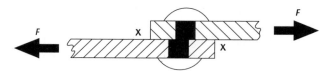

Fig. 12.8

(12.14) Shear and shear stress

Figure 12.8 shows a riveted connection, in which two plates overlap to form a lap-joint. Now consider what happens to the rivet when the joint is subjected to load. Two equal and opposite forces F applied to the plates as shown tend to move the top half of the rivet relative to the bottom half. A section of the rivet tends to slide across the interface X–X of the two plates. The rivet is said to *shear* and the material of the rivet is subject to a *shear stress*.

The shear stress τ is defined as the *load per unit area resisting shear*, i.e.

$$\text{shear stress} = \frac{\text{load } (F)}{\text{area } (A) \text{ resisting shear}}$$

or

$$\tau = \frac{F}{A}$$

This formula is similar to that for direct stress (tension or compression), but with two important differences:

1. The load F is made up of two equal and opposite forces whose lines of action are displaced, Fig. 12.8.
2. The area A resisting shear is measured on a surface *parallel* to the line of action of the shearing force.

For the rivet shown in Fig. 12.8, the area resisting shear is the area of cross-section of the rivet, i.e.

$$A = \frac{\pi}{4} d^2$$

where d = diameter of the rivet.

Since the rivet is sheared across one plane only it is said to be in *single shear*.

(12.15) Ultimate shear strength

To determine the allowable shear load that may be applied to a rivet or a pin, we need to know the greatest shear stress the material can carry before failure. This is called the *ultimate shear strength* and is the shear stress at which the rivet or pin will completely shear into two pieces.

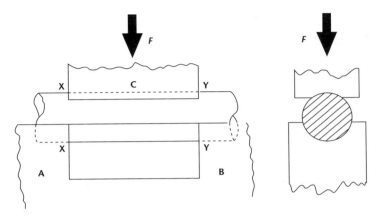

Fig. 12.9

The ultimate shear strength is usually determined by the test shown in Fig. 12.9. A bar of circular section is placed on a cradle AB and sheared by the punch C. The maximum punch load required to shear the bar into three pieces is then measured. The ultimate shear strength of the material is then the ratio

$$\frac{\text{maximum punch load}}{\textit{total}\ \text{area resisting shear}}$$

Note that shear occurs across both sections X–X and Y–Y; hence the total area resisting shear is now *twice* the cross-sectional area of the test specimen. A bar or rivet sheared in this manner is said to be in *double shear*.

Typical values of ultimate shear strength are as follows:

	MN/m²
mild steel	360
brass	150
cast-iron	150
hardened tool steel	680

The *allowable working stress* for design purposes is calculated as for direct stress, thus

$$\text{allowable shear stress} = \frac{\text{ultimate shear stress}}{\text{factor of safety}}$$

Example

A double-cover butt-joint is shown in Fig. 12.10. It is to carry a pull of 200 kN. Ten rivets are to be used in the arrangement shown. The ultimate shear strength of the rivet material is 360 MN/m². Allow a factor of safety of 4. Find the rivet diameter.

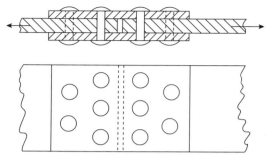

Fig. 12.10

Solution

It is usual to assume that the load is taken equally by all the rivets. Since five rivets in each plate take the full load,

$$\text{load per rivet} = \frac{200 \times 1000}{5} = 40\,000\,\text{N}$$

$$\text{Allowable shear stress} = \frac{\text{strength in shear}}{\text{factor of safety}}$$

$$= \frac{360 \times 10^6}{4}$$

$$= 90 \times 10^6\,\text{N/m}^2$$

Let the rivet diameter be d metres. Each rivet is in double shear and the area per rivet that resists shear is $2 \times \dfrac{\pi d^2}{4} = \dfrac{\pi d^2}{2}$. Therefore

$$\text{load per rivet} = \text{shear area} \times \text{stress}$$

$$40\,000 = \frac{\pi d^2}{2} \times 90 \times 10^6$$

i.e. $$d^2 = 0\cdot000\,283\,\text{m}^2$$

so that

$$d = 0\cdot017\,\text{m} = \textbf{17\,mm}$$

Example

Two coaxial shafts are to be connected by a flanged coupling having four 15 mm diameter bolts equally spaced around a 320 mm diameter pitch circle, Fig. 12.11. Determine the shear stress in the bolts when 150 kW is transmitted at 300 rev/min (a) assuming the bolts equally loaded and (b) assuming one bolt to be ahead of the others and to carry the whole torque.

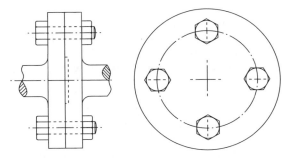

Fig. 12.11

Solution

$$\text{Power} = \frac{2\pi n T}{1000} \text{ kW}$$

Therefore torque transmitted,

$$T = \frac{\text{power}}{2\pi n}$$

$$= \frac{150 \times 1000 \times 60}{2\pi \times 300}$$

$$= 4770\,\text{N m}$$

Therefore tangential (shear) force at bolt circle

$$= \frac{\text{torque}}{\text{pitch circle radius}}$$

$$= \frac{4770}{0 \cdot 16}$$

$$= 29\,800\,\text{N}$$

(a) Shear stress in each bolt × area of bolt × no. of bolts = total shear force
    Therefore

$$\tau \times \frac{\pi}{4} \times (0 \cdot 015)^2 \times 4 = 29\,800$$

$$\tau = 42 \cdot 1 \times 10^6\,\text{N/m}^2 = \mathbf{42 \cdot 1\,MN/m^2}$$

(b) Shear stress in one bolt × area of bolt = total shear force
    Therefore

$$\tau \times \frac{\pi}{4}(0 \cdot 015)^2 = 29\,800$$

$$\tau = 168 \cdot 5 \times 10^6\,\text{N/m}^2 = \mathbf{168 \cdot 5\,MN/m^2}$$

12.17 Two plates overlap and are drilled for four 25 mm diameter steel rivets. The greatest allowable shear stress in the rivets is 60 N/mm². Calculate the maximum pull which can be carried at the joint formed by inserting the rivets.

(118 kN)

12.18 A 15 mm diameter steel bar is tested in double shear by a punch. It is found that the bar fails at a load of 90 kN. Calculate the ultimate strength in shear of the material.

If a bar of the same steel is used for making 25 mm diameter rivets, calculate the maximum allowable shear load (in single shear) on each rivet if a factor of safety of 6 is to be used.   (255 MN/m²; 20·9 kN)

12.19 The coupling shown, Fig. 12.12, carries a load of 10 kN, and the connecting pin is 6 mm in diameter. Calculate the shear stress in the pin. If the ultimate strength in shear is 380 MN/m², what is the factor of safety for the pin?

(177 MN/m²; 2·15)

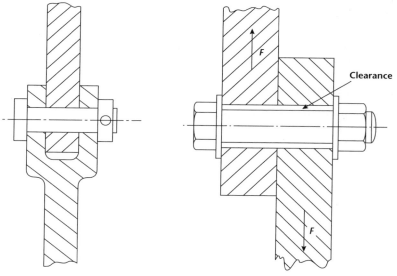

Fig. 12.12   Fig. 12.13

12.20 Two plates loaded as shown in Fig. 12.13 are fastened by a high tensile steel bolt. The load is transmitted by friction between the plates and the tensile load in the bolt due to initial tightening.

If the coefficient of friction is 0·7 and the bolt shank diameter is 20 mm, calculate the initial tensile stress in the bolt due to tightening if the load F to be transmitted is 80 kN.

If the bolt is replaced by a simple riveted joint, calculate the shear stress on a single rivet of 20 mm diameter.   (364 MN/m²; 255 MN/m²)

12.21 How many 12 mm diameter rivets are required for a single lap-joint which is to carry a load of 300 kN? The ultimate strength in shear is 360 N/mm² for the rivets and a factor of safety of at least 5 is to be used.   (37)

12.22 Two coaxial shafts are to be connected by a flanged coupling having six 6 mm diameter bolts equally spaced on a pitch circle diameter of 200 mm. If the maximum permissible shear stress in the material of the bolts is 180 N/mm², find the factor of safety for the bolts if the coupling transmits a torque of 2 kN m.   (1·5)

(12.16) Shearing and punching of flat plate

The shear stress required to cut or guillotine a plate is usually taken as the ultimate shear strength of the plate material. The area resisting shear depends on the shape of the cut-out and on whether the guillotine cuts straight or obliquely. For a straight cut, the area resisting shear is measured by the product

perimeter of cut-out × plate thickness

In Fig. 12.14 the area resisting shear by a punch of diameter d is shown with vertical shading. The perimeter of the hole is πd and therefore the area resisting shear is

$$A = \pi dt$$

where t is the thickness of the plate.

For a guillotine the area A resisting shear is equal to the plate width w multiplied by the plate thickness, i.e.

$$A = wt$$

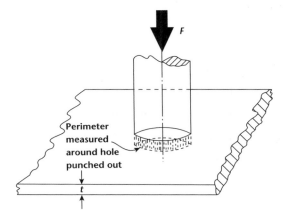

Fig. 12.14

For punching a rectangular cut-out the perimeter is $2(l + w)$, where l and w are lengths of the two sides, and the area is

$$A = 2(l + w)t$$

The *punch load F*, or force on the punch, required to cut the plate is equal to the ultimate shear stress τ multiplied by the area A resisting shear, i.e.

$$F = \tau A$$

The direct compressive stress σ on the punch is equal to the punch load divided by the area a of the punch normal to the load, i.e.

$$\sigma = \frac{F}{a}$$

For a rectangular punch

$$a = w \times l$$

For a circular punch

$$a = \frac{\pi d^2}{4}$$

Note that the thickness of plate that can be punched is limited by the strength of the punch as well as that of the plate.

Example

Calculate the greatest thickness of brass sheet that can be punched out by a steel punch 50 mm in diameter if the greatest available punch load that can be exerted by the press is 200 kN. The ultimate strength in shear for brass is 150 MN/m².

If the working stress in the steel punch is limited to 75 MN/m², what is the greatest thickness of sheet that can be punched?

Solution
Let t = thickness of brass sheet in millimetres.

$$\text{Area resisting shear} = \pi dt$$

$$= \pi \times 50 \times t$$

$$= 157t\,\text{mm}^2 = 157t \times 10^{-6}\,\text{m}^2$$

Load required to punch hole = shear area × shear strength

$$= 157t \times 10^{-6} \times 150 \times 10^6$$

$$= 23\,550t\,\text{N}$$

This must equal the available punch load. Therefore

$$23\,550t = 200 \times 1000$$

and

$$t = \mathbf{8 \cdot 5\,mm}$$

If stress in steel punch is $75\,\text{MN/m}^2$, then

maximum permissible punch load $= 75 \times 10^6 \times \dfrac{\pi d^2}{4}$

$$= 75 \times 10^6 \times \frac{\pi \times 0.05^2}{4}$$

$$= 147\,300\,\text{N}$$

Hence

$$23\,550t = 147\,500$$

and

$$t = 6.3\,\text{mm}$$

(**Example**)

A shearing machine has a flywheel of mass $100\,\text{kg}$. Assuming the mass of the flywheel to be concentrated at a radius of $2\,\text{m}$, find the greatest width of steel plate $12\,\text{mm}$ thick that can be guillotined when the reduction in speed of the flywheel is from 3 to $1.5\,\text{rev/s}$. The ultimate shear strength of the plate is $370\,\text{N/mm}^2$.

Solution

The work done in shearing the plate is equal to the kinetic energy given up by the flywheel (*see* sections 9.11 and 10.4).

Initial kinetic energy of flywheel $= \frac{1}{2}m\omega^2 r^2$

$$= \tfrac{1}{2} \times 100 \times (2\pi \times 3)^2 \times 2^2$$

$$= 71\,000\,\text{J}$$

Final kinetic energy of flywheel $= \tfrac{1}{2} \times 100 \times (2\pi \times 1.5)^2 \times 2^2$

$$= 17\,770\,\text{J}$$

Work done by shear force = loss of kinetic energy of flywheel; therefore

shear force $\times$ distance moved by force $= 71\,000 - 17\,770\,\text{J}$

or

shear force $\times$ plate thickness $= 53\,230\,\text{J}$

therefore

shear force $= \dfrac{53\,230}{0.012}$

$$= 4.44 \times 10^6\,\text{N}$$

Ultimate shear strength × shear area = shear force

i.e. $370 \times 10^6 \times 0.012 \times (\text{width}) = 4.44 \times 10^6$

so that

greatest width = **1 m approx.**

(**Problems**)

12.23 Calculate the load required to punch out a rectangular hole 100 mm by 50 mm in 2 mm thick aluminium sheet. The ultimate strength in shear is 125 MN/m².

If the maximum load that can be exerted by the press is 45 kN, what is the greatest thickness that can be punched out?   (75 kN; 1·2 mm)

12.24 Compare the load required to punch out a 75 mm diameter disc in sheet metal with that required to punch out an equilateral triangle of the same area and thickness.   (0·78:1)

12.25 Calculate the greatest thickness that can be punched in mild-steel sheet if the punch diameter is 25 mm: the greatest allowable compressive stress in the hardened steel punch is 450 N/mm² and the ultimate strength in shear of mild steel is 350 N/mm².   (8 mm)

12.26 A hole is to be punched out in steel plate 5 mm thick. The ultimate strength in shear of the plate is 360 MN/m² and the greatest allowable compressive stress in the punch is 180 MN/m². Calculate the diameter of the largest hole that can be made.   (40 mm)

12.27 A hole 50 mm square is to be punched out of 1·5 mm thick mild-steel sheet. If the ultimate strength in shear is 360 N/mm², calculate (a) the punch load required and (b) the compressive stress in the punch.

If the punch is known to fracture in compression under a stress of 240 N/mm²,what is the factor of safety during punching?
(108 kN; 43·2 MN/m²; 5·56)

12.28 Calculate the maximum width of 3 mm steel plate that can be cut by a guillotine. The ultimate strength in shear of the plate is 320 MN/m² and the greatest load the guillotine can exert is 1·2 MN.   (1·25 m)

12.29 A punching machine used for punching holes in sheet metal 10 mm thick has a flywheel rotating at 2 rev/s when the machine is not punching. If the speed of the flywheel is not to drop below 1 rev/s, calculate the greatest diameter of hole that can be punched. The mass of the flywheel is 300 kg assumed concentrated at 600 mm radius and the ultimate shear strength of the plate is 300 MN/m².   (68 mm)

12.30 A flywheel of a shearing machine is rotating at 30 rad/s and at the end of a shearing stroke its speed falls to 24 rad/s. The kinetic energy of the flywheel at 30 rad/s is 250 kJ and 0·8 of the *total* loss of energy is absorbed during the shearing operation when plate 18 mm thick is being cut. Find the maximum width of plate that can be sheared if the ultimate shear strength of the metal is 240 MN/m^2.   (927 mm)

12.31 A flypress has two spherical masses each 5 kg attached to the ends of an arm 1·5 m long. The arm rotates about its central axis at 4 rev/s. When used to punch holes 8 mm in diameter in sheet metal 10 mm thick the speed falls to *n* revolutions per second at the end of the punching stroke. Find *n*, taking the ultimate shearing strength of the metal as 360 N/mm^2.   (2·81 rev/s)

Chapter 13

Shear force and bending moment

Having in the previous chapter shown how to ascertain the allowable load on a member of a structure or machine in relation to the stresses and strains it can bear, the further effects of shearing and bending due to the forces produced by the loading on a structure have to be considered. Shear force and bending moment are defined and several standard cases of beams of various types, supports and loading are described. The shear force and bending moment are found for a number of points on the beam in each case, leading to the construction of diagrams showing the variation in values along the beam.

13.1 The cantilever beam

Consider the horizontal beam shown in Fig. 13.1. The end A is rigidly *built-in* to the wall and the end B is free. Such a beam is called a *cantilever*.

Fig. 13.1 Cantilever beam

13.2 The simply supported beam

The beam shown in Fig. 13.2(a) is supported on a pair of knife edges and is said to be *simply* or *freely* supported. Other forms of support that may be often considered as *simple* are shown in Figs. 13.2(b) and (c).

The reaction exerted by a simple support on a beam is assumed to act at a point.

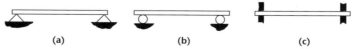

(a)   (b)   (c)

Fig. 13.2 Simply supported beams: (a) knife edge; (b) roller supports; (c) short bearings

13.3 Calculations of reactions

When a beam is fixed at some point or supported by props, the fixings and props exert reaction forces on the beam. To calculate these reactions the procedure is

- to equate the net transverse force to zero;
- to equate the total moment about any convenient point to zero.

For a horizontal cantilever beam, the fixed end is supported by a vertical reaction together with a moment about the same point, exerted by the fixing.

For a simply supported horizontal beam, the reactions may be found by taking moments about each support in turn and then checking that the net vertical force is zero.

13.4 Shear force

The cantilever beam shown in Fig. 13.3 carries a vertical transverse load of magnitude W at the free end. It will be assumed that the load is concentrated at a single point. Such a load is then called a *concentrated* or *point load*. For equilibrium of vertical forces acting on the beam AB the downward load W must be balanced by an equal and opposite vertical reaction W at the built-in end A (the weight of the beam is neglected here).

Consider now the effect of a transverse saw-cut at any section X of the beam. The effect of the load W would be to move the portion XB downwards. Before the saw-cut was made, therefore, there must have been an upward vertical force, equal and opposite to W, opposing the tendency of the load to move the portion XB downwards. This force must be exerted by the material of the beam itself at the section X. The effect of the applied load at any section therefore is to tend to *shear* the beam across at that section.

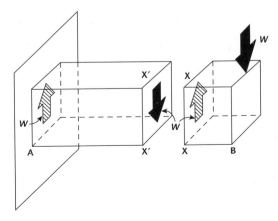

Fig. 13.3

We now define the *shear force* at any section as the force transverse to the beam tending to cause it to shear across the section.

13.5 Sign convention

The shear force at any section is taken *positive* if the portion of the beam to the *right* of the section tends to slide downwards relative to the left-hand portion, Fig. 13.4(a).

A *negative* shear force tends to cause the right-hand portion to slide *upwards* relative to the left-hand portion at the section considered, Fig. 13.4(b).

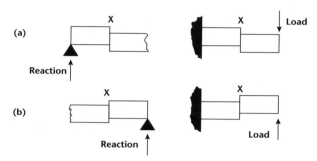

Fig. 13.4   (a) Positive shear force; (b) negative shear force

13.6 Resultant shear force

If several loads act on a beam, the *total*, *net* or *resultant* shear force at any section is equal in magnitude to the resultant of the transverse loads *on one side* of the section, due account being taken of sign. Thus

> **the resultant shear force at any section of a loaded beam is the algebraic sum of the loads to one side of the section.**

It does not matter which side of the section is considered provided that all loads on that side of the section are taken into account, including reactions exerted by fixings and supports. (The algebraic sum of the loads to the left-hand side of the section is, of course, equal to the algebraic sum of the loads to the right-hand side of the section.) Note that shear is caused only by loads *transverse* to the section of the beam.

13.7 Shear force diagram

The *shear force diagram* is a graph plotted along the axis of the beam showing the variation of shear force along the beam. For the beam shown in Fig. 13.5(a), the shear force is +W, and is uniform along the beam. Figure 13.5(b) shows the shear force diagram for this beam, the line O–O being the axis of zero shear force.

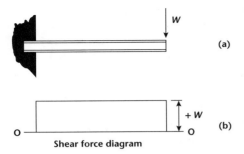

Fig. 13.5

Where the load W is a 'dead' load it may be given in units of mass, i.e. kilograms or tonnes, and it is necessary to calculate the *weight* of such loads in force units (newtons) before carrying out calculations.

Example

Draw the shear force diagram for the loaded cantilever beam shown in Fig. 13.6(a).

Solution

We first calculate the reaction at the built-in end A.

Net applied downward load on beam = $10 - 6 = 4\,\text{kN}$

The fixing reaction at A is equal and opposite to this and is therefore $4\,\text{kN}$ upwards, Fig. 13.6(b).

Shear force in AB: consider the force acting on the left-hand portion of any section X in AB. The only force *on* the beam to the left of X is the $4\,\text{kN}$ reaction and this tends to slide AX upward relative to XC. The shear force in AB is therefore $+4\,\text{kN}$ and is uniform along AB.

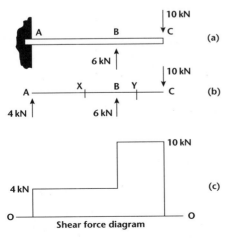

Fig. 13.6

Shear force in BC: consider section Y in BC. The force on the beam to the right-hand side of Y is 10 kN. This load tends to shear YC downward relative to AY. Hence the shear force in BC is uniform, positive, and equal to +10 kN.

As a check consider the forces to the left of Y. Their resultant is

$4 + 6 = 10$ kN, upward

This force tends to slide AY upward relative to YC. The shear force at Y is therefore +10 kN, as before.

The shear force diagram is shown in Fig. 13.6(c). It is seen that the magnitude of the shear force changes abruptly at each concentrated load and that the shear force is uniform on unloaded portions of the beam. The shear force diagram for a beam carrying only concentrated loads therefore consists of straight horizontal lines.

(**Example**)
_____

Draw the shear force diagram for the simply supported beam shown in Fig. 13.7(a). State the magnitude of the greatest shear force. What is the change of shear force at B?

Solution

We first calculate the reactions L and R of the simple supports at A and C respectively.

Taking moments about A,

$R \times 10 = 6 \times 6$

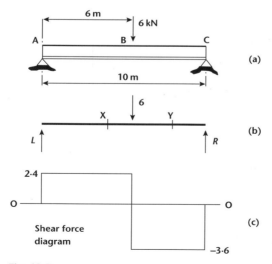

Fig. 13.7

therefore

$R = 3\cdot6\,\text{kN}$

Taking moments about C,

$L \times 10 = 6 \times 4$

therefore

$L = 2\cdot4\,\text{kN}$

Shear force in AB: the portion of beam to the left of section X in AB, Fig. 13.7(b), is subject to an upward force $L = 2\cdot4\,\text{kN}$. This force tends to slide AX upward relative to XC. Therefore, the shear force in AB = +2·4 kN.

Shear force in BC: the portion of beam to the right of section Y in BC is subject to an upward force $R = 3\cdot6\,\text{kN}$. This force tends to slide YC upward relative to AY. Therefore, shear force in BC = −3·6 kN. The shear force diagram is as shown in Fig. 13.7(c). The greatest shear force is **3·6 kN** at every section between B and C.

The change in shear force at B = 2·4 − (−3·6)

$$= \textbf{6 kN} \; (= \text{load at B})$$

<hr/>

Example

The beam shown in Fig. 13.8 is simply supported at B and D and carries at A and C concentrated masses of 1 tonne and 3 tonnes respectively. Determine the reactions L and R at B and D and draw the shear force diagram.

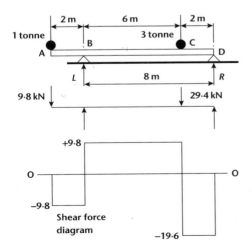

Fig. 13.8

Solution

The reaction at D is found by taking moments about B for *all* the loads on the beam, i.e. equating clockwise and anticlockwise moments about B. The weight of the 1 tonne mass is 9·8 kN and the weight of the 3 tonne mass is 29·4 kN. Thus

$$(9{\cdot}8 \times 2) + (R \times 8) = 29{\cdot}4 \times 6$$

Therefore

$$R = \textbf{19·6 kN}$$

Similarly, taking moments about D,

$$L \times 8 = (29{\cdot}4 \times 2) + (9{\cdot}8 \times 10)$$

Therefore

$$L = \textbf{19·6 kN}$$

Alternatively, equating the net vertical force to zero,

$$L + R = 29{\cdot}4 + 9{\cdot}8$$

so that

$$L = 39{\cdot}2 - 19{\cdot}6$$

$$= 19{\cdot}6 \text{ kN, as before}$$

Shear force diagram   The diagram is drawn by remembering that

- the shear force changes abruptly at a concentrated load;
- the shear force is uniform on an unloaded portion of the beam.

Using the given sign convention, we may start at the *left-hand end* and draw the diagram by following the arrows representing the loads and reactions; thus we draw

- at A, 9·8 kN down, then a horizontal line to B;
- at B, 19·6 kN up, then a horizontal line to C;
- at C, 29·4 kN down, then a horizontal line to D;
- at D, 19·6 kN up to the zero line again.

Note that in this method we have followed the *changes* in shear force along the beam.

Problems

13.1 Calculate, for the cantilever beam loaded as shown in Fig. 13.9(a), the shear force at A and the reaction at the built-in end.

Draw the shear force diagram for the beam.   (+2 kN; 3 kN; upward)

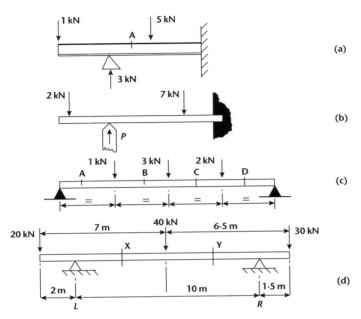

Fig. 13.9

13.2 A cantilever, Fig. 13.9(b), is loaded as shown and supported by a vertical prop. Calculate the load P in the prop for the reaction at the built-in end to be zero. Draw the shear force diagram for the beam.   (9 kN)

13.3 A simply supported beam of span 10 m between supports carries a mass of 5 t at a point 6 m from the left-hand support. Calculate (a) the shear force just to the right-hand side of the load, and (b) the change in shear force under the load.   (−29·4 kN; 49 kN)

13.4 A horizontal cantilever beam forming part of a wall crane carries the crane and its load totalling 3 t at the free end 3 m from the built-in end, and a vertical load of 1 t at 1 m from the wall. A tie-bar supports the beam and exerts an upward vertical force of 15 kN at the free end. Draw to scale the shear force diagram for the beam and state (a) the reaction at the built-in end and (b) the greatest shear force in the beam.   (24·2 kN upward; 24·2 kN)

13.5 Draw the shear force diagram for the simply supported beam shown in Fig. 13.9(c). State the value of the shear force at the points A, B, C and D.   (+2·75, +1·75, −1·25, −3·25 kN)

13.6 A beam is simply supported over a span of 6 m and carries a vertical concentrated load of 80 kN at a distance of 1·5 m from the left-hand support. The beam also carries a load of 100 kN at distance x from the right-hand support. Calculate the greatest value of x if the greatest shear force near the left-hand end is to be not greater than 80 kN.   (1·2 m)

13.7 An 0–6–0 diesel shunting locomotive has three axles, spaced 1·8 m apart. The mass of the locomotive is distributed as follows: front axle, 12 t; centre axle, 14 t; rear axle, 16 t. The locomotive rests on a bridge of 9 m span with the front axle 2·4 m from the left-hand support. If the bridge is simply supported, calculate the shear force at the midpoint of the span and draw the shear force diagram for the bridge.   (−43·1 kN)

13.8 The beam shown in Fig. 13.9(d) is simply supported but overhangs both supports. For the loading shown, calculate the reactions *L* and *R* at the supports and state the shear force at points X and Y.

(*L*, 39·5 kN; *R*, 50·5 kN; +19·5, −20·5 kN)

13.9 Draw to scale the shear force diagram for the simply supported beam loaded as shown in Fig. 13.10. State the magnitude of the reactions *L* and *R* and of the greatest and least shear forces on the beam.

(*L* = 6·3 kN; *R* = 3·7 kN; 6·3 kN along AB; 0·3 kN along CD)

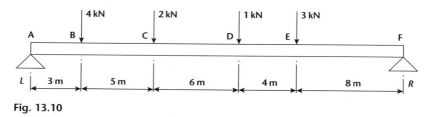

Fig. 13.10

13.8 **Bending moment**

Consider a cantilever beam loaded by a transverse point load *W*, Fig. 13.11. The effect of this load is to bend and deflect the beam downward. The *bending effect* at any section X is measured by the *moment* about the section of the load *W*. This moment is called the *bending moment*.

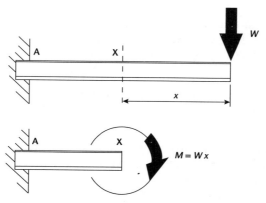

Fig. 13.11

If the perpendicular distance of the section X from the line of action of the load at B is x, the bending moment M at the section is given by

$M = W \times x$

The units of bending moment are Nm, kNm, MNm.

13.9 Sign convention

A bending moment is *positive* if its effect is to tend to make the beam *sag* at the section considered, Fig. 13.12. If the moment tends to make the beam bend upward at the section, or *hog*, it is *negative*.

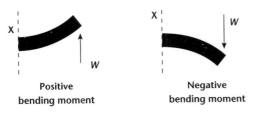

Positive
bending moment   Negative
                  bending moment

Fig. 13.12

13.10 Resultant bending moment

When more than one load acts on a beam, *the bending moment at any section is the algebraic sum of the moments due to all the forces on one side of the section.*

The moments of *all* the forces on *one side* of the section must be considered, including any moments exerted by end fixings and reactions. It does not matter which side of the section is considered, since the algebraic sum of the moments of all forces to the left of the section is equal to the sum of moments of all forces to the right of the section.

We must distinguish carefully between 'taking moments' and calculating a 'bending moment':

1. *The principle of moments* states that for equilibrium the algebraic sum of the moments of all the forces about any point is zero, i.e. when forces on *both* sides of a beam section are considered.
2. *The bending moment* is the algebraic sum of the moments of forces on *one* side of the section about that section.

13.11 Bending moment diagram

The *bending moment diagram* is a graph showing the variation along the axis of the beam of the applied bending moment.

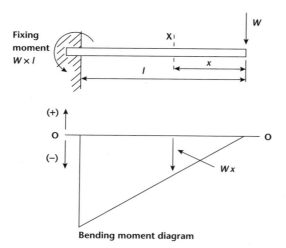

Fig. 13.13

For the beam shown in Fig. 13.13, the bending moment at section X is negative, since the beam tends to hog. Hence the bending moment M at section X is

$$M = -Wx$$

This expression for the bending moment applies to the whole length of the beam.

The moment M is proportional to x; hence the graph is a straight line. It is seen that the bending moment is zero at the free end where $x = 0$, and takes its greatest value at the built-in end where $x = l$. The bending moment diagram is therefore as shown in Fig. 13.13 and is everywhere negative.

The greatest bending moment is $-Wl$ and occurs at the built-in end. The fixing moment exerted by the wall on the beam at the built-in end is equal and opposite to the applied moment, therefore;

fixing moment = Wl, *anticlockwise*

Example

Draw the bending moment diagram for the simply supported beam shown in Fig. 13.14(a).

Solution
Reactions: we first calculate the reactions L and R at the simple supports A and C respectively. Taking moments about A,

$$R \times 10 = 5 \times 7$$

therefore

$$R = 3 \cdot 5 \, \text{kN}$$

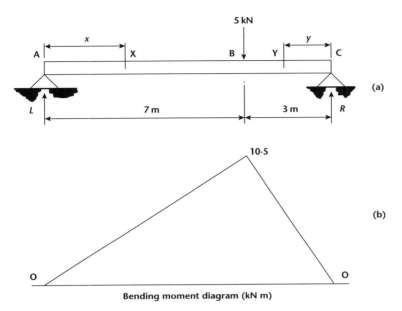

Fig. 13.14

Equating upward and downward forces,

$$L + R = 5\,\text{kN}$$

Hence

$$L = 1\cdot5\,\text{kN}$$

Consider section X, distance x from A.

Bending moment at X, $M_x = + L \times x$

$$= +1\cdot5x\,\text{kN m}$$

This moment is positive since it tends to make the beam sag at X. Since the moment M_x is proportional to x, the bending moment diagram for length AB is therefore a *straight line*. At point A, where $x = 0$, $M_x = 0$. At point B, where $x = 7$ m, $M_x = 10\cdot5$ kN m.

Consider now section Y, distant y from C. The bending moment at Y is $+R \times y$ $= +3\cdot5y$. This moment is proportional to y; hence the bending moment graph is a straight line between C and B. At C, where $y = 0$, the bending moment is zero, and at B, where $y = 3$ m, the bending moment is $+10\cdot5$ kN m, as already found.

To draw the bending moment diagram we plot the values of the moments at A, B and C and join the plotted points by straight lines, Fig. 13.14(b).

Note. This example shows that the bending moment diagram for a beam loaded only with concentrated point loads will consist of a series of straight lines. The slope of the diagram changes at each load. At the free end of a beam, the bending moment must be zero. Therefore, to draw the bending moment diagram for a beam carrying only point loads, it is only necessary to calculate the bending moment at each load, plot these values, and join the plotted points by straight lines. This method assumes that the weight of the beam itself is negligible.

Example

Draw the bending moment diagram for the propped cantilever loaded as shown in Fig. 13.15(a). The prop exerts an upward force of 6 kN. State the magnitude of the greatest bending moment and the fixing moment at the built-in end.

Solution

The bending moment diagram is made up of straight lines between the points A, B and C. It is sufficient, therefore, to calculate the moment at each of these points.

Bending moment at A due to load at $C = -4 \times 8 = -32\,\text{kN m}$

(negative, since it tends to make the beam hog at A)

Bending moment at A due to load at $B = +6 \times 5 = +30\,\text{kN m}$

(positive, since it tends to make the beam sag at A)

Resultant bending moment at $A = -32 + 30 = -2\,\text{kN m}$

(the resultant bending is negative and therefore tends to make the beam hog at A)

The fixing moment exerted at the built-in end is equal and opposite to the resultant bending moment exerted by the applied loads. The fixing moment is therefore **2 kN m, anticlockwise**, Fig. 13.15(b).

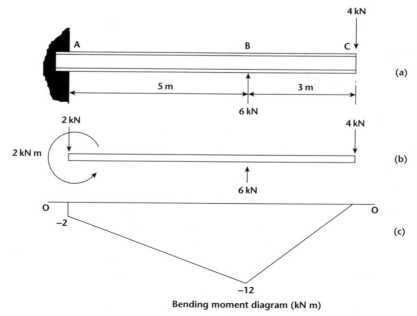

Bending moment diagram (kN m)

Fig. 13.15

To find the bending moment at B, consider the portion of beam to the right of B; then

bending moment at B = $-4 \times 3 = -12\,\text{kN m}$

Alternatively, considering the portion of beam to the left of B,

vertical reaction at A = $6 - 4 = 2\,\text{kN}$, downward

Therefore, bending moment at B

= moment of vertical reaction at A + fixing moment at A

= $-2 \times 5 - 2$

= $-12\,\text{kN m}$, as before

The bending moment at the free end C is zero.

The bending moment diagram is shown in Fig. 13.15(c). The greatest bending moment occurs at B and has magnitude **12 kN m**.

Example

Draw to scale the bending moment diagram for the simply supported beam loaded as shown in Fig. 13.16(a).

Solution
We first calculate the magnitude of the reactions L and R at A and E respectively. Taking moments about A and E in turn we obtain

$R = 4{\cdot}286\,\text{kN}$

$L = 3{\cdot}714\,\text{kN}$

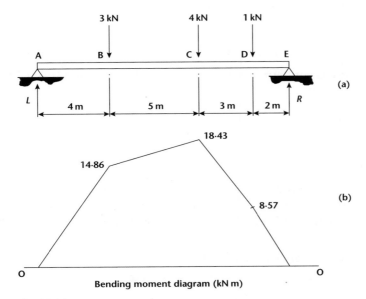

Fig. 13.16

The bending moment is zero at the free ends A and E.

Bending moment at B = +3·714 × 4 (considering loads on AB)

$$= +14·86\,kN\,m$$

Bending moment at C = +3·714 × 9 − 3 × 5 (considering loads on AC)

$$= +18·43\,kN\,m$$

Bending moment at D = +4·286 × 2 (considering loads on DE)

$$= +8·57\,kN\,m$$

The bending moment diagram is drawn by plotting the moments at A, B, C, D and E and joining the plotted points by straight lines, Fig. 13.16(b).

The greatest bending moment occurs at C and is of magnitude **18·4 kN m**.

$(\,$ **Example** $\,)$ _____

A horizontal joist carrying a travelling crane is simply supported over a span of 4 m. The mass of the crane is 150 kg. Determine the maximum load that the crane can support in kg, if the bending moment on the beam is not to exceed 5000 N m and the shear force is not to exceed 3 kN when the crane is positioned at midspan.

Solution

Let W newtons be the total weight of the crane and its load. The bending moment diagram is a triangle with its apex at the midpoint of the beam. The greatest bending moment occurs at the midpoint and is $Wl/4$, where $l = 4$ m is the span. Thus

$$\text{greatest bending moment} = \frac{Wl}{4} = 5000\,N\,m$$

Therefore

$$W = 5000\,N$$

The shear force is constant in magnitude along the beam and is equal to $W/2$; therefore

$$\frac{W}{2} = 3 \times 1000\,N$$

and

$$W = 6000\,N$$

The maximum permissible value of W is therefore 5000 N or 5 kN. For this value of W, both the bending moment and the shear force are kept within the given limiting values. This load is equivalent to the weight of a mass of 5000/9·8, or 510 kg. Therefore

$$\text{maximum load (mass) crane can carry} = 510 - \text{mass of crane}$$

$$= 510 - 150$$

$$= \mathbf{360\,kg}$$

Problems

13.10 Draw the bending moment diagram for the cantilever beam shown in Fig. 13.17(a) and state the magnitude of the fixing moment at the built-in end.

(90 kN m)

13.11 For the cantilever beam loaded as shown in Fig. 13.17(b), calculate the vertical load P in the prop if the fixing moment at the built-in end is to be zero. For this value of P, draw the shear force and bending moment diagrams and state the values of the bending moment and shearing force at a point 1 m from the built-in end.

(75 kN; 25 kN m; 25 kN)

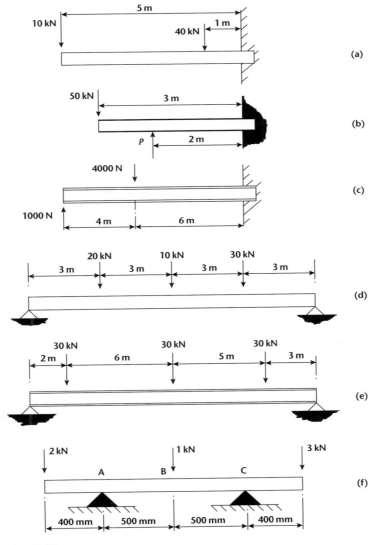

Fig. 13.17

13.12 Draw the shear force and bending moment diagrams for the loaded canti-lever shown in Fig. 13.17(c). State the greatest value of shear force and bending moment.   (3000 N; −14 000 N m)

13.13 Draw the shear force and bending moment diagrams for the simply sup-ported beam shown in Fig. 13.17(d). State the value of the bending moment at each load. What is the shear force at a section between the 20 kN and 10 kN loads?   (82·5, 105, 97·5 kN m; 7·5 kN)

13.14 Draw the shear force and bending moment diagrams for the simply sup-ported beam shown in Fig. 13.17(e). State (a) the reactions at the supports, (b) the bending moment at midspan and (c) the least and greatest shear forces.

(Left-hand reaction, 46·875 kN, right-hand, 43·125 kN; 195 kN m, 13·125 kN, 46·875 kN)

13.15 Draw the bending moment diagram for the beam shown in Fig. 13.17(f). The beam overhangs its supports at each end. State the magnitude of the reactions at the supports and the bending moment at points A, B and C.

(Left-hand reaction, 2·1 kN, right-hand, 3·9 kN; A, −0·8 kN m; B, −0·75 kN m; C, −1·2 kN m)

13.16 A wall crane is constructed from a horizontal cantilever beam. The crane mechanism has a mass of 180 kg and is mounted 2·5 m from the built-in end. Calculate the greatest mass which can be lifted if the maximum bend-ing moment in the cantilever is not to exceed 90 kN m.   (3·5 tonnes)

13.17 A four-wheel motor lorry has an unladen mass of $1\frac{1}{2}$ t (excluding wheels and axles) and its centre of gravity lies 1 m behind the front axle. The wheel base is 4 m. The lorry carries a load of 4 t having a centre of gravity midway between the axles. Calculate the load on the front axle and the bending moment on the chassis at a point midway between the axles.

(30·7 kN; 46·6 kN m)

13.18 A motor vehicle has a wheel base of 3 m. The load on the front axle is 12 kN and that on the rear axle is 16 kN. The vehicle rests on a simply supported bridge of span 8 m, the front axle being 2 m from one support. Calculate the reactions of the bridge supports and the bending moment under each axle.   (13, 15 kN; 39, 30 kN m)

13.19 A horizontal beam of span 6 m supports two equal masses M kg, one at midspan, the other at a point 2 m from one support. If the greatest shear force is limited to 10 kN and the greatest bending moment to 20 kN m, find the safe value for M.   (818 kg)

Chapter 14

Heat engines

....

Our study of heat engines in the following chapters is based on two definitions: (a) energy is the capacity to do work and (b) an engine is a machine that converts energy into useful work. We study, therefore, the release and control of energy to perform useful work and we should, perhaps, speak of an 'energy' engine rather than of a heat engine. First, we must further discuss energy in general and then heat energy in particular, which leads us in this chapter to deal with properties of matter, temperature and heat transfer.

14.1 Energy and its conversion

Energy is an abstract quantity, that is, it cannot be felt, tasted, smelt or weighed. Nevertheless, experience and experiment suggest that the ability to do work is best explained in terms of the idea of energy. Energy can take many forms, of which the following are important:

- chemical; nuclear (atomic);
- kinetic; potential, heat;
- electrical.

Kinetic and potential energy have already been studied in mechanics. Kinetic energy is measured by the speed and mass of a body. Potential energy is energy possessed by a body by virtue of its weight and its position (height). The measurement of heat energy will be considered in detail later. Its level, quality or intensity is measured by temperature but, as we shall see, temperature alone does not measure the quantity or total amount of heat energy possessed by a body.

Our personal experience of intense heat or intense cold is as a burn due to the flow of heat. Similarly our experience of electrical energy takes the form of a shock due to a flow of electrical energy. The flow of electrical energy is determined by a voltage or difference of potential in the same way as the flow of heat is determined by a temperature difference, or the flow of water by a difference in height. Electrical energy can be converted into heat, as in an electric fire. Again, electrical energy can be converted into mechanical work by means of the electric motor.

The study of energy is largely the study of its transformation from one form to another. For example, the burning of a match transforms chemical energy into heat and light energy. Also, in order to release the chemical energy the match is first 'struck'. That is, work is done against friction to produce heat. The temperature of the chemical in the match head rises until it starts to burn; the chemical then releases chemical energy in the form of heat. The energy released is sufficient to keep the temperature high enough for burning to continue, and the total energy released is usually much greater than that required to produce the initial spark. Note that we have assumed a connection between work, energy and temperature. These ideas will be developed later.

The 'energy engine' does useful work by a similar process. Fuel may be burnt by igniting it with a small initial spark. The chemical energy released is sufficient for burning to continue at a high temperature and the energy released is transformed into heat energy; this is shown by a rise in temperature of a gas or vapour in the engine. The gas expands when the temperature rises thus driving a piston or rotating a turbine wheel; hence useful work may be done.

The steps in the process of releasing energy to perform work suggest that we must study the relations between the following:

- burning and heat energy;
- heat energy and temperature;
- temperature and expansion;
- expansion and work.

Although it is not usually convenient to study these relationships in the order given, they should be kept in mind as a guide to the work that follows. Also, before embarking on a study of temperature we shall need to know a little more about the properties of matter.

(14.2) Matter and molecules

Matter may exist in one of three forms or phases: *solid*, *liquid* or *gas*. Whatever its form, matter is composed of *molecules*, a molecule being the smallest particle of any substance that can normally exist alone, i.e. not in chemical combination with any other similar particles. The number of molecules in a cubic millimetre of substance is very large indeed. Approximate figures for comparison are as follows:

air 0.085×10^{17} molecules per cubic millimetre

water 120×10^{17} molecules per cubic millimetre

iron 850×10^{17} molecules per cubic millimetre

Note that $10^{17} = 100\,000\,000\,000\,000\,000$.

The distance between molecules is exceedingly small. It is greater for air than for water and greater for water than for iron.

A gas is the least dense of all forms of matter. The molecules of a gas are in rapid and random motion and therefore possess kinetic energy. The kinetic energy of a gas molecule may be of the order of 300 J and its average speed of random motion 1500 m/s.

The molecules of a liquid also have a random motion but it is much less violent than that of a gas. In addition, the molecules of a liquid are in rapid vibration about a mean position and possess energy on this account. Although the molecules of a solid are relatively fixed in position, they also possess energy owing to their vibration.

14.3) Temperature

The kinetic energy of vibration of the molecules of a substance is apparent to the touch and is experienced as a degree of hotness which we call *temperature*. Temperature, for want of a better definition, may be defined as a measure of the *mean* kinetic energy possessed by all the molecules of a substance. A 'hot' body possesses more kinetic energy per molecule than a 'cold' body. As will be seen later, temperature may be thought of as a 'potential'; that is, for a given body, its temperature measures the level but not the quantity, of the energy available above a given datum, in the same way as the potential energy of a body is measured by its height.

Again, the molecules of a body at high temperatures are further apart than when the body is at a lower temperature. Thus a rise in temperature is, with certain exceptions, accompanied by an expansion. A measure of change of temperature can be obtained therefore by measuring the expansion of a body.

14.4) Expansion: measurement of temperature change

A change in temperature may be measured by the expansion of a body or a fluid. We might measure

- *linear expansion* – a change in length;
- *surface or superficial expansion* – a change in area;
- *volumetric or cubical expansion* – a change in volume.

The expansion of a solid is usually measured by observing the change in length of a bar. The expansion of a liquid or gas is generally measured by measuring the change in volume. Now, gases expand more readily than liquids or solids, and therefore temperature changes may be more accurately measured by the expansion of a gas. Gas thermometers are therefore used for accurate temperature measurement, but for the measurement of moderate temperatures with a reasonable degree of accuracy the ordinary mercury or alcohol thermometer is more convenient.

14.5) The scales of temperature: thermometers

The practical scale of temperature is chosen arbitrarily. A thermometer is placed in melting ice and the mercury level in the tube is marked. It is then placed in boiling water and the level again marked. These *fixed points* of the scale are determined at atmospheric pressure and are called the *ice point* and the *boiling* or *steam point* respectively.

On the *Celsius* scale of temperature, Fig. 14.1, the ice point is at zero, $0\,°C$, and the distance between the ice and steam points is divided into 100 equal divisions or *degrees*. Each division then represents an interval of one Celsius degree ($1\,°C$). The boiling point is therefore $100\,°C$. On this scale the freezing point of mercury is $-38.9\,°C$ and the boiling point $357\,°C$. Thus the use of mercury in a thermometer gives a working range of temperature for ordinary purposes.

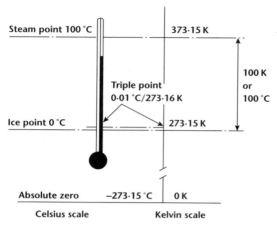

Fig. 14.1

Besides the Celsius scale the Kelvin scale has been adopted for scientific purposes. Chapter 16 deals with the gas laws and the gas thermometer which indicate the possibility of an absolute zero of temperature. From the laws of thermodynamics and the theoretical Carnot reversible heat engine cycle (*see* section 19.8) has developed a *thermodynamic* or *absolute scale of temperature* of which the Celsius scale is now a part. The absolute scale is independent of any particular substance or temperature scale. The unit (not degree) is called the *kelvin* (K) and an absolute temperature is denoted by *T*. The fixed point of the absolute scale is taken as the temperature of the *triple point* of *water* ($0.01\,°C$), where ice, water and water vapour exist together in equilibrium at one particular pressure (4·6 mmHg). The triple point is assigned the temperature 273·16 K thus setting the ice point temperature ($0\,°C$) at 273·15 K and the steam point at 373·15 K, Fig. 14.1. The absolute zero temperature is $T_0 = 0\,K = -273·15\,°C$. For practical purposes the figure 273 is used. Thus if $t\,°C$ is the temperature on the Celsius scale, the temperature on the absolute scale is

$$T = (t + T_0)\,K = (t + 273)\,K = (t + 273)\,°C\ abs.$$

The temperature difference, or interval, is the same on both scales and may be expressed in °C but preferably by the unit symbol K. For example, a temperature difference such as ($70\,°C - 20\,°C$) may be stated as $50\,°C$ or $50\,K$.

As well as mercury thermometers and other versions of the liquid thermometer, many other types of thermometer are in use. Platinum resistance thermometers are employed to measure temperatures as low as $-200\,°C$ and up to and beyond about $1000\,°C$. Above $1100\,°C$ various temperature measuring devices such as thermocouples utilize the high melting points of platinum and platinum–rhodium. Infra-red thermometers with a range $-20\,°C$ to $260\,°C$ mounted *in situ* or hand-held, are now used with electronic and electrical equipment to give instant measuring response and for holding close to an object so as not to affect the temperature of the object. For all these thermometers, including the triple point of water at which temperature they all agree, there are ten fixed points ranging from the triple point of hydrogen (13·8 K) to the freezing point of gold (1337·5 K).

(14.6) Linear expansion of solids

The change in length x of a solid bar when heated or cooled thro
ture change Δt is given by the experimental relation

$x = \alpha \Delta t l$

where l is the original length of the bar and α is a constant known as the *coefficient of linear expansion*. The value of this coefficient depends on the material of the bar. The change in temperature Δt will be the same on both the Celsius and Kelvin scales, i.e. $\Delta t\,^{\circ}C = \Delta t\,K$.

If unit length of a bar is raised in temperature one degree then $\Delta t = 1$, $l = 1$ and $x = \alpha$. Thus the coefficient of linear expansion may be defined as *the extension of a bar per degree rise in temperature per unit of original length*. To be exact, the temperature at which the coefficient is measured should be stated, but ordinarily we may neglect changes in the value of α with temperature.

The numerical value of α is the same whatever the unit of length. The coefficient is very small for metals. For steel, a bar 1 m in length extends 12×10^{-6} m per 1° rise in temperature, i.e. $\alpha = 12 \times 10^{-6}/K$. For Invar it is practically zero at ordinary temperatures.

The property of linear expansion and contraction of a substance due to temperature change applies to all linear dimensions of a body. For example, the radius r of a solid sphere increases by $\alpha \Delta t r$ when heated. Again, the diameter of a hole in a plate or sphere changes with temperature in exactly the same proportion as the other linear dimensions.

Typical values of α for different materials are given in Table 14.1.

Table 14.1 Coefficients of linear expansion

Substance	α/K
Aluminium	24×10^{-6}
Iron, carbon-steel	12×10^{-6}
Cast-iron	10×10^{-6}
Copper	17×10^{-6}
Brass, bronze	19×10^{-6}
Glass	8×10^{-6}
Invar	$1{\cdot}6 \times 10^{-6}$

(Example)

Two bars, one bronze, the other steel, are both nominally 150 mm long. When placed side by side at 20 °C, the bronze bar is found to be shorter than the steel by 0·25 mm. To what temperature must both bars be raised to make them exactly the same length?

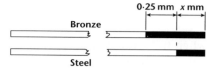

Fig. 14.2

Solution

Figure 14.2 shows the two bars. When they are the same length, the steel has extended a distance x mm (exaggerated in the diagram) and the bronze has therefore extended $(x + 0.25)$ mm.

Let Δt be the temperature rise. Appropriate values of the coefficient of linear expansion α for steel and bronze are taken from Table 14.1. For the bronze,

$$x + 0.25 = \alpha\Delta tl = 19 \times 10^{-6} \times \Delta t \times 150$$

For the steel,

$$x = \alpha\Delta tl = 12 \times 10^{-6} \times \Delta t \times 150$$

From these two equations,

$$x = (19 \times 10^{-6} \times \Delta t \times 150) - 0.25 = 12 \times 10^{-6} \times \Delta t \times 150$$

therefore

$$\Delta t = \frac{0.25}{(19 - 12) \times 10^{-6} \times 150}$$

$$= 238 \, \text{K} \quad \text{or} \quad 238 \, ^{\circ}\text{C}$$

so that

$$\text{final temperature} = 238 + 20 = \mathbf{258 \, ^{\circ}\text{C}}$$

Example

A ring of aluminium–bronze alloy has internal diameter of 300 mm and is 50 mm wide. For a temperature rise of 600 °C find (i) the final internal diameter and (ii) the change in width of the ring. Coefficient of linear expansion of the alloy = 17×10^{-6}/K.

Solution

Let d_1, d_2 be the initial and final internal diameters of the ring. When the ring is heated through Δt the change in length of the internal circumference is $\alpha\Delta t(\pi d_1)$; thus

$$\text{final circumference} = \pi d_2 = \pi d_1 + \alpha\Delta t(\pi d_1)$$

i.e. $$d_2 = d_1 + \alpha\Delta t d_1$$

The diameter of the hole has increased by $\alpha\Delta t d_1$, i.e. in exactly the same proportion as the increase in internal circumference. Thus the diameters of holes can be treated in the same way as any other linear dimension when calculating changes in length due to heating or cooling. Therefore, since $\Delta t = 600\,°C$ or $600\,K$, and $\alpha = 17 \times 10^{-6}/K$,

final internal diameter, $d_2 = 300 + (17 \times 10^{-6} \times 600 \times 300)$

$$= 303\cdot06\,mm$$

Similarly,

final outside diameter $= 400 + (17 \times 10^{-6} \times 600 \times 400)$

$$= 404\cdot08\,min$$

final width $= \frac{1}{2}(404\cdot08 - 303\cdot06) = 50\cdot51\,mm$

Therefore

change in width $= 50\cdot51 - 50 = \mathbf{0\cdot51\,mm}$

(**Example**)

A 75 mm diameter steel shaft runs in a bronze bush. At room temperature, 15 °C, the diametral clearance between shaft and bush is 0·15 mm. When running at speed the temperature of both shaft and bush is 100 °C. What is then the working radial clearance?

Solution
The diametral clearance is the difference in diameter between the shaft and its bush, Fig. 14.3(a).
The temperature rise is 85 °C or 85 K and, owing to this, both shaft and bush increase in diameter, thus altering the clearance.
The required values of α are taken from Table 14.1. Let d be the shaft diameter.

Increase in shaft diameter $= \alpha \Delta t d$

$$= 12 \times 10^{-6} \times 85 \times 75$$

$$= 0\cdot0765\,mm$$

Increase in bush diameter $= 19 \times 10^{-6} \times 85 \times 75$

$$= \mathbf{0\cdot1213\,mm}$$

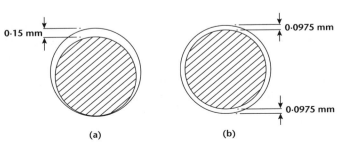

Fig. 14.3 Clearances: (a) diametral clearance at 15 °C; (b) radial clearance at 100 °C (both are exaggerated).

The increase in diameter of the bush is greater than that of the shaft by an amount 0·1213 − 0·0765, i.e. 0·0448 mm. The diametral clearance increases by this amount. Hence at 100 °C the total diametral clearance is 0·15 + 0·0448, i.e. 0·195 mm.

The radial clearance is the gap between shaft and bush when the shaft is concentric with the bush, Fig. 14.3(b). The radial clearance is therefore one-half the diametral clearance, i.e.

$$\text{radial clearance} = \frac{0 \cdot 195}{2} = 0 \cdot 0975 \text{ mm}$$

Problems

(Relevant values of α should be taken from Table 14.1.)

14.1 At the standard measuring temperature of 20 °C, a steel gauge is 250 mm long. What will be its actual length at 25 °C?   (250·015 mm)

14.2 A 50 mm steel gauge block is accurate when checked in the laboratory at 15 °C. Calculate the error when used in the workshop at 26 °C.

(0·0066 mm)

14.3 The spindle of an instrument is made of bronze. What rise in temperature will cause 1 per cent increase in the length of the spindle?   (526·3 K)

14.4 Calculate the increase in length of a 20 m length of steel rail track when the temperature changes from −1 °C (winter) to 32 °C (summer).   (7·9 mm)

14.5 A measuring tape is made of Invar. It is accurate at 30 °C. What is the error when a length between two points is measured as 120 m at 43 °C?

(2·5 mm)

14.6 The control rod in a thermostat is a thin metal bar 150 mm long, held fixed at one end and free to expand at the other end. The thermostat is assembled at 18 °C with a gap of 0·06 mm between the free end of the rod and the mechanism operating the shut-off valve. Find the temperature at which the valve acts. Coefficient of linear expansion of the metal = 20×10^{-6}/K.   (38 °C)

14.7 A copper rod is 50 mm diameter at 15 °C. What must be the diameter of a hole in a glass plate at 15 °C if the rod is to fit the hole exactly when both are placed in boiling water?   (50·038 mm)

14.8 Two tubes, one of copper, the other of steel, are nominally 250 mm long but the copper is 0·5 mm shorter than the steel. Both bars are raised in temperature by 330 K. What is the final difference in length between the bars?   (0·088 mm)

14.9 A brass rod is 200 mm long and a steel rod 200·76 mm long, both measured at 18 °C. At what temperature will their lengths be equal?   (561 °C)

14.10 A brass sphere 40 mm in diameter rests on an aluminium ring of internal diameter 39·90 mm at 150 °C. At what temperature will the sphere just pass through the ring? (650 °C)

14.11 An aluminium alloy piston, 100 mm in diameter at 15 °C is used in an oil pump. If the maximum permissible piston diameter for safe running is 100·10 mm, find the highest temperature at which the pump may run. The coefficient of linear expansion of the alloy is 25×10^{-6}/K. (55 °C)

14.12 The steel race of a 60 mm diameter ball bearing is machined 0·12 mm in diameter smaller than the steel shaft. In order to mount the bearing on the shaft, it is first heated in oil. What is the lowest temperature to which the bearing must be heated before it can be slipped on to the shaft? Room temperature, 18 °C. (185 °C)

14.13 A turbine steel casing flange is to be accurately machined to 1200 mm diameter at room temperature 15 °C. The diameter was checked and taken as correct when the casing was hot after machining. If the casing temperature was 120 °C, what is the diameter when the casing cools to room temperature? (1198·49 mm)

14.14 A 75 mm diameter aluminium piston fits into a cast-iron cylinder with a radial clearance of 0·75 mm at room temperature 20 °C. Calculate the working radial clearance when the piston is at 520 °C and the cylinder at 420 °C.

If the cast-iron cylinder is replaced by an aluminium one, what is then the working radial clearance at the same temperature?

(0·453 mm; 0·667 mm)

14.15 A steel bush 50 mm diameter at 15 °C is to be shrunk on to a steel shaft 50·3 mm diameter at 15 °C by heating, placing in position, and allowing to cool. Find the minimum temperature to which the bush must be heated. (515 °C)

14.16 A steel ring has a bore of 57·83 mm and is to be a shrink fit on a steel shaft 57·90 mm, both ring and shaft being measured at 18 °C. Allowing 0·015 mm on diameter for passing over the shaft, calculate the minimum temperature to which the ring must be heated. (141 °C)

Density

The *density* ρ of a body of mass m and volume V is defined as the *mass per unit volume*, i.e.

$$\text{density} = \frac{\text{mass}}{\text{volume}}$$

or

$$\rho = \frac{m}{V}$$

The SI unit of *volume* is the *cubic metre* (m³). Other units used are mm³ and dm³. The name *litre* is used for a volume of 1 dm³, thus

$$1 \text{ litre} = 1 \text{ dm}^3 = 10^{-3} \text{ or } \frac{1}{1000} \text{ m}^3$$

or

$$1 \text{ m}^3 = 1000 \text{ litres}$$

The SI unit of density is the kilogram per cubic metre (kg/m³). However, a convenient unit is Mg/m³, i.e. 1 tonne or 1000 kg per cubic metre. Temperature and pressure affect the density of any solid, liquid or gas, but for solids and liquids the variation is slight over a wide range. Water, however, is unusual in the way its density varies with temperature and its maximum density occurs at 4 °C and atmospheric pressure. The density of water is important and should be remembered, i.e.

$$\text{density of water} = 1000 \text{ kg/m}^3 = 1 \text{ Mg/m}^3 = 1 \text{ t/m}^3 = 1 \text{ kg/litre}$$

Table 14.2 gives some typical values of densities for various substances. The density of gases is dealt with in Chapter 16.

Table 14.2 Density and relative density of substances

Substance	Density (kg/m³)	Relative density
Steel	7 870	7·87
Cast-iron	7 200	7·2
Aluminium	2 700	2·70
Ice	920	0·917
Water	1 000	1·0 (at 4 °C)
Mercury	13 600	13·6
Alcohol	810	0·81
Zinc	7 000	7·0
Copper	8 500	8·5

(14.8) Relative density

The *relative density* of a substance is defined as the ratio of its mass to that of an equal volume of water at 4 °C and atmospheric pressure. More simply, it is the ratio of the density of a substance to that of water; it is a number, independent of any sets of units. Since mass is proportional to weight, relative density may be written as

$$\frac{\text{weight of substance}}{\text{weight of equal volume of water}}$$

This ratio was called *specific gravity* and is often used instead of the term relative density. In this text, only relative density will be used. For example, if the relative density of aluminium is given as 2·7 then its density is 2·7 times that of water, which is $1\,\text{Mg/m}^3$. The density of aluminium is therefore $2.7\,\text{Mg/m}^3$ and its weight is $2.7 \times 9.8\,\text{kN/m}^3$, i.e. 26.5kN/m^3.

14.9 Cubical and superficial expansion of solids

Consider a cube of unit side which is raised in temperature by an amount Δt, Fig. 14.4. Let the extension of each side (length) be x; then

$$x = \alpha\Delta t$$

new length of each side $= 1 + x$

new volume of cube $= (1 + x)^3$

$$= 1 + 3x + 3x^2 + x^3$$

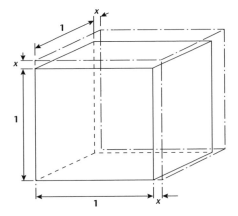

Fig. 14.4

The extension x is a very small quantity, hence the terms x^3 and $3x^2$ may be neglected. Therefore

new volume $\simeq 1 + 3x$

original volume $= 1$

Therefore

increase in volume $\simeq 3x$

$$= 3\alpha\Delta t$$

This is the change in volume per unit volume. If the original volume is V, then

change in volume $= 3\alpha\Delta t V$

The change in volume ΔV per unit rise in temperature is

$\Delta V = 3\alpha V$

The *coefficient of cubical expansion* γ is defined as the ratio

$$\frac{\text{change in volume per unit temperature rise } (\Delta V)}{\text{original volume } (V)}$$

i.e. $\gamma = \dfrac{\Delta V}{V}$

$\qquad = \dfrac{3\alpha V}{V}$

$\qquad = 3\alpha$

Thus the coefficient of cubical expansion γ is 3α, that is, three times the coefficient of linear expansion.

You should now satisfy yourself by similar reasoning that the *coefficient of superficial expansion* β is twice the coefficient of linear expansion, i.e.

$\qquad \beta = 2\alpha$

Hence, if A is the original area of a body, then for a rise in temperature Δt,

$\qquad$ change in area $= \beta \Delta t A$

$\qquad\qquad\qquad = 2\alpha \Delta t A$

(**Example**)
_____

A metal sphere is raised in temperature by 500 K. The coefficient of linear expansion of the metal is $20 \times 10^{-6}/\text{K}$. Find the ratio of the final density to the original density.

Solution

Let V = initial volume of sphere.

$\qquad$ Coefficient of cubical expansion $\gamma = 3\alpha$

$$= 3 \times 20 \times 10^{-6}/\text{K}$$

$\qquad$ Change in volume $= \gamma \Delta t V$

$$= 3 \times 20 \times 10^{-6} \times 500 \times V$$

$$= 0.03\, V$$

$\qquad$ Final volume $= V + 0.03 V$

$$= 1.03\, V$$

$\qquad$ Initial density $\rho_1 = \dfrac{\text{mass}}{\text{volume}} = \dfrac{m}{V}$

$\qquad$ Final density $\rho_2 = \dfrac{m}{1.03\, V}$

$\qquad$ Ratio of densities $= \dfrac{\rho_2}{\rho_1} = \dfrac{m/1.03\, V}{m/V} = \mathbf{1:1.03}$

(**Example**)

An 80 mm diameter aluminium piston fits into a cast-iron cylinder with a uniform radial clearance of 0·75 mm at room temperature, 15 °C. Determine the percentage increase in area of the gap between piston and cylinder at 130 °C. For aluminium, $\alpha = 24 \times 10^{-6}$/K; for cast-iron $\alpha = 10 \times 10^{-6}$/K.

Solution

$$\text{Area of piston} = \pi r^2 = \pi \times 40^2 = 5027 \text{ mm}^2$$

$$\text{Area of cylinder} = \pi r^2 = \pi \times 40\cdot 75^2 = 5217 \text{ mm}^2$$

Initial area of gap = area of cylinder − area of piston

$$= (\pi \times 40\cdot 75^2) - (\pi \times 40^2)$$

$$= 190 \text{ mm}^2$$

Coefficient of superficial expansion for aluminium is

$\beta = 2\alpha = 2 \times 24 \times 10^{-6}$/K

$\Delta t = (130 - 15)\,°C = 115\,K$

Increase in area of piston = $\beta \Delta t A$

$$= 2 \times 24 \times 10^{-6} \times 115 \times 5027$$

$$= 27\cdot 8 \text{ mm}^2$$

Similarly

increase in area of cylinder = $2 \times 10 \times 10^{-6} \times 115 \times 5217$

$$= 12 \text{ mm}^2$$

Change in area of gap = $27\cdot 8 - 12 = 15\cdot 8 \text{ mm}^2$

Percentage increase in area of gap = $\dfrac{15\cdot 8}{190} \times 100$

$$= \textbf{8·3 per cent}$$

(**Example**)

A copper sphere has a diameter of 14 cm at 20 °C. If the temperature of the sphere rises to 320 °C, find the percentage increase in external surface area. For copper $\alpha = 17 \times 10^{-6}$/K.

Solution

Coefficient of superficial expansion for copper is

$\beta = 2\alpha = 2 \times 17 \times 10^{-6}$/K

Surface area of the sphere = $4\pi r^2 = 4\pi \times 70^2 = 61\,600 \text{ mm}^2$

Increase in area of the sphere $= \beta \Delta t A$

$$= 2 \times 17 \times 10^{-6} \times 300 \times 4\pi \times 70^2$$

$$= 628 \, mm^2$$

Percentage increase in area $= \dfrac{increase \; in \; area}{original \; area} \times 100$

$$= \dfrac{628}{61\,600} \times 100$$

$$= 1 \cdot 02 \; \textbf{per cent}$$

Alternatively let r_1, r_2 be the initial and final radii of the sphere; then

percentage increase in area $= \dfrac{(4\pi r_2^2 - 4\pi r_1^2)}{4\pi r_1^2}$

$$= \dfrac{(r_2 - r_1)(r_1 + r_2)}{r_1^2} \times 100$$

$$= \dfrac{\alpha \Delta t r_1 \times 2r_1}{r_1^2} \times 100$$

$$= 2 \times \alpha \Delta t \times 100$$

$$= 1 \cdot 02 \; \text{per cent as before}$$

Note that $(r_2 - r_1)$, although a difference of two nearly equal quantities, is a significant quantity because it is a *factor* in the expression, whereas $(r_1 + r_2)$ is equal to $2r_1 + \alpha \Delta t r_1$ and the latter small quantity has no significance when multiplying. Thus we take $(r_1 + r_2) = 2r_1$.

Problems

(Take relevant values for coefficient of linear expansion from Table 14.1.)

14.17 Calculate the volume of an aluminium casting, of mass 300 kg, given that its density is $2 \cdot 7 \, Mg/m^3$. $(0 \cdot 11 \, m^3)$

14.18 Calculate the percentage change in density of an aluminium casting due to a temperature rise of 220 °C. (1·6 per cent, decrease)

14.19 Calculate the mass of a steel alloy casting of volume 10 litres and density $7 \cdot 85 \, Mg/m^3$. What is the change of density due to a rise in temperature of 500 K? The coefficient of linear expansion of the material is $12 \times 10^{-6}/K$. $(78 \cdot 5 \, kg; \; 0 \cdot 14 \, Mg/m^3)$

14.20 A hollow metal sphere 30 cm outside diameter and 26 cm inside diameter is raised in temperature by 820 K. The coefficient of linear expansion of the metal is $14 \times 10^{-6}/K$. Find (i) the ratio of the final and initial densities, (ii) the change in thickness and (iii) the percentage increase in external surface area. $(1{:}1 \cdot 034; \; 0 \cdot 23 \, mm; \; 2 \cdot 3 \; \text{per cent})$

14.21 A Duralumin piston slides in a cast-iron cylinder at 20 °C. At this temperature there is a uniform radial clearance of 0·13 mm and the piston diameter is 100 mm. Determine the area of the gap at 20 °C and also when the temperature has been raised to 120 °C. For Duralumin $\alpha = 22\cdot5 \times 10^{-6}$/K.

(40·8 mm^2; 21·2 mm^2)

14.22 A silver sphere has a diameter of 50 mm at room temperature. If the temperature is increased at the rate of 0·5 °C per second, find the increase in surface area after 15 min. Coefficient of linear expansion of silver $= 19 \times 10^{-6}$/K.

(135 mm^2)

Chapter 15

Temperature and heat

....

The basic information given here concerns the measurement and units of heat energy and the connections between temperature, heat and mechanical energy. This information is then applied to the capacity of bodies and gases to store heat, with the problems dealt with being limited to calculations on heating-rates and elementary heat transfer. The transfer of heat by conduction, convection and radiation is described together with the ability of the surfaces of bodies to emit, absorb and transfer radiant heat energy.

15.1 Quantity of heat

We have seen that the kinetic energy of vibration of the molecules of a body is experienced as a degree of hotness, and that temperature may be considered a measure of the mean kinetic energy per molecule. It is not practicable to measure the energy per molecule, but the temperature of a body is determined in practice by the level of liquid in a thermometer.

Now, a large quantity of water contains more molecules than a small quantity and the larger quantity therefore requires a greater amount of energy to increase the average energy per molecule by a given amount, i.e. to increase its temperature by one degree. Similarly, if the same amount of energy (heat) is supplied to different quantities of water, the smaller quantity will attain the higher temperature.

Thus heat is energy – the energy of motion and the quantity of heat flowing into or out of a body are accounted for by a change in the temperature of the body. As we have seen, this change in temperature depends on the number of molecules and hence on the mass of the body. We shall see that the temperature change depends also on the nature of the body. Heat flow therefore is measured by (a) a temperature change and (b) the mass of the body. This may be compared with the method by which we measure the potential energy of a body, i.e. by (a) the height or change in height and (b) the weight of the body. Temperature plays the same part in heat measurement as does height in the measurement of potential energy.

15.2 Units of heat energy

Heat is a form of energy and the SI unit of heat is the same as for other forms of energy, i.e. the *joule* (J). The joule is defined in terms of mechanical energy (*see* section 5.1) and is the work done when the point of application of a force of 1 N is displaced through a distance of 1 m in the direction of the force. The basic heat unit* is therefore independent of any particular substance, temperature scale or range of temperature. The joule is rather a small unit of heat and the larger multiples kJ, MJ and GJ are more often used. It is advisable, however, in most calculations to revert to the basic unit.

15.3 Relation between heat and mechanical energy

The relation between heat energy and mechanical energy is derived from experimental work and is dealt with in Chapter 19. Experiment shows that approximately 4·2 kJ of mechanical energy is required to raise the temperature of 1 kg of water through 1 °C (or 1 K). The exact value depends on the position on the temperature scale at which the experiment is carried out. The amount of heat therefore that is required by burning fuel or from electrical energy in order to raise 1 kg of water through 1 °C is 4·2 kJ approximately.

15.4 Specific heat capacity

If 1 kg of water and 1 kg of steel receive the same quantity of heat the resulting changes in temperature of the two substances are quite different. The reason is that the kinds of molecules making up steel are different from those of water; the number of molecules per kg of each substance is different, and, finally, the different molecules have different masses. The capacity of a body for taking up heat therefore depends on the nature of the body. The quantity of heat required to raise 1 kg of any substance through 1 °C or 1 K is called the *specific heat capacity*[†] of the substance and is denoted by c. The SI unit of c is the *joule per kilogram per kelvin* or *per °C*, i.e. **J/kg K** or **J/kg °C**. Note again that the Celsius scale is part of the Kelvin scale of temperature so that a temperature interval is the same on both scales. However, the kelvin unit is used now for temperature difference so that the form J/kg K is preferred.

Typical values of specific heat are given in Table 15.1.

---

* The basic heat unit was formerly defined in terms of the heat required to raise unit mass of water through one degree of temperature. Various authorities defined the unit using different masses of water, different temperature scales and different positions on the temperature scale. Such units were the British thermal unit, the centigrade heat unit and the calorie.

† The word 'specific' before a quantity such as 'heat capacity' or 'volume' means 'per unit mass'.

Table 15.1 Specific heat capacities

Substance	Specific heat capacity (kJ/kg K)
Metals	
Aluminium	0·92
Iron, carbon-steel	0·46
Cast-iron	0·54
Copper, brass	0·39
Lead	0·13
Non-metals	
Ice	2·1 at 0 °C
Glass	0·7
Steam	2·1 approx.
Mercury	0·14
Alcohol	2·3
Water (pure)	4·2174 at 0 °C
	4·2045 at 4 °C
	4·1855 at 15 °C
	4·2145 at 99 °C

Specific heat capacity of water

The specific heat capacity of water is particularly important. Although, as shown in Table 15.1, there is a slight variation in the value with temperature, the figure of 4.2 is sufficiently accurate for our purposes and will be used here. This value should be remembered. Thus

specific heat capacity of water = 4·2 kJ/kg K

Specific heat capacities of a gas

The heat taken in by a gas depends on the conditions under which heating now takes place. The pressure may be maintained constant or if the gas is in a closed vessel then the volume is constant. However, both pressure and volume may vary during heating or cooling, and it is also possible for the temperature to remain constant (*see* section 16.14).

The specific heat capacities of gases have been determined for the two particular conditions of constant pressure and constant volume. Thus the specific heat capacity of a gas is defined as *the heat required to raise 1 kg of gas at constant pressure (or constant volume) through 1 K (or 1 °C).*

The specific heat capacity at constant pressure is denoted by c_p and that at constant volume by c_v. The units are the same as for the specific heat capacity of a substance (kJ/kg K).

At constant volume and increasing pressure all the heat required to raise the temperature of a gas by 1 degree goes to increase the internal energy of the gas. At constant pressure, as for example when a gas is heated in a cylinder and expands against a movable piston, the gas must do external work against the piston. *Additional* heat, therefore, has to be applied for a rise in temperature of 1 degree, i.e. the difference $c_p - c_v$ is the heat required per 1 degree rise in temperature to expand the gas under constant pressure. The ratio of these two specific heats will be found to be important later and is denoted by γ, i.e.

$$\gamma = \frac{c_p}{c_v}$$

The most important gas is normal atmospheric air for which $c_p = 1\,\text{kJ/kg K}$ and $c_v = 0.72\,\text{kJ/kg K}$. For a mixture of gases such as the flue gases from a boiler or the exhaust gases from an engine a *mean* specific heat capacity is used. For example, for boiler flue gases c_p is taken as about $1.1\,\text{kJ/kg K}$.

Table 15.2 gives values of c_p and c_v for some common gases. These values are assumed to be constant for 'perfect' gases, but there is a variation with temperature.

Table 15.2

Gas	Specific heat capacities (kJ/kg K)	
	Constant pressure, c_p	Constant volume, c_v
Air	1.00	0.72
Nitrogen	1.04	0.74
Oxygen	0.92	0.66
Hydrogen	14.20	10.10
Carbon dioxide	0.85	0.63
Carbon monoxide	1.04	0.74

15.5   Thermal capacity and heat flow

The *thermal* or *heat capacity* C of a body is defined as the heat required to raise its temperature by 1 K. Since the specific heat capacity c gives the heat required to raise unit mass of the body's substance through 1 K (or 1 °C) then for a body of mass m its thermal capacity is given by

thermal capacity C = specific heat capacity × mass

$$= cm$$

The unit of c is J/kg K and if m is in kg then the unit of C is J/K (or J/°C). More often it is convenient to work in kilojoules.

If the body is now heated through Δt the total quantity of heat flowing into the body is given by

Q = specific heat capacity × mass × temperature rise

$$= cm\Delta t$$

$$= C\Delta t$$

and the unit of Q will be an energy unit, i.e. the kilojoule if c is in kJ/kg K and m is in kg.

15.6 Water equivalent

It is possible to compare the thermal capacity of any substance with that of water. For example, it can be shown experimentally that, when the same amount of heat is supplied to 1 kg of copper and 0·093 kg of water, both the copper and water rise in temperature by the same amount. The *water equivalent* of copper is said to be 0·093 kg. In general, if an amount of heat Q produces a rise in temperature Δt in a body of mass m and specific heat capacity c and the same heat produces the same rise in temperature in a mass m_w of water, then m_w is the water equivalent of the body; thus

$$Q = cm\Delta t = c_w m_w \Delta t$$

i.e. equivalent mass of water,

$$m_w = \frac{c}{c_w} m$$

where c_w = specific heat capacity of water.

15.7 Heating-rate calculations

If the temperature of a body of mass m rises at the rate of $\dot{t}$ degrees per unit time, then the heat flow rate required to do this is given by

$\dot{Q}$ = specific heat capacity × mass × *rate* of temperature rise

or

$$\dot{Q} = cm\dot{t}$$

$\dot{Q}$ is the *rate* at which heat must be supplied to the body. In terms of the basic units of c (J/kg K), m (kg) and $\dot{t}$ (°C/s or K/s), the unit of $\dot{Q}$ is the joule per second (J/s) or watt (W). The unit for heat flow rate is therefore the same as the unit of power. The watt is a small quantity and the higher multiples are usually used, i.e. kW, MW and GW.

Example

If heat is supplied to a 5 kg mass of oil of specific heat capacity 1·6 kJ/kg K at the rate of 4 kW, find the time taken to raise the temperature of the oil from 15 °C to 50 °C.

Solution

$$\dot{Q} = cm\dot{t}$$

i.e. $4 \times 1000 = 1 \cdot 6 \times 1000 \times 5 \times \dot{t}$

Therefore

$$\dot{t} = 0 \cdot 5 \, \text{K/s} \quad \text{or} \quad 0 \cdot 5 \, ^\circ\text{C/s}$$

so that

time taken to raise temperature $35\,^\circ\text{C} = \dfrac{35}{0 \cdot 5} = 70\,\text{s}$

(Example)

A Lancashire boiler is $7 \cdot 5$ m long, 2 m diameter, and is half-full of water at $20\,^\circ$C. Heat is supplied to the boiler at a constant rate of 3 MW. Calculate the time taken to raise the temperature of the water to $150\,^\circ$C if the thermal capacity of the boiler metal is neglected.

Solution

Volume of boiler $= \dfrac{\pi \times 2^2}{4} \times 7 \cdot 5 = 23 \cdot 6 \, \text{m}^3$

Volume of water $= \frac{1}{2} \times 23 \cdot 6 = 11 \cdot 8 \, \text{m}^3$

Mass of water $= 11 \cdot 8 \times 1000 = 11\,800 \, \text{kg}$

since density of water is $1000 \, \text{kg/m}^3$.

If t is the time in seconds taken to raise the temperature from $20\,^\circ$C to $150\,^\circ$C then

heat supplied per second $= 3\,\text{MW} = 3 \times 10^6 \, \text{J/s}$

heat supplied in time $t = 3 \times 10^6 \times t\,\text{J}$

Therefore

$3 \times 10^6 \times t = cm\Delta t$

$\qquad = 4 \cdot 2 \times 10^3 \times 11\,800 \times (150 - 20)$

so that

$t = 2150\,\text{s} \quad \text{or} \quad 35 \cdot 8 \, \text{min}$

(Example)

Steel alloy components of total mass 6 t at a temperature of $25\,^\circ$C are heated in a gas-fired furnace to $825\,^\circ$C prior to quenching. If only 25 per cent of the heat available from the gas is absorbed by the components, find the total heat supplied in the gas. The specific heat capacity of the steel alloy is $0 \cdot 5 \, \text{kJ/kg K}$.

Solution

$$6\,t = 6000\,\text{kg}$$

Heat absorbed by the components $= cm\Delta t$

$$= 0.5 \times 6000 \times (825 - 25)$$

$$= 2.4 \times 10^6\,\text{kJ}$$

$$= 2.4\,\text{GJ}$$

This quantity represents only 25 per cent of the total heat supplied, Q; therefore

$$0.25 \times Q = 2.4\,\text{GJ}$$

and

$$Q = 9.6\,\text{GJ}$$

Example

A mass of 40 g of oxygen is cooled in a closed vessel from 16 °C to −44 °C. The same mass of oxygen is cooled through the same temperature range in a cylinder so that the pressure is maintained constant. Compare the heat removed in the two cases. For oxygen, $c_v = 0.66\,\text{kJ/kg K}$, and $c_p = 0.92\,\text{kJ/kg K}$.

Solution
At constant volume,

heat removed $= c_v m\Delta t$

$$= 0.66 \times (40 \times 10^{-3}) \times [16 - (-44)]$$

$$= 1.58\,\text{kJ}$$

At constant pressure,

heat removed $= c_p m\Delta t$

$$= 0.92 \times (40 \times 10^{-3}) \times [16 - (-44)]$$

$$= 2.21\,\text{kJ}$$

Problems

15.1 Find the heat required to raise the temperature of 15 litres of water from 10 °C to 80 °C. If heat is supplied in 6 min, what is the heat flow rate?
(4.41 MJ; 12.3 kW)

15.2 Calculate the heat required to raise the temperature of 2 t of water in a boiler from 15 °C to the boiling-point, 100 °C. (714 MJ)

15.3 If 66·5 kJ are required to raise the temperature of 2·25 kg of a metal by 55 °C, what is the specific heat capacity of the metal? (0·54 kJ/kg K)

15.4 Calculate the average rate at which heat must be supplied to a kettle of water in order to heat its contents from a temperature of 15 °C to the boiling point in 2 min. Mass of kettle = 0·25 kg; specific heat capacity of the metal = 0·84 kJ/kg K; mass of water = 0·7 kg. (2·23 kW)

15.5 The hot gases from a combustion chamber enter a heat exchanger at 295 °C and leave at 15 °C. 50 kg of gas pass through the exchanger per minute. If the mean specific heat capacity of the gases is 1·05 kJ/kg K, calculate the heat transferred per hour. (882 MJ)

15.6 The cooling system of a diesel engine pumps 270 kg of water per hour through the radiator where it is cooled from 68 °C to 38 °C. Calculate the rate at which heat is given up to the cooling air blown past the radiator. (9·45 kW)

15.7 120 metal bushes, each of mass 0·5 kg and specific heat capacity 0·5 kJ/kg K, are to be raised in temperature by 500 K prior to shrink-fitting on shafts. Find the total heat necessary. (15 MJ)

15.8 A Lancashire boiler is 9 m long, 2 m diameter, and is two-thirds full of water at 20 °C. Heat is supplied to the boiler at the uniform rate of 1·8 MW. Calculate the time to raise the temperature of the water to 200 °C if the thermal capacity of the boiler itself is neglected. (2·2 h)

15.9 The cylinder of a gas engine contains 1 dm^3 of a mixture of town gas and air at 90 °C. The mean density of the gas mixture is 8 kg/m^3. The gas is burnt and the temperature of the burnt gases rises to 870 °C. If the average specific heat capacity of the mixture during burning is 1·1 kJ/kg K, calculate the heat energy which must be supplied. (6·86 kJ)

15.10 The water in a boiler of the Calder Hall nuclear reactor plant is converted to steam by heating with carbon dioxide gas. The gas enters the boiler at 350 °C and leaves at 180 °C. The mass of gas flowing through the boiler is 200 kg/s. Calculate the rate at which heat is given up to the boiler if the specific heat capacity of carbon dioxide at constant pressure is 0·85 kJ/kg K. (28·9 MW)

15.11 A mass of 12 g of nitrogen in a closed vessel is heated from 26 °C to 50 °C. Find the heat required. If the same quantity of heat is supplied under constant pressure conditions, what would be the final temperature of the gas? For nitrogen, c_p = 1·04 kJ/kg K, c_v = 0·74 kJ/kg K. (0·213 kJ; 43·1 °C)

15.12 A cylinder contains 6 g of the rare gas argon. Careful experiment showed that the gas absorbed 18·72 J of heat for a temperature rise of 10 °C when the volume was kept constant. A second experiment showed that the gas absorbed 50·16 J of heat for a temperature rise of 16 °C when the pressure was maintained constant. Find the specific heat capacities of argon at constant volume and constant pressure. (0·312, 0·522 kJ/kg K)

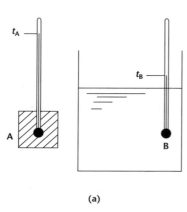

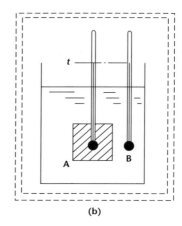

(a)   (b)

Fig. 15.1

15.8 Transference of heat; mixtures

When cold water is added to hot water the final temperature of the mixture lies somewhere between that of the cold and that of the hot. When two quantities of water at the same temperature are mixed there is no change in temperature. These are facts that are known to everyone. In order to investigate more closely the transfer of heat consider a body A immersed in a liquid B, Fig. 15.1. Suppose, before immersion, that the body is at temperature t_A and the liquid at temperature t_B, and assume that the body A has the higher temperature. When the body is immersed, observation of thermometers placed in the body and liquid would show a fall in temperature of the body and a rise in temperature of the liquid. This continues until both are at the same temperature t after which no further change takes place.

Since A has fallen in temperature from t_A to t, it will have given up an amount of heat

Q_A = specific heat capacity × mass × temperature change

$$= c_A m_A (t_A - t)$$

Similarly the rise of temperature of B = $t - t_B$. Therefore, heat gained by B is

$$Q_B = c_B m_B (t - t_B)$$

We now say that, owing to the temperature difference between A and B, heat flows from A to B. Also the heat flowing out of A is equal to the heat flowing into B, i.e.

heat given out by A = heat taken in by B

i.e. $Q_A = Q_B$

i.e. $c_A m_A (t_A - t) = c_B m_B (t - t_B)$

From this equation the final temperature t of the mixture may be found. The same reasoning applies if A consists of several bodies having different heat capacities. The product $c_A m_A$ then represents the total thermal capacity of all the masses constituting A. Note that we must not add the masses but we add their thermal capacities or water equivalents.

In the above discussion A and B might also represent two masses brought into contact, or two liquids being mixed together.

Note that we have assumed that all the heat Q_A given up by A is received by B; that is, we are assuming no heat to be transferred to the surroundings. Under these conditions A and B together are said to be *isolated*, or *thermally insulated*. This isolation is symbolized by the broken line, drawn around A and B together. Isolation means that whatever changes take place in the temperature outside the broken line no effect can be measured on thermometers in A and B. In practice, however, isolation is never perfect; some heat is lost to the surroundings if they are at a temperature lower than that of A and B, or heat is gained by A and B if the surroundings are at a higher temperature.

Example

A steel tank of mass 50 kg contains 200 kg of water at 15 °C. A further 100 kg of water is poured into the tank at 95 °C. If the specific heat capacity of the steel is 0·45 kJ/kg K, calculate the final temperature of the water in the tank.

Solution
Let t°C be the final temperature of the mixture.

Heat given up by the water poured into the tank $= cm\Delta t$

$$= 4\cdot2 \times 100 \times (95 - t)\,\text{kJ}$$

Heat gained by steel tank $= cm\Delta t$

$$= 0\cdot45 \times 50 \times (t - 15)\,\text{kJ}$$

$$= 22\cdot5(t - 15)\,\text{kJ}$$

Heat gained by water in the tank $= cm\Delta t$

$$= 4\cdot2 \times 200 \times (t - 15)\,\text{kJ}$$

Equating heat given up by the water poured in, to the heat taken up by the tank and the water, then

$$4\cdot2 \times 100 \times (95 - t) = [22\cdot5(t - 15)] + [4\cdot2 \times 200(t - 15)]$$

$$= 862\cdot5(t - 15)$$

Therefore

$$t = \mathbf{41\,°C}$$

Example

The specific heat of a brass specimen is to be determined by the method of mixtures. The specimen, of mass 32 g, is placed in water at 100 °C until it reaches the same temperature as the water. It is then quickly transferred to a copper calorimeter containing water at 24 °C. The final temperature in the calorimeter is found to be 32·8 °C. Calculate the specific heat capacity of the brass using the following data: water equivalent of calorimeter, 4·5 g; mass of water, 20 g.

Solution

The initial temperature of the brass is 100 °C and its final temperature 32·8 °C. If c is the specific heat capacity of the brass, then

heat given out by brass = specific heat capacity × mass × temperature drop

$$= c \times 32 \times 10^{-3}(100 - 32 \cdot 8)$$

$$= c \times 2 \cdot 15 \, J$$

The initial temperature of the water and calorimeter is 24 °C; the final temperature 32·8 °C. The equivalent mass of water is 20 + 4·5 = 24·5 g; hence

heat taken in by water and calorimeter = $4 \cdot 2 \times 10^3 \times (24 \cdot 5 \times 10^{-3}) \times (32 \cdot 8 - 24)$

$$= 906 \, J$$

Equating heat given out by brass to that taken in by water and calorimeter,

$c \times 2 \cdot 15 = 906$

$c = \mathbf{422 \, J/kg\,K}$

Example

In a hardening process a metal rod of mass 3 kg is heated to a temperature of 700 °C and then quenched in oil initially at a temperature of 20 °C. The quantity of oil is 25 litres, its specific heat capacity is 1·47 kJ/kg K, and its density is 0·94 Mg/m³. Find the temperature to which the oil will rise during quenching.

If the oil temperature must not rise above 42 °C, what is the minimum quantity of oil in litres required? The specific heat capacity of the metal is 0·5 kJ/kg K.

Solution

$$25 \, \text{litres} = 25 \times 10^{-3} \, m^3$$

Mass of oil = volume × density = $25 \times 10^{-3} \times (0 \cdot 94 \times 10^3) \, kg$

$$= 23 \cdot 5 \, kg$$

(note that 0·94 Mg/m³ = 0·94 kg/litre, i.e. 1 litre of the oil has a mass of 0·94 kg).

Let t °C be the final temperature of the oil after the quenching operation; then

heat given out by rod = $cm\Delta t = 0 \cdot 5 \times 10^3 \times 3 \times (700 - t) \, J$

and

heat taken in by oil = $1.47 \times 10^3 \times 23.5 \times (t - 20) \, \text{J}$

Equating the heat given out by the rod to that taken in by the oil, we have

$$0.5 \times 10^3 \times 3 \times (700 - t) = 1.47 \times 10^3 \times 23.5 \times (t - 20)$$

$$t = 48\,^\circ\text{C}$$

If the final temperature of the oil is limited to $42\,^\circ\text{C}$, then $t = 42$, and if m kg is the mass of oil required,

$$0.5 \times 10^3 \times 3 \times (700 - 42) = 1.47 \times 10^3 \times m \times (42 - 20)$$

therefore

$$m = 30.5 \, \text{kg}$$

Hence,

$$\text{volume of oil} = \frac{\text{mass}}{\text{density}} = \frac{30.5 \,(\text{kg})}{0.94 \times 10^3 (\text{kg}/\text{m}^3)}$$

$$= 0.0325 \, \text{m}^3$$

$$= 32.5 \, \text{litres}$$

Problems

15.13 2 kg of iron is first placed in water at $100\,^\circ\text{C}$, then quickly removed and placed in 4 kg of water at $15\,^\circ\text{C}$. What is the final temperature of the water and iron? Take specific heat capacity of iron as $0.45 \, \text{kJ/kg K}$. $(19.3\,^\circ\text{C})$

15.14 1.2 kg of lead shot is taken out of water at $100\,^\circ\text{C}$ and poured into 0.75 kg of water at $20\,^\circ\text{C}$. Find the final temperature of the mixture. The specific heat capacity of lead is $0.13 \, \text{kJ/kg K}$. $(23.8\,^\circ\text{C})$

15.15 How much water at $15\,^\circ\text{C}$ must be mixed with 1 kg of water at $100\,^\circ\text{C}$ in order to reduce the temperature to $50\,^\circ\text{C}$? $(1.43 \, \text{kg})$

15.16 A small cylinder block has a mass of 40 kg; its specific heat capacity is $0.55 \, \text{kJ/kg K}$ and original temperature $205\,^\circ\text{C}$. 4.5 litres of water at $5\,^\circ\text{C}$ are poured into the empty cooling passages in the block and rest there. What is the final temperature? $(113\,^\circ\text{C})$

15.17 32 cutting tools each of mass 300 g are removed from a heat treatment furnace at $1200\,^\circ\text{C}$ and plunged in a quenching bath containing 225 litres of oil at $25\,^\circ\text{C}$. What is the final temperature of the oil and cutters? The specific heat capacities of the oil and steel are respectively 1.7 and $0.46 \, \text{kJ/kg K}$. The relative density of the oil is 0.86. $(40.7\,^\circ\text{C})$

15.18 A steel specimen of mass 30 g is first placed in boiling water. It is then quickly transferred to a copper calorimeter containing 20 g of water at $20\,^\circ\text{C}$.

The calorimeter's mass is 30 g and its specific heat capacity is 0·42 kJ/kg K. It is observed that the temperature of the water in the calorimeter rises to 30·9 °C. Calculate the specific heat capacity of steel.   (0·51 kJ/kg K)

15.19 The specific heat capacity of an aluminium specimen of mass 16 g is to be determined by the method of mixtures. The copper calorimeter used has a water equivalent of 2 g and contains 26 g of water at 22 °C. When the specimen has been raised in temperature to 98 °C, it is then quickly placed in the calorimeter. The final temperature of the water is found to be 28 °C. (a) Calculate the specific heat capacity of aluminium. (b) If the mass of the copper calorimeter is neglected, what would be the calculated value of the specific heat capacity?   (0·63, 0·59 kJ/kg K)

15.20 An experiment is to be carried out to measure the specific heat capacity of an oil by the method of mixtures. The copper calorimeter of mass 28 g contains 32 g of water at 21 °C. The specific heat capacity of copper is 0·4 kJ/kg K. The oil is first heated to 200 °C before being mixed with the water. In order that the heat losses shall be as small as possible, the temperature rise of the water must not be too great. For accuracy of measurement, however, the temperature rise must be about 20 °C. For this temperature rise, what mass of oil would be required, assuming the specific heat capacity of the oil to be about 1·3 kJ/kg K.

If this mass of oil is used and the final temperature is observed to be 39 °C, what is the correct value of the specific heat capacity of the oil?   (14·1 g; 1·17 kJ/kg K)

(15.9) Heat transfer

Heat transfer is *energy in transit* and heat flows because of a temperature difference; only when heat flows from one body to another can it be detected by a change of temperature or a change of state. The direction of flow is from the higher temperature to the lower. A temperature difference is the 'thermal potential' or 'height' which determines the direction of heat flow and its rate. The greater the temperature difference, the more rapidly does the exchange of heat take place.

Heat transfer takes place in one or more of three ways: *conduction, convection* and *radiation*.

Conduction

Close-packed molecules in a body assist the direct transport of energy from molecule to molecule. In general, the more closely packed the molecules, the better the conduction of heat. Thus solids are better conductors of heat than liquids, and liquids are better than gases. Among the solids, the metals are the best conductors.

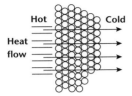

Fig. 15.2   Close-packed solid molecules conduct heat directly.

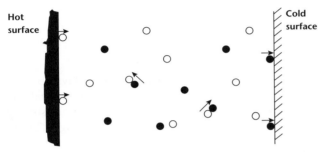

Fig. 15.3   Conduction in gases. Gases conduct heat by collisions between (a) hot (●) and cold (○) molecules, (b) cold molecules and a hot surface, (c) hot molecules and a cold surface.

A good conductor is a material that requires a small temperature difference to transfer a given amount of heat per unit time. An *insulator* or poor conductor requires a larger temperature difference to transfer the same amount of heat in the same time.

If the molecules at one end of a *solid* are heated, they are set into more rapid vibration and this increase in the energy of vibration is communicated to neighbouring molecules. The energy supplied in the form of heat is therefore transmitted along the solid as energy of vibration by direct contact between the molecules, Fig. 15.2.

Liquids are not usually good conductors of heat. Exceptions are mercury and liquid metals at high temperatures. Water is a poor conductor but is better than most other liquids.

The molecules of a *gas* are not in contact except when they collide during random motion. Conduction then takes place during momentary contact between molecules or when molecules impinge on a solid surface, Fig. 15.3. Gases, therefore, are very poor conductors of heat. Trapped, still air serves as a good insulator since it is a very poor conductor. Similarly, any porous or honeycombed material which can trap air acts as a good insulator. Typical materials are straw, fibrous asbestos, wool, mineral wools such as glass or rock, crumpled aluminium foil, cork and expanded polystyrene. The perfect insulator is a complete vacuum, empty of all material. For example, the Thermos flask is an effective insulator because the space between the inner and outer containers is almost empty of air.

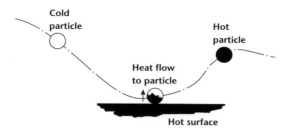

Fig. 15.4

Convection

In the conduction of heat the mass movement of material plays no part. *Convection*, on the other hand, is the transport of heat by movement of material, as in a flowing liquid or a current of air. Fluid particles moving past a hot surface pick up heat by random contact between fluid and surface molecules, Fig. 15.4. If the fluid then travels on to a cold surface it will give up its heat by conduction from its hot molecules to the cold surface.

Natural and forced convection

It is usual to describe the convection effect of a fluid according to whether the fluid flows naturally due to a temperature difference or is mechanically pumped. If parts of a fluid are at different temperatures the densities differ, a hot fluid being less dense than a cold fluid. The denser cold fluid tends to fall and the lighter hot fluid to rise. The two effects act together to set up convection currents in the fluid. This is called *natural convection*. Fluid that is blown or pumped past a surface or along a pipe gives rise to heat transfer by *forced convection*.

Figure 15.5(a) shows a U-tube filled with water. If the right-hand arm of the tube is heated the water in it will rise; cold water, being denser than hot, will fall in the left-hand arm to fill the space vacated by the hot water. A flow of water in the anticlockwise direction is then set up. This arrangement is called a thermosyphon and illustrates the principle of natural convection.

Figure 15.5(b) shows diagrammatically a simple hot-water central heating system. Water, heated in the boiler, rises to the radiator system in the floor above, where it is cooled and flows back to the lower level. Note that cold water enters the boilers at the lowest position to assist the natural convection in the boiler itself.

Figure 15.5(c) shows the principle of the ordinary hot water 'radiator' or hot-plate convector for room heating. Air in contact with the hot surface of the 'radiator' rises, its place being taken by cold air. Thus the heater sets up a steady circulation of warm air in the room. The most suitable position for a radiator is beneath a window where it may warm incoming air.

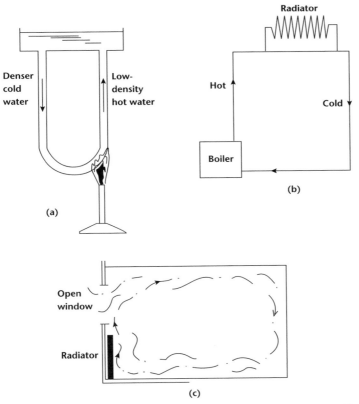

Fig. 15.5 Natural convection (a) thermosyphon (b) hot water system (c) convection from heater.

Radiation

Radiation is the transfer of energy through empty space. If a hot and a cold body were placed together in a vacuum, but not in contact, heat would be transmitted from one to the other by radiation. Radiation can take place through air or any other medium but the effect does not require the presence of any matter whatsoever. The transfer of heat and light energy from the sun across empty space is the simplest example of radiation. A magnifying glass can 'focus' heat as well as light energy from the sun to produce a considerable heating effect.

The heat energy transmitted by radiation is of the same nature as light and travels with the velocity of light, i.e. 299 800 km/s in empty space.

All bodies radiate heat energy and, at a sufficiently high temperature, this energy is observed as light. The heat treatment of steel at one time depended on a proper assessment of temperature from the colour of the heated metal. As the temperature increases, the steel turns a dull red followed by a bright orange and yellow, until, at very high temperatures, all the colours of the spectrum are emitted, the radiation then appearing 'white hot'.

The transfer of heat by radiation is usually important at high temperatures or when conduction and convection are negligible.

(15.10) Emission; absorption; reflection

As we have seen, radiant heat energy is of the same nature as light. It is emitted by all bodies and may be absorbed or reflected by other bodies. A material transparent to radiant energy allows radiant heat to pass freely and is called *diathermous*. Bodies that reflect or absorb radiant heat and do not allow it to pass through are called *athermous*. If the energy is wholly transmitted or reflected the temperature of the body does not change. If the radiant heat is absorbed the temperature of the body rises, Fig. 15.6.

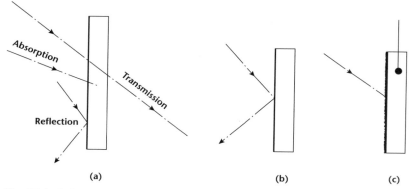

Fig. 15.6 Reflection and absorption. (a) Reflection, (b) light smooth surfaces *reflect* heat radiation, (c) black rough surfaces *absorb* heat radiation.

The ability of a body to emit, absorb or reflect heat depends on the colour and nature of its *surface*. The best absorbers of radiant heat are black rough surfaces. The best reflectors are light coloured and have a smooth polished surface, e.g. a metal mirror.

It is found experimentally that the best absorbers of heat are also the best emitters or radiators of heat. Also, a good reflector is both a poor absorber and a poor emitter. For example, an aluminium kettle has a bright polished surface and therefore tends to retain the heat of the water. The bottom of the kettle, on the other hand, should have a dull surface to absorb heat quickly. Aluminium foil is used as a heat insulator since it traps still air, reflects heat, and acts as a poor emitter and absorber.

A bright metal steam-pipe is a poor radiator but allows heat loss by conduction, through the metal and by natural convection from the surface. A thin layer of asbestos or wool-like lagging reduces heat loss by conduction but the dull rough surface of the lagging is often a better emitter than bright steel. Hence a thin layer of lagging may be worse than no lagging at all. One of the most efficient methods of insulation is the use of such lagging covered by a smooth polished metal casing.

Asbestos was hitherto a most common form of lagging but is now greatly restricted in use because of risk to health. A typical lagging material utilizing the properties of trapped air is a mineral wool formed from fine flexible fibres. Such lagging usually has a long life. An example of a short-life thermal insulating material employed in a one-off situation at very high temperatures is the *ablative* tile used to provide a heat-shield for capsules returning from outer space. Ablation is the carrying away of heat by a gas stream causing erosion of a material similar to the erosion of rocks by weathering. The tile is a composite plastic which may include asbestos, cork or glass fibre. When the capsule impacts on the earth's atmosphere aerodynamic friction forces cause the tiles to melt or vaporize (i.e. change their state) and since the tiles have very low thermal conductivity the heat does not pass through to the capsule before the liquid or vapour is swept away from the surface. Expendable tiles are lost but the inner capsule shield is maintained intact. The heat absorbed when there is a change of state of a substance is dealt with in Chapter 18. Space vehicles may be subject to temperatures as high as 2500 K and heat transfer rates of the order of 30 000 kW per square metre of external surface, even if only for a few seconds. A large range of materials and techniques have been brought into use besides the tiles already mentioned, e.g. coatings of plastic ablatives, silicon carbide and specially treated nylon as well as fibreglass, carbon-fibre cloth and metal or graphite plating shields.

Chapter 16

Expansion and compression of gases

We begin here with the definition of pressure and its units, derived from the SI base units, then go on to the meaning of absolute, atmospheric and gauge pressures. Further work on temperature is necessary, bringing in the absolute zero of temperature before dealing with the characteristic equation of an ideal gas and the laws governing the expansion and compression of gases.

16.1 Pressure exerted by a gas

The molecules of a gas are in rapid and random motion, and when placed in a container, a gas always expands to fill the container completely. The rapidly moving molecules continually bombard the surface of the container and the effect of their impact is to produce a force normal to the surface. The number of impacts per unit time is so large that it appears that the force is continuous and uniform.

The force normal to unit area of surface is the *pressure* acting on the surface. Therefore

$$\text{pressure} = \frac{\text{force}}{\text{area}}$$

The derived SI unit of pressure is the same as that for stress, i.e. the *newton per square metre* (N/m^2). The multiple forms kN/m^2, MN/m^2 and GN/m^2 are also used. A further variation of the basic form is N/mm^2 which is the same as MN/m^2. Other units for pressure are

1 bar $= 10^5\,N/m^2 = 100\,kN/m^2$

1 millibar (mb) $= 100\,N/m^2 = 1/1000\,\text{bar}$

1 hectobar (hbar) $= 10^7\,N/m^2 = 10\,MN/m^2 = 100\,\text{bar}$

The newton per square metre is also called the *pascal* (**Pa**), whose multiples are: *kilopascal* (**kPa**), *megapascal* (**MPa**), *gigapascal* (**GPa**). In this book, however, for the units of pressure we shall use only the names newton per square metre and bar, and the names for their multiples.

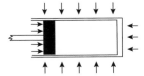

Fig. 16.1

16.2 Absolute pressure

Figure 16.1 shows a cylinder containing a close-fitting piston. Suppose the cylinder is completely emptied of all gas. Then the piston would be driven into the cylinder. Since the cylinder is empty of gas there can be no pressure exerted on the inside face of the piston and the pressure of the atmospheric air on the outer face produces an external driving force on the piston.

The space within the cylinder is said to be a *perfect vacuum* and is at zero pressure. A pressure measured above this zero is called an *absolute pressure*. For example, the pressure of atmospheric air at the earth's surface is 101.3 kN/m^2 (1.013 bar) absolute (abs.) and is due to the weight of the earth's atmospheric air.

Unless otherwise stated all pressures used in this book are absolute pressures.

16.3 Atmospheric pressure; the barometer

A practical method of measuring the pressure of the atmosphere is by means of the *mercury barometer*, Fig. 16.2. The space at the top of a closed vertical tube dipped in an open bath of mercury is emptied of air as far as possible. The pressure in this space is then equal only to the *vapour pressure* of the mercury vapour (*see* Ch. 18). However, this vapour pressure is low and may often be neglected. The top space is therefore almost at zero pressure.

The pressure of the atmosphere on the mercury in the open bath forces the liquid up the tube until the weight of the mercury column is sufficient to balance the pressure of the atmosphere.

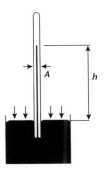

Fig. 16.2

Let h be the height of the mercury column above the level of mercury in the bath; that is, h is the 'height' of the mercury barometer. Let ρ be the density of mercury and A the cross-sectional area of the tube; then

volume of mercury column $= A \times h$

mass of mercury column $= \rho \times A \times h$

Pressure exerted by the mercury column at the level of the bath

$$= \frac{\text{weight of column}}{\text{area of tube}}$$

$$= \frac{\rho.Ah.g}{A}$$

$$= \rho gh$$

This is the pressure exerted by the atmosphere on the mercury surface in the bath. The height h of the barometer is therefore directly proportional to the pressure of the atmosphere.

The density of mercury may be taken as $13 \cdot 6 \times 10^3 \, \text{kg/m}^3$; therefore if h is in metres:

atmospheric pressure $= \rho gh = 13 \cdot 6 \times 10^3 \times 9 \cdot 8h \, \text{N/m}^2$

$$= 133 \cdot 3h \, \text{kN/m}^2$$

Thus one metre of mercury, $h = 1$, corresponds to a pressure of $133 \cdot 3 \, \text{kN/m}^2$.

If the atmospheric pressure is $101 \cdot 3 \, \text{kN/m}^2$, then

$$h = \frac{101 \cdot 3}{133 \cdot 3} = 0 \cdot 76 \, \text{m}$$

An atmospheric pressure of $101 \cdot 3 \, \text{kN/m}^2$ therefore corresponds to a barometric height of $760 \, \text{mm}$ of mercury (mm Hg). Again, for example, a pressure of $800 \, \text{mm Hg}$ corresponds to a pressure of $0 \cdot 8 \times 133 \cdot 3 = 106 \cdot 6 \, \text{kN/m}^2$ or

$$\frac{800}{760} \times 101 \cdot 3 = 106 \cdot 6 \, \text{kN/m}^2.$$

The atmospheric pressure may also be related to an equivalent height or 'head' of water. The density of water is $1000 \, \text{kg/m}^3$ (at $4 \, °\text{C}$ and atmospheric pressure) and the pressure due to a column of water of height h is ρgh, i.e. $1000 \times 9 \cdot 8h \, \text{N/m}^2$. The height of a column of water to balance an atmospheric pressure of $101 \cdot 3 \, \text{kN/m}^2$ is therefore about

$$\frac{101 \cdot 3 \times 10^3}{1000 \times 9 \cdot 8} \quad \text{i.e. } 10 \cdot 4 \, \text{m}$$

For large pressures, the unit used for the atmospheric pressure is the *atmosphere*. Thus

1 atmosphere (atm) = $101 \cdot 3 \, \text{kN/m}^2$

$= 1 \cdot 013 \, \text{bar}$

$= 760 \, \text{mm of mercury}$

$= 10 \cdot 4 \, \text{m of water}$

In this text, unless otherwise stated, the value of $101 \, \text{kN/m}^2$ ($1 \cdot 01$ bar) for atmospheric pressure will be used in calculations.

(16.4) Gauge pressure

The pressure of gas in a cylinder is usually measured by a *pressure gauge*. The simplest form of gauge is the *manometer*. This is a simple U-tube containing a liquid, Fig. 16.3. One limb of the tube is open to the atmosphere and the other is connected to the space containing gas under pressure.

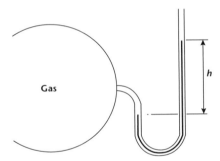

Fig. 16.3

If the pressure of the gas is greater than that of the atmosphere, the liquid in the U-tube is forced down in the left-hand limb and up in the right-hand limb. If the difference in the levels is h then the difference in pressure is $\rho g \times h$, where ρ is the density of the liquid in the tube. The manometer thus measures the *difference* between the pressure of the gas and the pressure of the atmosphere. This difference is called the *gauge pressure*. It is the pressure measured by a pressure gauge and is expressed as N/m^2 *gauge*.

Thus to convert gauge pressure to absolute pressure we must add the pressure of the atmosphere at the place where the pressure is measured, Fig. 16.4. Thus

absolute pressure = gauge pressure + atmospheric pressure

Where the term pressure is used, therefore, care should be taken to ascertain whether absolute pressure or gauge pressure is meant. In this book the term 'pressure' alone always means absolute pressure.

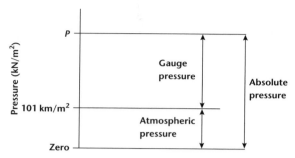

Fig. 16.4

(16.5) The Bourdon gauge

Another practical method of measuring gauge pressure is the *Bourdon* gauge. This consists essentially of a hollow curved tube of approximately elliptical cross-section, Fig. 16.5. If the pressure of the gas or liquid inside the tube is greater than that of the atmosphere, the sides of the tube tend to become circular. The tube tends to straighten out and the deflection of the end of the tube measures the gauge pressure to some scale.

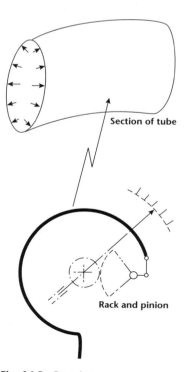

Fig. 16.5 Bourdon gauge

Vacuum gauges

Pressure gauges used for measuring pressures less than atmospheric are calibrated to read from 0 to $101 \, kN/m^2$ (and from 0 to 760 mm Hg). Such a *vacuum* gauge gives the amount the pressure is *below* that of the atmosphere and the gauge reading must be *subtracted* from atmospheric pressure. Zero on the gauge indicates atmospheric pressure. A reading of $101 \, kN/m^2$ indicates zero absolute pressure, i.e. a perfect vacuum. A reading of 100 mm Hg, for example, is referred to as 100 mm Hg of vacuum and corresponds to an absolute pressure of $760 - 100 = 660$ mm Hg or $133 \cdot 3 \times 0 \cdot 66 = 88 \, kN/m^2$.

16.6 Relation between pressure and volume of a gas: Boyle's law

Consider a cylinder of gas closed by a piston at one end, Fig. 16.6. If the piston is moved to the left, the volume of gas is decreased. The mass of gas, however, remains the same and therefore the number of molecules in the cylinder is unaltered. The molecules have a shorter distance to travel in a smaller space and therefore rebound more often against the walls of the cylinder and piston. *The pressure therefore increases.* In addition, the molecules may move more rapidly due to the compression, i.e. the gas has a higher temperature. If, however, the piston is moved slowly and we wait for a sufficient length of time, the temperature will return to its original value but the pressure will still be greater than it was originally.

Suppose, therefore, when the gas in the cylinder is at constant temperature we measure the gauge pressure for each value of the volume and then convert each

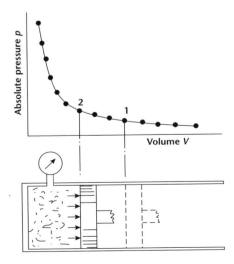

Fig. 16.6

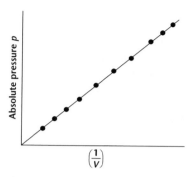

Fig. 16.7

pressure gauge reading to absolute pressure. If the absolute pressure p is plotted against the volume V, a curve of the form shown in Fig. 16.6 will be obtained. However, if we plot pressure against the reciprocal of the volume, i.e. p against $1/V$, then the straight line of Fig. 16.7 will be obtained. The line passes through the origin, hence showing that the pressure is directly proportional to $1/V$; thus

p is proportional to $\dfrac{1}{V}$

i.e. $p \propto \dfrac{1}{V}$

therefore

pV = constant, C

or

$$pV = C$$

This is known as *Boyle's law* which may be stated thus:

The absolute pressure of a given mass of gas varies inversely as its volume if the temperature is maintained constant, or in other words, *for a given mass of gas at constant temperature,*

pressure × volume = constant

Boyle's law is not followed exactly by any gas but it is obeyed very closely by gases at high temperatures and low pressures.

If points 1 and 2, Fig. 16.6, denote two values of pressure and volume then, since pV = constant, we may write

$$p_1 V_1 = p_2 V_2 \text{ at constant temperature}$$

Note that Boyle's law holds only for *absolute* pressures.

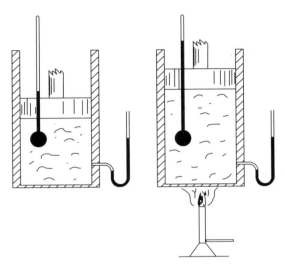

Fig. 16.8

(16.7) Relation between volume and temperature of a gas: Charles's law

Boyle's law gives only the relation between pressure and volume of a given mass of gas when the temperature is kept constant. Now we investigate the relation between volume and temperature *when the pressure is kept constant*. Consider the cylinder and piston shown in Fig. 16.8. The piston traps a given mass of gas in the cylinder. If friction is neglected, the weight of the piston is balanced by the gas pressure acting upwards on the inner face of the piston. Since the weight of the piston is constant, the pressure in the gas will be constant also.

Now let us raise the temperature of the gas in the cylinder. As the temperature rises the piston will be found to rise also. Thus the volume of gas has increased, the pressure remaining constant.

The *coefficient of volumetric* or *cubical expansion* γ of a gas is defined as the increase in volume per unit volume per degree rise in temperature, i.e.

$$\gamma = \frac{\text{increase in volume per unit volume}}{\text{temperature rise}}$$

i.e. $\gamma = \dfrac{\Delta V}{\Delta t}$

or

$\Delta V = \gamma \Delta t$

If V_0 is the original volume of the gas, then the increase in volume for a rise in temperature Δt is $\Delta V \times V_0$, and the final volume is therefore

$V = V_0 + \Delta V V_0$

$\quad = V_0 + \gamma \Delta t V_0$

$\quad = V_0(1 + \gamma \Delta t)$

This relation is known as *Charles's (or Gay-Lussac's)* law. The law may be stated as follows:

> *For a given mass of gas at constant pressure a given increase of temperature always produces the same fractional increase in volume.*

Experiment shows that the coefficient of volumetric expansion is substantially constant for *all* gases and is equal to 1/273, i.e. the increase in volume per unit volume per 1 degree rise in temperature is 1/273; thus

$$\gamma = \frac{1}{273} \Big/ {}^{\circ}C$$

Again, if V_0 is the volume of a gas at $0\,^{\circ}C$, the volume V at temperature $t\,^{\circ}C$ at the same pressure is

$$V = V_0\left(1 + \frac{t}{273}\right)$$

since $\Delta t = t$, and $\gamma = 1/273$.

(16.8) The absolute zero of temperature

Consider a volume of gas, $273\,\text{mm}^3$, at temperature $0\,^{\circ}C$. Charles's law states that the volume is reduced by $1/273\,\text{mm}^3$ for each degree reduction in temperature at constant pressure. Therefore, if the temperature were reduced by $273\,^{\circ}C$, the volume of the gas would be zero. The temperature when the volume is zero is therefore $-273\,^{\circ}C$.

Figure 16.9 shows the straight-line relation between volume V and temperature $t\,^{\circ}C$ for a gas cooled at constant pressure. This is the graphical representation of

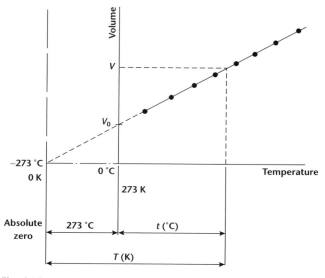

Fig. 16.9

Charles's law. The straight line cuts the axis of temperature at a temperature of −273 °C. This temperature is known as the *absolute zero*. Of course, no gas would remain a gas at such low temperatures; it would become either liquid or solid.

In terms of our molecular picture, all the molecules have come to rest. They have no motion whatsoever at absolute zero temperature. The absolute zero of temperature is very difficult to attain but by special methods a temperature within one ten-millionth part of a degree of absolute zero has been reached.

16.9 The absolute scale of temperature

It is convenient to measure temperature from the absolute zero of −273 °C. A temperature measured from the absolute zero is an *absolute* temperature on the *Kelvin, absolute* or *thermodynamic* temperature scale and the unit is the *kelvin* (**K**). If t °C is the temperature of a body measured above 0 °C then the corresponding temperature measured above absolute zero is $T = (t + 273)$ K = $(t + 273)$ °C abs. The units of Celsius and Kelvin temperature intervals are identical but the Kelvin is used to indicate a temperature difference. Further notes on the Celsius and Kelvin scales will be found on page 313.

Thus:

To convert a temperature in °C to the absolute scale add 273.

Now from Charles's law, if V_0 is the volume of a gas at 0 °C and V the volume at t °C then

$$V = V_0\left(1 + \frac{t}{273}\right)$$

$$= V_0\frac{(273 + t)}{273}$$

But $273 + t$ is the absolute temperature T corresponding to temperature t °C; therefore, denoting 273 °C by T_0, the absolute temperature at 0 °C, we may write

$$V = V_0\frac{T}{T_0}$$

or

$$\frac{V}{T} = \frac{V_0}{T_0} = \text{constant}$$

since both V_0 and T_0, the original volume and temperature respectively, are constants.

Hence Charles's law may be restated as follows:

> *The volume of a given mass of gas at constant pressure varies directly as its absolute temperature.*

or

$$\frac{V}{T} = \text{constant, when } p \text{ is constant}$$

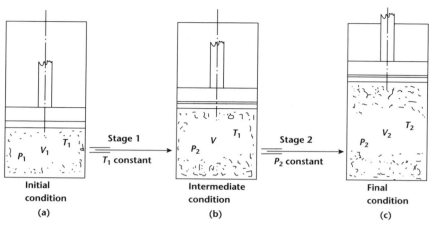

Fig. 16.10

(16.10) The ideal-gas rule; characteristic equation of an ideal gas

Suppose the condition of a given quantity of gas of mass m changes from initial conditions p_1, V_1, T_1 to final conditions p_2, V_2, T_2. Irrespective of how the changes are effected we may assume the final conditions to be arrived at in two stages, Fig. 16.10.

Stage 1

A change in pressure from p_1 to p_2 while the volume changes from V_1 to some intermediate value V. The temperature T_1 is kept constant; that is, the change is according to Boyle's law. Thus

$$p_2 V = p_1 V_1$$

i.e. $V = \dfrac{p_1}{p_2} V_1$ [1]

Stage 2

A change in volume from V to the final volume V_2 while the temperature changes to T_2. The pressure p_2 is kept constant; that is, the change is according to Charles's law. Thus

$$\frac{V_2}{T_2} = \frac{V}{T_1}$$

$$= \frac{p_1}{p_2} \frac{V_1}{T_1} \text{ from equation [1]}$$

Hence

$$\frac{p_1 V_1}{T_1} = \frac{p_2 V_2}{T_2}$$

or, in general, for a *given mass of gas,*

$$\frac{pV}{T} = \text{constant}$$

Since the mass of gas m is constant we may write

$$\frac{pV}{T} = m \times \text{constant}$$

$$= mR$$

R is a constant for that particular gas, and is called the *characteristic gas constant.* Thus

$pV = mRT$

This is known as the *ideal gas rule* or the *characteristic equation of an ideal gas.* This rule is not obeyed exactly by any real gas but describes the behaviour of gases well enough for practical purposes.

A gas that is assumed to obey the ideal gas rule and also has a constant specific heat is called a *perfect* gas.

Note that we may write the ideal gas rule as

$$\frac{p_1 V_1}{T_1} = \frac{p_2 V_2}{T_2} = mR = \text{constant}$$

In applying the gas rule absolute pressures and absolute temperatures must be used.

The value of the gas constant R depends on the nature of the gas. Its *numerical* value depends on the units in which p, V, m and T are measured.

The *units* of R may be deduced from the ideal gas rule

$$R = \frac{pV}{mT} = \frac{(\text{N/m}^2) \times \text{m}^3}{\text{kg} \times \text{K}}$$

$$= \text{N m/kg K} \quad \text{or} \quad \text{N m/kg}\,^{\circ}\text{C, i.e. } \mathbf{J/kg\,K}$$

since the intervals on the Celsius and Kelvin scales are equal and since $1\,\text{N m} = 1\,\text{J}$. R may also be expressed as kJ/kg K.

Values of the gas constant for some important gases are shown in Table 16.1.

Table 16.1 Values of the gas constant for important gases

Gas	Gas constant (R)	
	J/kg K	kJ/kg K
Air	287	0·287
Oxygen	260	0·260
Hydrogen	4150	4·150
Nitrogen	297	0·297
Carbon dioxide	189	0·189
Dry steam	462	0·462

(Example)
_____

Air at $101\,kN/m^2$ and $0\,°C$ has a density of $1·293\,kg/m^3$. Calculate the gas constant for air.

Solution

$$p = 101\,kN/m^2 = 101 \times 10^3\,N/m^2$$

$$V = 1\,m^3$$

$$m = 1·293\,kg$$

$$T = 273\,K$$

$$R = \frac{pV}{mT} = \frac{101 \times 10^3 \times 1}{1·293 \times 273}$$

$$= 286\,J/kg\,K$$

(Example)
_____

$0·06\,m^3$ of hexane gas in a combustion chamber has a pressure of $170\,kN/m^2$ gauge and a temperature $150\,°C$. The mass of gas is $0·4\,kg$. What is the gas constant for hexane?

Solution

$$V = 0·06\,m^3$$

$$m = 0·4\,kg$$

$$T = 150 + 273 = 423\,°C\ (abs.) = 423\,K$$

$$p = 170 + 101 = 271\,kN/m^2\ (abs.)$$

$$= 271 \times 10^3\,N/m^2\ (abs.)$$

$$R = \frac{pV}{mT}$$

$$= \frac{271 \times 10^3 \times 0·06}{0·4 \times 423}$$

$$= 96\,J/kg\,K$$

(Example)
_____

Air is compressed in an engine cylinder. The ratio of initial to final volume is 15 to 1. The initial temperature and pressure are $15\,°C$ and $1·01$ bar. The final pressure is 34 bar. Calculate the final temperature.

Solution

Initial temperature $T_1 = 15 + 273 = 288\,K$.

If the initial volume V_1 is taken as 15 then the final volume V_2 is unity. Applying the gas rule,

$$\frac{p_1V_1}{T_1} = \frac{p_2V_2}{T_2}$$

Therefore

$$\text{final temperature } T_2 = \frac{p_2}{p_1} \times \frac{V_2}{V_1} \times T_1$$

$$= \frac{34}{1\cdot01} \times \frac{1}{15} \times 288$$

$$= 646\,\text{K}$$

$$= 373\,°\text{C}$$

The increase of temperature due to compression is made use of in the oil engine in order to attain a high enough temperature to ignite the fuel. The fuel is injected into the air in the cylinder at the end of the compression. Similarly, many explosives are porous and contain tiny pockets of air or gas. When struck, gas near the surface is greatly compressed, the temperature rises very rapidly and ignites the explosive.

Example

A cylinder has a volume of 56 litres and contains air at a pressure of $1\cdot4\,\text{MN/m}^2$ and temperature 38 °C. Calculate the mass of air in the cylinder.

Solution

$$T = 38 + 273 = 311\,\text{K}$$

$$V = 56\,\text{litres} = 0\cdot056\,\text{m}^3$$

$$p = 1\cdot4\,\text{MN/m}^2 = 1\cdot4 \times 10^6\,\text{N/m}^2$$

$$R = 287\,\text{J/kg\,K (from Table 16.1)}$$

Since $pV = mRT$

$$m = \frac{pV}{RT}$$

$$= \frac{1\cdot4 \times 10^6 \times 0\cdot056}{287 \times 311}$$

$$= 0\cdot88\,\text{kg}$$

Example

A gas–air fuel supplied to an engine has a gas constant of 700 J/kg K and its pressure is measured as 80 mm water gauge when the barometer is standing at 765 mm Hg. The temperature of the fuel is 30 °C. Find the volume per kg of fuel. Relative density of mercury = 13·6.

Solution

$$80 \text{ mm of water} = \frac{80}{13 \cdot 6} = 5 \cdot 9 \text{ mm Hg}$$

Hence

$$p = 765 + 5 \cdot 9 = 770 \cdot 9 \text{ mm Hg}$$

$$= 770 \cdot 9 \times \frac{101 \cdot 3}{760}$$

$$= 103 \text{ kN/m}^2$$

From $pV = mRT$,

$$103 \times 10^3 \times V = 1 \times 700 \times (273 + 30)$$

Therefore

$$V = 2 \cdot 06 \text{ m}^3/\text{kg}$$

Problems

(Where necessary, appropriate values of the gas constant should be taken from Table 16.1.)

16.1   (a) Express 15 bar in newtons per square metre, hectobars and metres of water.
        (b) Express 0·8 mm Hg in millibars and newtons per square metre.
        (c) Express 100 mm water gauge in millimetres of mercury and kilonewtons per square metre.
                    ((a) 15×10^5 N/m²; 0·15 hbar; 153 m; (b) 1·065 mbar; 106·5 N/m²;
                                                        (c) 7·35 mm; 0·98 kN/m²)

16.2   (a) Express a vacuum gauge reading of 30 mm Hg in N/m² if the barometer reading is 764 mm Hg. What is the corresponding absolute pressure in kN/m²?
        (b) Convert 200 kN/m² to bars, 80 mbar to kN/m², 0.7 kN/m² to water gauge.
                    ((a) 4530 N/m²; 97·8 kN/m²; (b) 2 bar; 8 kN/m²; 72 mm H₂O))

16.3   (a) What is the vacuum gauge reading in kN/m² corresponding to an absolute pressure of 80 kN/m² when the barometer reading is 745 mm Hg?
        (b) A U-tube containing mercury records a pressure by showing a difference in level of 78 mm. What is the pressure in kN/m²?
        (c) A manometer to measure gauge pressure is connected to a pressure vessel. The level of the fluid in the manometer is 380 mm and the atmospheric pressure is 1 bar. What is the pressure reading in kN/m² if the relative density of the fluid is 1·4?
                    ((a) 19·3 kN/m²; (b) 104 kN/m²; (c) 5·2 kN/m²))

16.4 60 litres of gas, mass 0·14 kg are contained in a cylinder at a pressure of 200 kN/m^2 and temperature 21 °C. Find the value of the gas constant.

(291 J/kg K)

16.5 The cylinder of a car engine has a volume of 1000 cm^3 and contains air at a pressure of 110 kN/m^2 and temperature 21 °C. Calculate the mass of air contained in the cylinder.

(0·0013 kg)

16.6 Air is to be compressed in a diesel engine cylinder. The initial pressure and temperature are 105 kN/m^2 and 15 °C respectively and the final pressure is to be 7 MN/m^2. In order to ignite the fuel charge injected at the end of compression, the temperature required is 1100 °C. Calculate the ratio of initial to final volume for this to be possible.

(14:1)

16.7 The total volume of the gas circulating in the coolant circuit of a gas-cooled nuclear reactor is 3000 m^3. If the gas is carbon dioxide at a pressure of 1·1 MN/m^2 and an average temperature of 300 °C, calculate the total mass of gas.

(30·5 tonnes)

16.8 A cylinder contains air at 85·4 °C and a pressure 150 kN/m^2. This is compressed to one-sixth of its original volume, the final temperature being 134 °C. Calculate the final pressure in the cylinder.   (1·025 MN/m^2)

16.9 A closed cylinder, 150 mm diameter, contains gas at a pressure of 140 kN/m^2 *gauge* and temperature 15 °C. The gas is heated to a temperature of 1100 °C. What is then the total force on the end of the cylinder, due to *internal* pressure?

(20·3 kN)

16.10 A compressor delivers 90 m^3 of air per hour at 101 kN/m^2 and 15 °C. How many cubic metres of air would the compressor deliver per hour at 7 MN/m^2 and 50 °C?   (1·46 m^3/h)

16.11 A closed vessel of capacity 0·03 m^3 contains 70 g of hydrogen. If the temperature in the vessel is 20 °C, find the pressure. At what temperature would the pressure in the vessel be 1·4 MN/m^2?   (2·84 MN/m^2; −128·5 °C)

16.12 A spherical vessel of 100 mm inside diameter contains helium at −183 °C. If the mass of the gas is 5 g and R for helium is 2080 J/kg K, what is the pressure in the vessel?   (1·79 MN/m^2)

16.13 Air entering a nozzle has a temperature of 17 °C and a pressure of 140 kN/m^2 *gauge*. What is its volume per kg?   (0·345 m^3)

16.14 A cylinder, 200 mm in diameter, is fitted with a spring-controlled piston as shown in Fig. 16.11. What is the force in the spring when space A is occupied by 0·014 m^3 of air at 200 °C? The mass of air is 50 g and the pressure on the spring side of the piston is 1 bar.   (12·1 kN)

Fig. 16.11

16.15 An aircraft flying at a height of 16 km is surrounded by air at a pressure of 100 mbar and a temperature of −57 °C. The air actually drawn into the engine cylinder has a volume of 1800 cm³, pressure 126 kN/m³ and temperature 100 °C. What is the volume of air drawn in at the surrounding conditions?

(0·013 m³)

16.16 The gaseous fuel consumption per minute of an engine is 1·5 m³ at 16 °C and 70 mm of water. The barometer reading is 740 mm Hg. If R for the fuel is 800 J/kg K, find the mass flow rate of fuel in kg per minute.

(0·65 kg/min)

(16.11) Density of a gas

Consider a volume of gas V having mass m. The *density* ρ of the gas is its mass per unit volume, i.e.

$$\text{density} = \frac{\text{mass}}{\text{volume}}$$

i.e. $\rho = \dfrac{m}{V}$

The SI unit of density is **kg/m³**. Other forms of unit that may be met with are Mg/m³ = tonne/m³ or kg/litre.

The density of a given gas is determined by its pressure and temperature. To compare the densities of gases, it is necessary to specify certain standard conditions of pressure and temperature.

(16.12) Standard conditions

Standard temperature and pressure (STP) means a temperature of 0 °C and pressure 101·3 kN/m² or *one standard atmosphere*.

Normal temperature and pressure (NTP) means a temperature of 16 °C and pressure 101·3 kN/m².

Unfortunately there is no recognized term for either of the above 'standards' and the definitions given above will be used here. Standards are specified in particular fields of work for all quantities and tests where values depend on ambient conditions, for example, BS. *Standard Reference Conditions* for engine testing (*see* section 21.16). The use of a standard temperature and pressure is necessary to compare, for example, tests on the ability of different engines to induce air into the cylinder. In air-compressor work, the term *free air* is often used. This corresponds to air at the conditions on site, unaffected by the compressor.

Example

A large ship's diesel engine induces a volume of 200 litres of air into its cylinder at a pressure of $130 \, kN/m^2$ and temperature $32\,°C$. To what volume does this correspond at NTP?

Solution
At NTP,

$p = 101·3 \, kN/m^2$ and $T = 273 + 16 = 289 \, K$

Denoting NTP conditions by suffix n and applying the ideal gas rule,

$$\frac{pV}{T} = \frac{p_n V_n}{T_n}$$

Therefore

$$V_n = \frac{p}{p_n} \times \frac{T_n}{T} \times V$$

$$= \frac{130}{101·3} \times \frac{289}{(273 + 32)} \times (200 \times 10^{-3})$$

$$= 0·24 \, m^3$$

Example

The density of nitrogen is $1·25 \, kg/m^3$ at STP. Calculate the gas constant for nitrogen.

Solution
At STP,

$p = 101·3 \, kN/m^2$, $T = 273 \, K$

Consider unit volume of gas, then

$V = 1 \, m^3$, $m = 1·25 \, kg$

Therefore, from the gas rule,

$$pV = mRT$$

$$101·3 \times 10^3 \times 1 = 1·25 \times R \times 273$$

Hence

$$R = 297 \, J/kg \, K$$

Problems

(Where necessary, appropriate values of the gas constant should be taken from Table 16.1.)

16.17 A volume of 60 litres of air is drawn into a cylinder at a temperature of 50 °C and pressure 130 kN/m². What are the corresponding volumes at NTP and STP?   (69; 65·2 litres)

16.18 A jet-engined aircraft flies at a height where the air pressure is 490 mbar and the temperature −7 °C. The engine consumes 35 kg of air per second. What is the air consumption in cubic metres per second at these conditions and at the conditions of STP?   (54·4; 27·1 m³/s)

16.19 During a test on a gas engine the measured gas consumption was 3·8 m³/h at 20 °C and 1·04 bar. What is the consumption referred to NTP?

(3·85 m³/h)

16.20 A closed vessel has a volume of 150 litres and contains methane at 1·5 MN/m² and 120 °C. Taking the density of methane as 0·71 kg/m³ at STP find the mass of methane in the vessel.   (1·1 kg)

16.21 Hydrogen peroxide used in augmenting the power developed by a helicopter has a density of 1·44 kg/m³ at NTP. Estimate the gas constant for hydrogen peroxide and its density at 3·5 MN/m² and 100 °C.

(243 J/kg K; 38·6 kg/m³)

16.22 At a height of 12 200 m above the earth's surface the temperature and pressure of the air are −56·3 °C and 187 mbar, respectively. Calculate the ratio of the density at this height to the density at ground level. At ground level the pressure is 101 kN/m² and the temperature is 15 °C.   (0·25)

16.23 The carbon dioxide gas coolant circuit of a nuclear reactor is at a pressure of 700 kN/m² *gauge* and a temperature 140 °C at the gas circulator outlet. What is the density of the gas at this point?   (10·3 kg/m³)

(16.13) ## The index law of expansion

The ideal gas rule $pV = mRT$ relates the four variables p, V, m and T for a given gas. Given any three of the four variables, the fourth one may be found using this rule. For a given mass of gas, to calculate a change in pressure, both the change in volume and the change in temperature must be known. In practice, when a gas undergoes an expansion or compression the ideal gas rule alone cannot provide all the information required. It is necessary to know more about the way in which the expansion or compression takes place. Experiment shows that the relation between pressure and volume of a gas during any expansion or compression may be stated as

pV^n = constant

where the index n is a constant which depends on the nature of the gas and on the way in which the expansion is carried out. This is known as the *index law of expansion*.

It is possible for an expansion to take place without any change in the temperature of the gas or without change in the pressure. Again, both pressure and temperature may change without heat flow into or out of the gas. Finally, in the most general case, there might be changes in pressure and temperature together with heat flow.

The possible types of expansion will now be considered in detail.

(16.14) Isothermal expansion and compression

An *isothermal* expansion or compression is one which takes place at constant temperature, i.e. T = constant. This is possible if the change takes place very slowly. The heat exchanges that take place have then sufficient time to allow the temperature to remain constant. However, if a gas expands rapidly with a consequent change in temperature, and if at the end of expansion the temperature, after a lapse of time, returns to its original value, then the term *isothermal change* may be applied to the initial and final conditions of the gas. For example, when air is compressed in a compressor it leaves the compressor at a very high temperature. It may then be cooled back to its original temperature in a storage vessel. The term isothermal change then applies to the air entering the compressor and finally in the storage vessel.

Since T is constant in an isothermal change we have, from the ideal gas rule,

$$\frac{pV}{T} = \text{constant}$$

i.e. $pV = \text{constant}$

This is the law of isothermal expansion or compression and is of course Boyle's law.

It can be seen that an isothermal change obeys the index law where the index n has the value unity, i.e. $n = 1$. Figure 16.12 shows the curve representing an isothermal expansion or compression. The curve is a hyperbola.

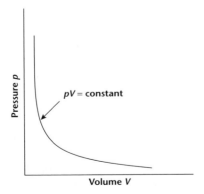

Fig. 16.12

16.15 Adiabatic expansion and compression

Consider a gas under pressure in a perfectly lagged cylinder. The lagging prevents the flow of heat into or out of the gas in the cylinder. Any expansion or compression of the gas therefore takes place without any heat flow from or to it, during the process. Such a change is said to be *adiabatic*. Insulation is never perfect, but if an expansion takes place very rapidly, approximately adiabatic conditions may be achieved. For example, the compression of a gas in the cylinder of a high-speed engine is approximately adiabatic even though the cylinder may be cooled by a water jacket.

Experimentally it is found that for an adiabatic expansion or compression a gas obeys the relation

$$pV^\gamma = \text{constant}$$

where γ is a constant for a particular gas, known as the *adiabatic index*. Adiabatic expansion is therefore a special case of the index law where n has the special value γ. γ can be shown to be equal to the ratio of the specific heat capacities at constant pressure (c_p) and constant volume (c_v), i.e. $\gamma = c_p/c_v$.

Typical values of γ are

air 1·4
carbon dioxide 1·25
steam 1·3

For a perfect gas γ is a constant. Gases are not perfect, however, and γ is found to vary slightly with temperature and also very slightly with pressure.

16.16 Polytropic change

A process that obeys the general relation $pV^n = \text{constant}$ is called a *polytropic* process. When n has the value unity we use the special name *isothermal*. When n has the value γ we use the name *adiabatic*.

In practice, for air compressed in an engine cylinder the value of n lies between 1·3 and 1·35. For expansion of burnt gases in a petrol engine n may lie between 1·25 and 1·3.

Figure 16.13 shows the curves $pV^n = \text{constant}$ for various values of the index n. The higher the value of n, the steeper the curve, i.e. the greater the rise in pressure for a given change in volume.

The two equations $pV^n = \text{constant}$ and $pV = mRT$ are together sufficient to allow us to calculate any change in p, V or T for a given mass of a particular gas that obeys Boyle's and Charles's laws.

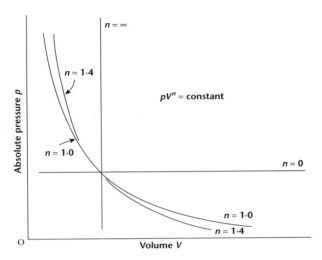

Fig. 16.13

(**Example**)

Air at $100 \, \text{kN/m}^2$ enters the cylinder of an oil engine. The inlet valve closes and the air is compressed by the piston according to the law $pV^{1.33} = $ constant. The ratio of initial to final volume is $12:1$. Calculate the pressure in the cylinder when the air is fully compressed.

Solution

Let suffixes 1 and 2 denote the initial and final conditions respectively; then the ratio of the volumes is

$$\frac{V_1}{V_2} = 12$$

Also

$$p_1 V_1^{1.33} = p_2 V_2^{1.33}$$

Hence

$$p_2 = p_1 \times \left(\frac{V_1}{V_2}\right)^{1.33}$$

$$= 100 \times 12^{1.33} \, \text{kN/m}^2$$

$$= 100 \times 27 \cdot 23 = 2723 \, \text{kN/m}^2$$

$$= 2 \cdot 72 \, \text{MN/m}^2$$

(**Example**)

Carbon dioxide gas, initially at a temperature of 200 °C and pressure 1·2 MN/m², expands adiabatically to a pressure of 240 kN/m². Calculate the final temperature if the adiabatic index for carbon dioxide is 1·25.

Solution

Consider 1 kg of gas. For adiabatic expansion, pV^γ = constant, and since γ = 1·25,

$$p_1 V_1^{1\cdot25} = p_2 V_2^{1\cdot25}$$

Therefore

$$\left(\frac{V_2}{V_1}\right)^{1\cdot25} = \frac{p_1}{p_2} = \frac{1\cdot2 \times 10^3}{240} = 5$$

so that

$$\frac{V_2}{V_1} = 5^{1/1\cdot25} = 5^{0\cdot8} = 3\cdot624$$

From the ideal gas rule,

$$\frac{p_2 V_2}{T_2} = \frac{p_1 V_1}{T_1}$$

Hence the final temperature

$$T_2 = \frac{p_2}{p_1} \times \frac{V_2}{V_1} \times T_1$$

$$= \frac{240 \times 10^3}{1\cdot2 \times 10^6} \times 3\cdot624 \times (200 + 273)$$

$$= 342\cdot8 \,\text{K}$$

$$= 69\cdot8 \,°\text{C}$$

(16.17) The constant-volume gas thermometer

From the ideal gas rule,

$$pV = mRT$$

Therefore

$$T = \left(\frac{V}{mR}\right) \times p$$

and if the volume V is constant,

$$T = \text{constant} \times p$$

Hence, by measuring changes in pressure at constant volume, the corresponding

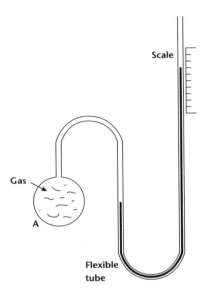

Fig. 16.14   Constant-volume gas thermometer

changes in absolute temperature may be measured. This forms the basis of the constant-volume gas thermometer, Fig. 16.14.

The bulb A contains a gas (e.g. hydrogen, nitrogen or oxygen) and is connected to a flexible tube containing a liquid, e.g. mercury. The pressure of gas in the reservoir is measured by the difference in level of the two columns of liquid.

The bulb is placed in a fluid the temperature of which is to be measured. The vertical right-hand limb of the U-tube is raised or lowered until the liquid in the left-hand limb regains a *fixed* level; by doing so the volume of gas is kept constant. The height of liquid in the right-hand limb is measured against a scale fixed to the tube. The scale is marked (calibrated) in exactly the same manner as for a mercury thermometer, i.e. by marking the levels of the ice point and boiling point of water. When the scale is calibrated, the gas thermometer can be used to measure temperature from pressure changes in the reservoir.

The coefficient of expansion of a gas is much greater than that of mercury or other liquids. Therefore a gas is more sensitive to temperature change. The gas thermometer is much more accurate than a mercury one and serves as a standard against which the more convenient and practical mercury thermometer may be checked periodically in the same way as workshop gauges are checked against more accurate standards.

(16.18) Constant-pressure gas thermometer

The constant-pressure gas thermometer is almost identical with the constant-volume thermometer described above. It is operated, however, so as to keep the

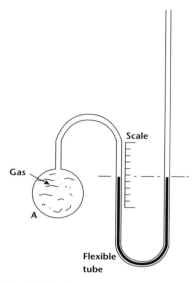

Fig. 16.15   Constant-pressure gas thermometer

pressure constant instead of the volume, Fig. 16.15. At every reading the right-hand limb of the U-tube is raised or lowered to keep the liquid in both limbs at the same level. The pressure of gas in the reservoir is then always constant. Changes in volume of the gas in the reservoir are measured on the scale on the left-hand limb.

Since the relation between volume and temperature at constant pressure is not a linear one, i.e. volume is not directly proportional to temperature, the scale divisions are not evenly spaced. This thermometer is not so convenient, therefore, as the constant volume apparatus.

Problems

(For air $R = 287$ J/kg K and $\gamma = 1 \cdot 4$.)

16.24 Air expands adiabatically from a pressure of 1 hbar and the ratio of final to initial volume is 10 to 1. Calculate the final pressure. If the expansion had been isothermal, what would then be the final pressure?

$(400\,\text{kN/m}^2;\ 1\,\text{MN/m}^2)$

16.25 The compressor of a turbo-jet engine compresses air from 105 kN/m² and 15 °C to 420 kN/m². If the compression is adiabatic, calculate the ratio of the final and initial volumes of a given mass of air. Hence, using the ideal gas rule, find the final temperature. $(1{:}2{\cdot}69;\ 155\,°\text{C})$

16.26 An air compressor takes in air at 105 kN/m² and 15 °C and compresses it to a pressure of 420 kN/m² *gauge*. Calculate (a) the initial volume of air per kg and (b) the final air temperature. Assume the compression to be adiabatic.

$(0{\cdot}79\,\text{m}^3/\text{kg};\ 182\,°\text{C})$

16.27 At the start of expansion in an engine cylinder the burnt gases are at a temperature of 650 °C and pressure 3·2 MN/m^2. The gases expand to ten times their original volume before being exhausted from the cylinder. Calculate the pressure and temperature of the exhaust gases, assuming the expansion to take place according to the law $pV^{1·25}$ = constant.

(180 kN/m^2; 246 °C)

16.28 A gas mixture is expanded in a cylinder from a volume of 2·5 litres to a volume of 10 litres. If the initial pressure is 3 MN/m^2, what is the pressure at the end of expansion, assuming a law $pV^{1·34}$ = constant? Calculate also the initial temperature if the final temperature is 120 °C.

(468 kN/m^2; 356 °C)

16.29 Air occupies a volume of 80 litres at a pressure of 3 MN/m^2 and expands to a volume of 500 litres and pressure 315 kN/m^2. If the law of expansion is pV^n = constant, what is the value of the index n? (1·23)

16.30 0·1 m^3 of gas at 5·25 MN/m^2 and 2200 °C expands to 420 kN/m^2 and 1094 °C. Using the ideal gas rule find the final volume. If the expansion is assumed to obey the law pV^n = constant what is the value of the index n?

(0·69 m^3; 1·31)

16.31 An engine compresses the air–fuel intake charge through a volume ratio of 7·25 to 1. The inlet pressure is 105 kN/m^2 and temperature 27 °C. The law of compression is $pV^{1·3}$ = constant. Calculate the pressure and temperature at the end of compression. (1380 kN/m^2; 270 °C)

16.32 A motor cycle engine has an 11 to 1 compression ratio, i.e. the ratio of initial to final volume during compression is 11 to 1. If the inlet pressure and temperature are 96 kN/m^2 and 37 °C respectively, calculate the final pressure and temperature. Assume compression according to the law $pV^{1·32}$ = constant. (2·27 MN/m^2; 394 °C)

16.33 The piston of an air motor is driven by compressed air. The air enters the cylinder of the motor at pressure 2·8 MN/m^2 and temperature 39 °C. The air expands to twice its original volume. Calculate the final pressure and temperature of the air, assuming the expansion to be adiabatic.

(1·06 MN/m^2; −37 °C)

16.34 Oxygen is drawn from a gas holder at 105 kN/m^2 and compressed according to the law $pV^{1·3}$ = constant, to a pressure of 525 kN/m^2. What is the ratio of the initial and final specific volumes of the oxygen. If the initial temperature is 15 °C, calculate the final specific volume. For oxygen, R = 260 J/kg K.

(3·45; 1; 0·207 m^3/kg)

16.35 In a refrigerating plant air is expanded in a motor from 525 kN/m^2 and 20 °C to a pressure of 105 kN/m^2 according to the law $pV^{1·35}$ = constant. What is the final temperature of the air? (−80 °C)

Chapter 17

Fuels; combustion; energy release

A critical factor regarding all heat engines is the fuel, the source of energy. Discussion of the common fuels entails explanations of the fundamental properties of matter and the chemistry of combustion; this involves study of the formation of chemical equations, analysis of fuels and products of combustion as well as the heat generated during a chemical reaction. The coverage is extended with a brief section on chemical and nuclear energy.

17.1 Fuel

A *fuel* is a substance that releases heat energy when ignited at a sufficiently high temperature. A fuel may be a solid, liquid or gas. Nearly all fuels contain hydrogen and carbon; those that do are known as *hydrocarbons*. Typical hydrocarbon fuels are given in Table 17.1.

Table 17.1

Solid	Liquid	Gas
Coal	Petrol	Hydrogen
Wood	Paraffin	Methane (marsh gas or firedamp)
Peat	Alcohol	Acetylene
Coke	Diesel oil	Coal gas
		Blast furnace gas
		North Sea (natural) gas

Fossil-fuels are those of organic origin produced over long periods in the earth, i.e. coal, oil, petrol and gas. Vegetable matter used as fuel is termed *biomass*.

The main types of coal are:

- bituminous (hard, over 70% carbon);
- lignite (soft, brown, 70% carbon);
- anthracite (very hard, 94% carbon).

The release of heat energy by burning can take place only if (a) a supply of oxygen is available and (b) combustion is started by a sufficiently high temperature. Atmospheric air is the usual source of oxygen since it contains 23 per cent of air by mass of oxygen.

Ignition may be by a spark, as in a petrol engine, or by the rise in temperature of the air due to compression, as in a diesel engine cylinder. The ignition temperature required to initiate combustion varies widely with the different types of fuel. For example, the ignition temperature of hydrogen is approximately 610 °C and a rough value for carbon is 430 °C.

The engineer's problems are to estimate the quantity of air required to burn a given fuel and the amount of heat energy released as well as to control and deal with the products of combustion. To do this requires a further knowledge of the properties of matter and the chemistry of combustion.

17.2 The elements

A chemical element is the fundamental building block of chemistry from which all substances are formed by chemical combination. There are now 103 elements, about 90 of which are known in nature and the others are produced artificially. Some of the more common elements are: *oxygen, nitrogen, hydrogen, helium, carbon, sulphur, iron, copper, tin, gold, silver, lead* and *mercury*. Of these, hydrogen is the element most generally found in the universe, followed by helium.

17.3 The atom; relative atomic mass

The smallest amount of any element is the *atom*. Chemistry is the study of the formation of substances by the combination of atoms of the elements.

Every atom has a mass. The lightest atom is that of ordinary hydrogen which has a mass of about 1.66×10^{-24} g. It is more convenient, however, to measure the mass of an element by its *atomic mass* (often referred to as *atomic weight*) which is the mass of its atom relative to that of a certain type of carbon called carbon-12, taken as 12.* On this scale, hydrogen has an atomic mass of 1.008, Fig. 17.1. Many elements have relative atomic masses varying slightly from whole numbers but, for convenience, the values are rounded off to give whole numbers. The relative atomic masses of elements met with in combustion calculations are shown in Table 17.2 and this list should be memorized. An atom of each element is described by a chemical symbol, e.g. N represents *one* atom of nitrogen.

---

* The unit for an amount of chemical substance (elements or compounds) is defined by fixing the corresponding mass of carbon 12. By international agreement this mass has been fixed at 0.012 kg and the unit of this quantity of substance is called the *mole*, the symbol for which is *mol*. The mole is a base SI unit.

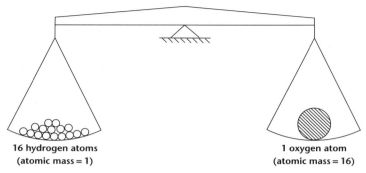

16 hydrogen atoms
(atomic mass = 1)

1 oxygen atom
(atomic mass = 16)

Fig. 17.1

Table 17.2

Element	Chemical symbol (representing 1 atom)	Relative atomic mass
Hydrogen	H	1
Oxygen	O	16
Nitrogen	N	14
Carbon	C	12
Sulphur	S	32

(17.4) The molecule; relative molecular mass

The *molecule* is the smallest particle of any substance that can normally exist alone, i.e. not in chemical combination with any other similar particle. A molecule may contain one or more atoms. For example, the molecule of carbon contains a single atom and is represented by C. The molecule of oxygen is formed by the combination of *two* oxygen atoms and is thus represented by O_2.

The *relative molecular mass* (often referred to as *molecular weight*) is the total relative atomic mass of the atoms making up the molecule. Thus the relative molecular mass of oxygen is twice the relative atomic mass of the oxygen atom. Table 17.3 gives a list of useful masses and symbols.

Table 17.3

Element	Molecular symbol	Relative molecular mass
Hydrogen	H_2	2
Oxygen	O_2	32
Nitrogen	N_2	28
Carbon	C	12
Sulphur	S	32

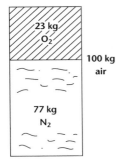

Fig. 17.2 Air is a mixture of mainly oxyen and nitrogen.

17.5 Mixtures and compounds

Substances may exist together either as a *mixture* or as a chemical *compound*. In a mixture the substances exist side by side and do not combine to form a new chemical substance. A mixture can often be separated out into its component substances. For example, oil may be mixed with water. If allowed to stand the mixture will separate out, the oil rising to the surface. Similarly, atmospheric air is a mixture containing mainly oxygen and nitrogen, the proportions being 23 per cent oxygen, 77 per cent nitrogen, by mass, Fig. 17.2.

A compound is a chemical combination of two or more elements to form a molecule of a new substance, different in kind and properties, from the elements of which it is made up. A water molecule, for example, is a compound of hydrogen and oxygen but has properties quite different from either hydrogen or oxygen. In the formation of compounds, oxygen is the most active element.

The atmosphere

Engineers' proportions, *by volume*, for oxygen and nitrogen in air or atmosphere are taken as 21 per cent and 78 per cent respectively. More exactly, for *dry* air they are 20.94 and 78.08 per cent. The proportions in the atmosphere differ slightly throughout day and night and are variable in the lower atmosphere, particularly because of the presence of water vapour below an average height globally of about 12 km; water vapour content of the atmosphere is usually below 4 per cent. Of the remaining 1 per cent of the atmosphere, the most significant volume of gas is argon (0.93 per cent) followed by carbon dioxide (0.03 per cent, but variable), then in order of measurable quantities, neon, helium, ozone, hydrogen and finally traces of krypton, xenon, methane and nitrogen oxides.

Thus nitrogen, oxygen, carbon dioxide and argon account for all but 0.04 per cent of the dry atmosphere. These gases, with water vapour, may be considered the permanent constituents of air.

In the lower atmosphere there are also gases with relatively short lives and particles (*aerosols*) such as dust, ash, sea-salt, ice and sulphur dioxide. The lighter gases such as hydrogen and helium have a greater presence in the upper atmosphere.

(17.6) Relative molecular mass of a compound: molecular symbol

The *relative molecular mass* of a compound is the sum total of the relative atomic masses of its constituent elements. For example, the water molecule is made up of two atoms of hydrogen (H_2) and one atom of oxygen (O), Fig. 17.3. The relative molecular mass of water is therefore the sum of the relative masses of two atoms of hydrogen and one of oxygen, i.e. ($2 \times 1 + 16$), or 18.

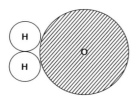

Fig. 17.3   The water molecule; water is a compound of hydrogen and oxygen.

The molecular symbol for a compound shows the constituent elements of which it is made up and shows also the number of atoms of each element present. Thus the symbol for one molecule of water is H_2O. The student should check by calculation the relative molecular masses of the compounds given in Table 17.4.

Table 17.4

Compound	Molecular symbol	Relative molecular mass
Water	H_2O	18
Carbon dioxide	CO_2	44
Sulphur dioxide	SO_2	64
Carbon monoxide	CO	28
Methane (marsh gas)	CH_4	16

(17.7) Reaction

A *reaction* is said to occur when a chemical compound is formed by the combination of elements or when a compound is split up (decomposed) into its separate elements. For example, hydrogen burnt in oxygen forms water vapour. At very high temperatures, on the other hand, water vapour may be split up into the separate elements, hydrogen and oxygen.

17.8   Combustion

Most elements, under suitable conditions, will take part in a chemical reaction. The type of reaction with which we are concerned is the reaction with oxygen which also liberates heat. This is called *combustion*. It should be noted that, if the oxygen is supplied from atmospheric air, the nitrogen in the air takes no part in the reaction. As far as combustion is concerned, nitrogen is an *inert* gas and does not burn with oxygen. It can, however, form chemical compounds with oxygen or other elements, but this need not concern us at this stage. Also, nitrogen serves to slow down the rate of combustion and the peak temperature reached.

In practice the most important reactions are those involving hydrogen, carbon and sulphur. The compounds formed by *complete* combustion of these elements in oxygen are as follows:

hydrogen forms water vapour (H_2O)
carbon forms carbon dioxide (CO_2)
sulphur forms sulphur dioxide (SO_2)

17.9   The chemical equation

The relation between the numbers of atoms (or molecules) and the atomic (or molecular) relative masses taking part in a chemical reaction is expressed as a *chemical equation*. Any reaction must obey the following rules:

1. The total number of atoms of each kind is the same before and after the reaction.
2. Before the reaction, each separate element usually (or normally) exists as a complete molecule.

The application of these rules to the combustion of hydrogen and carbon will now be shown.

17.10   Formation of chemical equations

Consider the combustion of hydrogen, illustrated diagrammatically in Fig. 17.4. Before the reaction takes place, the oxygen exists as a complete molecule (O_2) composed of two atoms. Similarly, the hydrogen exists as hydrogen molecules (H_2). At the end of the reaction, each oxygen atom has combined with two hydrogen atoms to form water (H_2O). Since there are two oxygen atoms available per molecule of oxygen the total number of hydrogen atoms required is four, made up of two hydrogen molecules, each containing two atoms. At the end of the reaction there will be two molecules of water, each molecule having three atoms, giving a total of six atoms as expected.

The chemical equation is written down as follows: on the left-hand side we have two hydrogen molecules ($2H_2$) together with one oxygen molecule (O_2) and on the right-hand side we have two molecules of water ($2H_2O$), i.e.

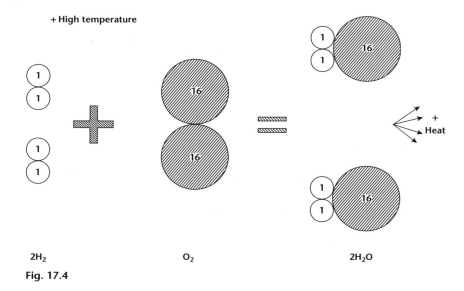

Fig. 17.4

$$2H_2 \quad + O_2 \qquad = 2H_2O$$

molecules	2	1	2
atoms	4	+ 2	= 6
relative atomic masses	4×1	$+ 2 \times 16$	$= 2(2 + 16)$
or	4	+ 32	= 36

Note:

1. The number of atoms is the same on both sides of the equation.
2. The total relative atomic mass is the same on both sides of the equation.
3. There is not necessarily the same number of molecules on the two sides of the equation.
4. The positive sign in a chemical equation means 'chemically combining with'.

As a second example of the formation of a chemical equation, consider the complete combustion of carbon with oxygen to form carbon dioxide. The reaction is illustrated in Fig. 17.5.

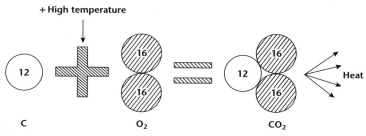

Fig. 17.5

A molecule of carbon (C) contains only one atom of carbon and combines with one molecule of oxygen (O_2) to form one molecule of carbon dioxide (CO_2). There are therefore two oxygen atoms and one carbon atom before and after the reaction. The equation is

$$C + O_2 = CO_2$$

molecules	1	1	1
atoms	1 +	2 =	3
relative atomic masses	12 +	32 =	44

Burning of hydrogen; air required

The combustion of hydrogen is represented by the equation

$$2H_2 + O_2 = 2H_2O$$

relative atomic masses $4 + 32 = 36$

We may choose unit mass as 1 kg. Hence

4 kg of hydrogen combines with 32 kg of oxygen to form 36 kg of water vapour.

The oxygen required is usually provided by atmospheric air and 100 kg of air contains approximately 23 kg of oxygen and 77 kg of nitrogen. Hence, the mass of air required is given by

(mass of oxygen required) $\times \frac{100}{23}$

For the burning of hydrogen in air we have, therefore, the following results:

mass of oxygen required per kg of hydrogen $= \frac{32}{4}$, or 8 kg

mass of air required per kg of hydrogen $= \frac{32}{4} \times \frac{100}{23}$, or 34·8 kg

mass of associated nitrogen $= \frac{32}{4} \times \frac{77}{23}$, or 26·8 kg

Burning of carbon; air required

The chemical equation is

$$C + O_2 = CO_2$$

relative atomic masses $12 + 32 = 44$

Hence

12 kg of carbon combines with 32 kg of oxygen to form 44 kg of carbon dioxide

Hence

mass of oxygen required per kg of carbon $= \frac{32}{12}$, or 2·67 kg

mass of air required per kg of carbon $= \frac{32}{12} \times \frac{100}{23}$, or 11·6 kg

mass of associated nitrogen $= \frac{32}{12} \times \frac{77}{23}$, or 8·93 kg

Burning of sulphur; air required

The chemical equation for the burning of sulphur to form sulphur dioxide (SO_2) is

$$S + O_2 = SO_2$$

atoms $\qquad$ $1 + 2 = 3$

relative atomic masses $\quad$ $32 + 32 = 64$

Hence

32 kg of sulphur combines with 32 kg of oxygen to form 64 kg of sulphur dioxide.

Hence

mass of oxygen required per kg of sulphur $= \frac{32}{32}$, or 1 kg

mass of air required per kg of sulphur $\quad = \frac{32}{32} \times \frac{100}{23}$, or 4·35 kg

mass of associated nitrogen $\qquad\qquad = \frac{32}{32} \times \frac{77}{23}$, or 3·35 kg

(17.11) Chemical analysis of a fuel

A chemical analysis of a fuel provides the percentage of each element present. This is also termed the *ultimate analysis*. The analysis of a solid or liquid fuel is usually expressed as a percentage by mass and is termed a *gravimetric analysis*. The analysis of a gaseous fuel is usually given as a percentage by volume and is termed a *volumetric analysis*.

The ultimate analysis is required for combustion calculations, but in practice a simpler and cheaper test of a solid fuel is often carried out. This latter test is termed a *proximate analysis* and determines merely:

- the mass of material which escapes when the fuel is heated without combustion (volatile matter);
- the residual carbon, or coke, remaining after heating;
- the moisture content;
- the ash in the unburnt fuel.

These items are then expressed as a percentage by mass.

We are restricting the calculations in this chapter to analysis by mass, i.e. gravimetric analysis.

(17.12) Minimum air for combustion

Combustion calculations for a fuel require that the ultimate analysis be known. If this is so, the mass of oxygen required for complete combustion of the given amount of each element per kg of fuel may be calculated. The total oxygen per kg of fuel is obtained by simple addition. Then, since air contains 23 per cent oxygen by mass, the total air required is found by multiplying the mass of oxygen by 100/23. The mass of air calculated in this manner is the theoretical air required, and, as will be shown later, is the *minimum* mass of air for complete combustion. This chemically correct amount of air for complete combustion is called the *stoichiometric air*.

Certain fuels contain small amounts of oxygen. If so, the oxygen required for combustion of the remaining elements of the fuel is first calculated. The mass of oxygen already present is then subtracted. The result is the mass of oxygen which must be supplied from atmospheric air. The above methods of calculation are illustrated in the following examples.

(**Example**)

A coal contains 90 per cent carbon, 6 per cent hydrogen, 2 per cent sulphur, and 2 per cent ash, by mass. Calculate the theoretical mass of air required for complete combustion per kg of fuel.

Solution

The calculation of the mass of oxygen required for 1 kg of fuel is set out in Table 17.5. The numbers shown below the reaction equations are the relative masses of each element entering into the reaction. Ash present in the fuel does not burn. The ash content therefore does not enter into the calculation.

Table 17.5

Element	Reaction	Fraction by mass	Oxygen required (kg)
Carbon	$C + O_2 = CO_2$ $12 + 32 = 44$	0·9	$0.9 \times \frac{32}{12} = 2.4$
Hydrogen	$2H_2 + O_2 = 2H_2O$ $4 + 32 = 36$	0·06	$0.06 \times \frac{32}{4} = 0.48$
Sulphur	$S + O_2 = SO_2$ $32 + 32 = 64$	0·02	$0.02 \times \frac{32}{32} = 0.02$
			Total oxygen = 2·9 kg

Total mass of air required for complete combustion per kg of fuel

$= \text{weight of oxygen required} \times \frac{100}{23}$

$= 2.9 \times \frac{100}{23}$

$= 12.6 \, \text{kg}$

(**Example**)

A dried wood contains carbon 50 per cent, hydrogen 6 per cent, and oxygen 44 per cent, by mass. What is the theoretical mass of air required for complete combustion?

Solution

Oxygen required for 0·5 kg of carbon $= 0·5 \times \frac{32}{12} = 1·33\,\text{kg}$

Oxygen required for 0·06 kg of hydrogen $= 0·06 \times \frac{32}{4} = 0·48\,\text{kg}$

Total oxygen required per 1 kg of fuel $= 1·33 + 0·48 = 1·81\,\text{kg}$

But 0·44 kg of oxygen is already present in each kg of fuel; therefore, actual oxygen required = 1·81 − 0·44 = 1·37 kg/kg of fuel; therefore,

air required $= 1·37 \times \frac{100}{23}$

$= \mathbf{6\,kg/kg\ of\ fuel}$

Problems

17.1 Dried peat, used as fuel in a gas turbine combustion chamber, has a percentage composition by mass: carbon 58, hydrogen 6·3, oxygen 30·8, nitrogen 0·9 and ash 4. Calculate the minimum mass of air required for complete combustion per kg of peat.   (7·6 kg)

17.2 A fuel has an analysis by mass: carbon 0·91, hydrogen 0·07, sulphur 0·01, the remainder being incombustible. Find the minimum amount of air required for complete combustion per kg of fuel.   (13 kg)

17.3 A boiler furnace consumes 200 kg of coal per hour. The composition of the coal by mass is: carbon 0·9, hydrogen 0·05, remainder ash. Air is supplied to the furnace by a forced-draught fan. What is the minimum mass of air to be delivered by the fan per hour for the complete combustion of the fuel?
                                                          (2434 kg/h)

17.4 Coal burned in a furnace has a composition by mass of carbon 0·88, hydrogen 0·06, remainder ash. If 50 per cent more air is supplied per kg of coal than the theoretical minimum required for complete combustion, what is the total air supplied per kg of fuel? How much nitrogen enters the furnace?
                                                     (18·4 kg; 14·2 kg)

17.5 A sample of coal showed a gravimetric analysis: C 60 per cent, H_2 15 per cent, O_2 15 per cent, N_2 4 per cent, S 3 per cent, remainder ash. Calculate the minimum mass of air required for complete combustion per kg of coal.
                                                          (11·7 kg)

17.6 Fuel oil for a diesel engine yields the following analysis: C 0·84, H_2 0·14, S 0·02. What is the minimum mass of air required for complete combustion per kg of oil? How much nitrogen is associated with this air?
                                                     (14·7; 11·3 kg)

17.7 Derive from first principles the chemical equation for the combustion of carbon to form carbon monoxide. Hence, calculate the minimum mass of air required for the combustion of carbon to carbon monoxide. If the carbon is completely burnt to carbon dioxide, how much air is required?
                                           $(2C + O_2 = 2CO; 5·8; 11·6\,\text{kg})$

17.8 Carbon monoxide (CO) may be burnt in the presence of oxygen to form carbon dioxide. The reaction takes place according to the equation

$$2CO + O_2 = 2CO_2$$

Calculate from first principles the theoretical air required for complete combustion, and the mass of carbon dioxide per kg of carbon monoxide.

(2·49; 1·57 kg)

(17.13) Excess air

So far our calculations have provided us with the *minimum* amount of air required for the complete combustion of a fuel. In practice, in order to ensure a sufficient supply of oxygen, an excess mass of air over and above the minimum is supplied.

The amount of air actually supplied to an engine cylinder or to a furnace is determined by many factors. For example, air supplied to an engine cylinder takes in approximately three times as much nitrogen as useful oxygen. Nitrogen, being an inert gas, does not contribute to the combustion process and takes up useful space in the cylinder. In addition, the exhaust of hot nitrogen from the cylinder represents heat energy which is not utilized to do useful work. Again, too much excess air will tend to chill a cylinder or furnace and may tend to reduce the temperature below that required for combustion of all the fuel. On the other hand, if only the minimum quantity of air is supplied, poor distribution in the cylinder or furnace may mean that not all the oxygen comes into contact with hot fuel. Incomplete combustion results. Finally, the rate at which fuel is burnt depends on the rate of air supply. In practice, a balance is struck between conflicting requirements of complete combustion and excessive chilling and heat loss to exhaust gases.

The excess air supply is generally expressed as a percentage of the minimum mass required for complete combustion. Thus, per kg of fuel,

percentage excess air

$$= \frac{\text{total air supplied} - \text{minimum air required}}{\text{minimum air required}} \times 100$$

Typical values of excess air supplied, expressed as a percentage of the quantity theoretically required, are as follows:

hand-fired boiler furnace	100
mechanically stoked furnace	40
petrol engine	20
oil engine	20

Problems

17.9 The percentage composition of the fuel used in an oil engine is: C 86, H_2 13. When the air supply is 100 per cent in excess of that theoretically required for complete combustion, calculate how much nitrogen leaves the engine per kg of fuel burned.   (22·3 kg)

17.10 A boiler uses fuel oil (gravimetric analysis: carbon 0·86, hydrogen 0·14) at the rate of 550 kg/h. The air supplied is 25 per cent in excess of the theoretical minimum required for complete combustion. What is the total mass of air supplied per hour?   (10 200 kg/h)

17.11 A petrol engine uses a petrol–air mixture, 17 kg of air per kg of petrol. The petrol analysis is 84 per cent carbon, 16 per cent hydrogen. What is the excess air supplied over the minimum required for complete combustion?   (11·1%)

17.12 A coal contains by percentage: C 84, H_2 5, S 3, O_2 6, ash 2. A fan supplies air for combustion of this coal at the rate of 20 kg per kg of coal. What is the percentage excess air supplied?   (76 per cent)

17.13 The fuel oil used in an engine has the following analysis: C 86 per cent, H_2 14 per cent. The air–fuel ratio by mass is 18:1. How much excess oxygen is supplied per kg of fuel?   (0·73 kg)

17.14 A boiler uses fuel oil (gravimetric composition: C 0·86, H_2 0·13) at the rate of 8·5 kg/min. The air supplied is 25 per cent in excess of the minimum required for complete combustion. An induced-draught fan delivers the flue gases to a chimney. Find the total mass of gas to be delivered by the fan per minute if losses are neglected.   (162 kg/min)

17.14 The products of combustion

When fuel is burnt with air, the *products of combustion* or *exhaust gases* are made up as follows:

carbon dioxide (CO_2) ⎫
water vapour (H_2O)   ⎬ the burnt gases
sulphur dioxide (SO_2) ⎭

nitrogen (N_2)   ⎰ from the air
                  ⎱ from the fuel

oxygen (O_2)   from the excess air

If insufficient air is supplied, or if the combustion process is inefficient, the carbon burns partly to carbon monoxide (CO) instead of carbon dioxide. In this case there will be carbon monoxide and possibly unburned carbon, as well as carbon dioxide present in the products. It is by a study of the analysis of the products of combustion that engineers determine whether the combustion is efficient.

The total mass of the products of combustion is the sum of the masses of fuel and air, when due allowance is made for losses, such as fuel falling through firebars, leakage, etc. Thus, if 20 kg of air is supplied per kg of fuel and losses are neglected, there should be 21 kg of products.

Nitrogen gas is normally inert but does not always remain so in high-temperature, high-pressure conditions. The nitrogen in the air and fuel may form oxides of nitrogen (NO_x) depending on flame temperatures, air–fuel mixture and flow conditions at some stage of combustion. Similarly, oxides of sulphur and hydrogen may form from the molecules of sulphur and hydrogen in the fuel. Another complication is that some molecules of an element such as nitrogen may *dissociate* into its atoms.

(17.15) Dry products of combustion

A distinction is often made between *wet* and *dry* products of combustion. The reason for this is that the apparatus commonly used in the analysis of exhaust gases contains water through which the gases pass. Any water vapour present in the gases condenses and is not recorded among the exhaust gases. Similarly, the sulphur dioxide dissolves in the water and does not appear in the recorded analysis. The apparatus (known as the Orsat apparatus) therefore gives the composition of the *dry* products, and it is a volumetric analysis (*see* section 17.11). However, we shall restrict ourselves to analysis by mass.

The analysis of the dry products of combustion therefore omits both water vapour and sulphur dioxide.

The method of calculation of the combustion products is illustrated in the following examples. Again, we are dealing only with analysis *by mass*.

(**Example**)
_____

A petrol has an analysis by mass of: carbon 90 per cent, hydrogen 10 per cent. It is burned with the theoretical minimum quantity of air for complete combustion. Estimate the percentage composition of the dry products of combustion.

Solution
For the carbon (0·9 kg/kg of fuel),

$$C + O_2 = CO_2$$

(masses) 12 + 32 = 44

By proportion

$$0 \cdot 9 + \tfrac{32}{12} \times 0 \cdot 9 = \tfrac{44}{12} \times 0 \cdot 9$$

i.e. $0 \cdot 9 + 2 \cdot 4 = 3 \cdot 3$

Therefore, 0·9 kg of carbon requires 2·4 kg of oxygen and produces 3·3 kg of carbon dioxide.

For the hydrogen ($0 \cdot 1$ kg/kg of fuel),

$$2H_2 + O_2 = 2H_2O$$

(masses) $4 + 32 = 36$

$$0 \cdot 1 + \tfrac{32}{4} \times 0 \cdot 1 = \tfrac{36}{4} \times 0 \cdot 1$$

$$0 \cdot 1 + 0 \cdot 8 = 0 \cdot 9$$

Therefore, $0 \cdot 1$ kg of hydrogen requires $0 \cdot 8$ kg of oxygen and produces $0 \cdot 9$ kg of water vapour.

Total oxygen required $= 2 \cdot 4 + 0 \cdot 8 = 3 \cdot 2$ kg/kg of fuel

Total air required $= 3 \cdot 2 \times \tfrac{100}{23}$ kg/kg of fuel

Therefore, nitrogen in the air supplied

$$= 3 \cdot 2 \times \tfrac{100}{23} \times \tfrac{77}{100}$$

$$= 3 \cdot 2 \times \tfrac{77}{23}$$

$$= 10 \cdot 71 \text{ kg/kg of fuel}$$

The analysis of the dry products of combustion (H_2O omitted) is set out as in Table 17.6.

Table 17.6

Constituent	kg/kg of fuel	Percentage composition
CO_2	$3 \cdot 3$	$\dfrac{3 \cdot 3}{14 \cdot 01} \times 100 = 23 \cdot 55$
N_2	$10 \cdot 71$	$\dfrac{10 \cdot 71}{14 \cdot 01} \times 100 = 76 \cdot 45$
Totals	**$14 \cdot 01$**	**$100 \cdot 0$**

Note that mass of fuel + mass of air = mass of wet products

i.e. $1 + (3 \cdot 2 + 10 \cdot 71) = 14 \cdot 01 + 0 \cdot 9$

$$= 14 \cdot 9 \text{ kg}$$

(Example)

An oil engine uses oil having gravimetric analysis: carbon $0 \cdot 86$, hydrogen $0 \cdot 13$, ash $0 \cdot 01$. The ratio of air supplied to fuel burnt was $30 : 1$. Estimate the masses of the various constituents of the wet products and the percentage composition of the dry products.

Solution
Per kg of fuel,

total air supplied $= 30$ kg

total oxygen supplied $= 30 \times \tfrac{23}{100} = 6 \cdot 9$ kg

total nitrogen supplied $= 30 \times \tfrac{77}{100} = 23 \cdot 1$ kg

The solution is set out in Table 17.7. Note that the oxygen not required for combustion appears as a product of combustion. It is given by the difference between the total oxygen supplied and the minimum oxygen required for complete combustion.

The analysis of the wet and dry products is set out in Table 17.8. Note that:

mass of wet products (30·99) + mass of ash (0·01) = mass of air (30) + mass of fuel (1)

Table 17.7

Constituent	Mass kg/kg of fuel	Equation	Minimum oxygen required	Products of combustion			
				CO_2	H_2O	N_2	O_2
C	0·86	$C + O_2 = CO_2$ (12 + 32 = 44)	$\frac{32}{12} \times 0.86$ = 2·29	$\frac{44}{12} \times 0.86$ = 3·15	–	–	–
H_2	0·13	$2H_2 + O_2 = 2H_2O$ (4 + 32 = 36)	$\frac{32}{4} \times 0.13$ = 1·04	–	$\frac{36}{4} \times 0.13$ = 1·17	–	–
Ash	0·01	–	–	–	–	–	–
Air	30·0	–	–	–	–	23·1	6·9 – 3·33 = 3·57
Totals	**31·0**		**3·33**	**3·15**	**1·17**	**23·1**	**3·57**

Table 17.8

Wet products		Dry products		
Constituent	kg/kg of fuel	Constituent	kg/kg of fuel	Percentage composition
CO_2	3·15	CO_2	3·15	10·55
H_2O	1·17			
N_2	23·1	N_2	23·1	77·5
O_2	3·57	O_2	3·57	11·95
Totals	**30·99**		**29·82**	**100·00**

Problems

17.15 A paraffin oil burnt in a gas turbine contains 84 per cent carbon and 16 per cent hydrogen by mass. Estimate the masses of the various products of combustion per kg of fuel if 20 per cent excess air is supplied.

(CO_2, 3·08; H_2O, 1·44; N_2, 14·15; O_2, 0·704 kg)

17.16 A sample of ship's fuel oil contains 84 per cent carbon, 14 per cent hydrogen, 1·5 per cent oxygen and 0·5 per cent nitrogen. Find the minimum mass of air required for compete combustion per kg of fuel. If 30 per cent excess air is supplied, find the mass of each constituent of the flue gases per kg of fuel. (14·6 kg; CO_2, 3·08; H_2O, 1·26; O_2, 1; N_2, 14·6 kg)

17.17 A sample of coal has a gravimetric analysis: C 0·86, H_2 0·05, O_2 0·04, ash 0·05. Find the minimum mass of air required per kg of coal. If 18 kg of air is supplied per kg of coal, determine the masses of the constituents of the wet waste gases produced per kg of coal.

(11·6 kg; CO_2, 3·15; H_2O, 0·45; N_2, 13·9; O_2, 1·49 kg)

17.18 A petrol engine uses a volatile fuel of the following composition by mass: carbon 85 per cent, hydrogen 15 per cent. If the air used is 10 per cent in excess of the minimum required, find the percentage composition by mass of the dry products of combustion.

(CO_2, 19·2; O_2, 2·13; N_2, 78·7 per cent)

17.19 A sample of boiler coal yields the following ultimate analysis by percentage: C 80, H_2 5, O_2 8, S 1·25, N_2 1·2, ash 4·55. Calculate the theoretical mass of air required per kg of coal for complete combustion. If 20 kg of air is actually supplied per kg of coal, determine the masses of the constituents in the wet flue gases.

(10·7 kg; CO_2, 2·94; H_2O, 0·45; SO_2, 0·025; O_2, 2·13; N_2, 15·4 kg)

17.20 A boiler furnace consumes 200 kg of coal per hour. The analysis of the coal by mass is: C 0·82, H_2 0·08, N_2 0·01, S 0·014, remainder ash. The furnace air is supplied by a forced-draught fan and the flue gases are delivered to the stack by an induced-draught fan. If 50 per cent excess air is supplied, find (a) the mass of air per hour passing through the forced-draught fan, (b) the mass of gas per hour passing through the induced draught fan, (c) the mass of carbon dioxide produced and (d) the mass of the dry products passing up the chimney per kg of coal.   (3710 kg/h; 3895 kg/h; 3, 18·7 kg)

(17.16) Heat of reaction

When a compound is formed from elements or split up into its elements, heat enters into the reaction; it may be given out or taken in during the reaction and is known as the *heat of reaction*. If the reaction is a rapid combination of an element with oxygen, heat is generated and is called the *heat of combustion*. The calculation of the heat of reaction is complex and we shall restrict ourselves here to simple and approximate calculations.

The heat of combustion is found experimentally and is usually stated 'per kg' of element burnt. Approximate values are as follows:

hydrogen	143 MJ/kg
carbon	33·7 MJ/kg
sulphur	9·3 MJ/kg

These values assume complete combustion of the element. The value for hydrogen assumes that all the water vapour formed by combustion is condensed to liquid.

17.17 Calorific value of a fuel

The heat liberated when 1 kg of fuel is completely burnt is known as its *calorific value*. The heating effects of the elements carbon, hydrogen and sulphur are known. Given, therefore, the ultimate analysis of a fuel, an estimate of its calorific value may be found by summing the heating effects due to its constituents. The result is an estimate only, since it assumes that every element of the fuel is free to react with the oxygen supplied. In practice, this may not be so. For example, heat may be required to break up any compound already formed between elements in the fuel before burning takes place. Typical figures of calorific values of fuels are:

coal	35 MJ/kg
wood	20 MJ/kg
petrol	45 MJ/kg
coal gas	20 MJ/m^3
natural gas	38 MJ/m^3

17.18 Incomplete combustion

If insufficient air is supplied, or if the distribution or mixing of oxygen and fuel is inefficient, then combustion may be incomplete. The total heat liberated will be reduced accordingly. Consider, for example, the incomplete combustion of carbon. If the air supplied is insufficient, part of the carbon reacts with oxygen to form carbon monoxide (CO) instead of carbon dioxide (CO_2). The chemical equation for the production of carbon monoxide from carbon is

$$2C + O_2 = 2CO$$

(masses) 24 + 32 = 56

i.e. 1 kg of carbon combines with 32/24 kg of oxygen to form 56/24 kg of carbon monoxide. The heat of combustion in this case is only approximately 9·3 MJ/kg of carbon as compared with 33·7 MJ/kg when carbon dioxide is formed. There is thus a loss of approximately 24·4 MJ/kg of carbon incompletely burnt.

Incomplete combustion may be due to incorrect adjustment of the fuel or air supply. Black smoke in the exhaust of a diesel engine is evidence of incomplete combustion due to bad mixing of air and fuel. Black smoke from the chimney stack of a boiler plant is usually evidence of insufficient air supply. It may also be due to stoking, when fresh fuel added to the furnace grate tends to pass unburnt up the stack as solid carbon particles. Emission of black smoke from a chimney stack is illegal.

Example

Estimate the calorific value of a light fuel oil having ultimate analysis by mass: C 0·85, H$_2$ 0·12, S 0·01, remainder ash.

Solution

Using the calorific values of the elements given on page 386, we obtain the approximate calorific value of the fuel as follows:

carbon $0.85 \times 33.7 = 28.7$
hydrogen $0.12 \times 143 \quad = 17.2$
sulphur $0.01 \times \quad 9.3 = \underline{0.093}$
                              Total = $\underline{45.99}$ MJ/kg

The calorific value of the oil is therefore approximately **46 MJ/kg**.

Problems

(For the calorific values of carbon, hydrogen and sulphur, *see* page 386).

17.21 A boiler burns coal containing by mass: C 83 per cent, H_2 16 per cent, ash 1 per cent. Estimate the calorific value of the coal.   (50.9 MJ/kg)

17.22 A dried wood has an analysis by mass: C 0.5, H_2 0.05, O_2 0.45. Estimate its calorific value. If the wood contains 40 per cent by mass of moisture, what is then its calorific value per kg of wet wood? What is the theoretical mass of air required per kg of wet wood for complete combustion?
                              (24 MJ/kg; 14.4 MJ/kg; 3.35 kg)

17.23 A coal contains by percentage: carbon 84, hydrogen 5, sulphur 2, oxygen 5, ash 4. Estimate the calorific value of the fuel and the theoretical mass of air per kg of coal required for complete combustion.
                              (35.6 MJ/kg; 11.4 kg)

17.24 A fuel oil has a relative density of 0.7. The ultimate analysis gives: carbon 83 per cent; hydrogen 12 per cent; ash 5 per cent. Estimate the heat available by combustion per litre of oil. Density of water = $1 \, Mg/m^3$.
                              (31.6 MJ/kg)

17.25 A fuel oil contains 85 per cent carbon, 14 per cent hydrogen. Estimate its calorific value. If the air–fuel ratio is 14:1, what is the heat generated per kg of air–fuel mixture?   (48.8 MJ/kg; 3.25 MJ/kg)

17.26 A boiler plant is to be converted from coal burning to oil burning. The calorific value of the coal used is 35 MJ/kg and of the oil 43 MJ/kg. When burning coal, the heat actually supplied to the steam per kg of fuel is 80 per cent of the calorific value: when burning oil, the percentage is 90 per cent. If oil costs x per cent more than coal per kg, calculate the greatest value of x if the cost of raising steam is to remain the same.
                              (38 per cent)

17.27 A coal contains hydrogen and carbon, with negligible ash. Estimate the minimum percentage of hydrogen that should be present if the fuel is to have a calorific value of at least 44 MJ/kg.   (9.4 per cent)

(17.19) Chemical and nuclear energy; the atomic nucleus

The nucleus of the atom

We have seen that the ultimate particles of compounds are the molecules, and that the ultimate particles of elements are the atoms. There are a vast number of chemical compounds that may be formed from the atoms of the basic elements. In addition the atoms themselves are compounded from still simpler 'particles' – the *proton*, the *neutron* and the *electron*.

Figure 17.6 is a diagrammatic representation of a hydrogen atom. It consists of a central nucleus containing a proton, together with a much smaller particle, an electron, which may be *imagined* to circle the nucleus as a planet circles the sun. Similarly, Fig. 17.7 illustrates an oxygen atom. This consists of a nucleus containing eight protons and eight neutrons, with eight electrons circling the nucleus.

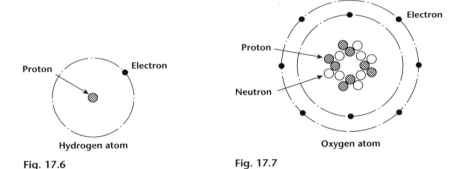

Fig. 17.6   Fig. 17.7

All other atoms are built up in a similar manner from a nucleus containing protons and neutrons and an outer shell of electrons. The proton and the neutron are of equal mass, but the proton carries a positive electrical charge and the neutron is uncharged. The electron carries a negative charge and, in an atom in its normal state, the number of electrons is equal to the number of protons in the nucleus and the atom is electrically *neutral*. Since the mass of the electron is less than one-thousandth part of the mass of a proton nearly all the mass of the atom resides in the nucleus. The total number of protons and neutrons in the nucleus is the *mass number* of the atom. The total mass of the atom is the *atomic mass*. The atomic mass of an element therefore differs slightly from the mass number. Typical values of the mass numbers are given in Table 17.9.

Table 17.9

Nucleus	Particles in nucleus	Symbol	Mass number
Hydrogen	1 proton	^{1}H	1
Helium	2 protons, 2 neutrons	^{4}He	4
Uranium	92 protons, 143 neutrons	^{235}U3	235
	92 protons, 146 neutrons	^{238}U6	238

The number of protons in the nucleus determines the chemical properties of the element, i.e. they fix the nature of the element – hydrogen, helium, carbon and so on. A nucleus of a given element will always contain the same number of protons but may contain a varying number of neutrons. For example, uranium occurring in its natural state contains 99·3 per cent of ^{238}U and 0·7 per cent of ^{235}U. Nuclei of the same element which contain different numbers of neutrons are called *isotopes* of the element.

We have so far explained that in the almost empty space enclosed by an atom, a tiny nucleus of protons and neutrons is orbited by a shell of electrons and that the nucleus accounts for nearly all the mass of the atom. In the past the atom was considered to be the fundamental constituent of matter but when its internal structure was discovered it was then thought that the electron, proton and neutron were the fundamental units. Now, as a result of the science of 'particle physics' and the use of *accelerator* machines producing beams of protons or electrons to bombard the nucleus of the atom, several hundred subatomic particles have been detected and classified according to their properties, such as mass and electrical charge. Most of these particles are unstable and short-lived, having an extraordinary 'fleeting' existence; they are found to decay and re-assemble into other particle forms and as information builds up, known ones are being reclassified. Many are hypothetical, arising from the theories governing particle physics and the theoretical models constructed to explain their relationships. The hypothetical particles include a number that have no internal structure and cannot be divided, i.e. they are among the *elementary particles* now considered to be the fundamental constituents of matter.

The elementary particles are:

- *Leptons*, of which there are six – electron, tau, muon and three associated neutrinos. The muon, an unstable particle with a fleeting lifetime, is becoming important and although similar to the electron has a mass over 200 times as great. For each lepton there is an equivalent 'anti-particle'.

- *Quarks*, of which there are six, have not been detected experimentally; they are categorized using a complex system involving 'flavours' and 'colours' and are given the odd names of *up, down, charmed, strange, top* and *bottom*. In the theory of quarks it is the quark that makes up the 'structured' sub-atomic particles called *hadrons*, in which there are two sub-classes, *baryons* (three quarks) and *mesons* (two quarks). The hadrons include the proton and neutron and other particles. There are also 'anti-quarks' with electric charges opposite to those of the quarks.

- *Gauge bosons*, of which there are four, are postulated in 'gauge theory' as carrying the four fundamental forces of nature* between particles in the atom, thus aiding the solution of the puzzle as to how the nucleus is held together. The

---

* The four fundamental forces in order of magnitude are gravity, weak nuclear interaction, electromagnetism and the strong nuclear force. Taking the gravity force as unity their strengths are respectively considered to be in the ratio $1:10^{34}:10^{38}:10^{40}$. As can be seen, the gravitational force is extraordinarily weak but differs greatly from the others in that it acts over vast distances.

four forces are gravitational attraction, the weak nuclear, radioactive decaying force, electromagnetism and the strong strong nuclear force binding quarks together. The corresponding bosons are respectively the *graviton, weakon, photon* and *gluon*. Again, there is only indirect evidence as to their existence.

Chemical reactions

When hydrogen is burnt in the presence of oxygen to form a water molecule, the result may be illustrated as in Fig. 17.8. Two hydrogen atoms combine with each oxygen atom and each hydrogen atom shares two electrons with the oxygen atom. The chemical process of combustion therefore involves only the electrons, leaving the nucleus of each atom unchanged. That is, the nature of each element in the resulting compound is unaltered by any chemical reaction. The chemical properties of the compound differ of course from those of the original separate elements.

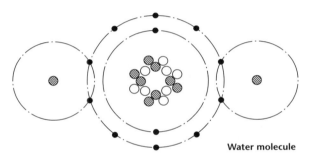

Water molecule

Fig. 17.8

Ionization: ions; ionosphere

An electrically neutral atom has its normal complement of electrons, i.e. it has an equal number of protons and electrons. *Ionization* is when an atom is stripped of an outer electron by an electrical discharge, solar X-rays and ultra-violet radiation or by an interchange of electrons in a chemical reaction as in a fuel cell (*see* section 20.17). An atom with unequal protons and electrons becomes electrically charged and is called an *ion*. When it loses an electron, an atom is left with a surplus positive ion and becomes a particle with a positive charge, i.e. a positive ion, and if it gains an electron it becomes a negative ion. The escaping electron becomes a negative ion. Hydrogen and metals form positive ions, whereas non-metals (e.g. chloride) form negative ions. Loosely bonded molecules are formed when ions with equal and opposite charges bond together.

Ions are used, for example, in fuel cells and for rocket propulsion. They occur naturally and exist in the atmosphere. The *ionosphere* is a region of the upper atmosphere, above about 80 km, where ionization starts to occur and the region extends upwards to 300 km or higher. This layer of ionized air is able to reflect low-frequency radio waves from earth back to earth.

The following terms are used in connection with ionization:

- *Electrolysis* is the use of electricity to break down a chemical compound into its elements.
- An *electrolyte* is a molten compound or acid which permits the conduction of currents through ionization.
- An *electrode* is a conductor for leading an electric current into or out of an electrolyte. A positive electrode is called the 'cathode' and a negative electrode the 'anode'.

Nuclear reactions

Fission

Certain elements in nature (e.g. radium) break up spontaneously, emitting fast moving particles in the process. These elements are called *naturally radioactive*. If a small quantity of radium is mixed with a fluorescent powder each particle given off by the radium causes a small flash of light to be emitted. This process is the basis of the luminous materials used for watches and instrument dials.

The nuclei of some elements of large mass number may also be caused to break up. This process is called *fission*, Fig. 17.9. If the mass of the original nucleus is compared with the sum of the masses of the resulting parts, it is found that a minute loss in mass has occurred. In addition a considerable amount of energy is released. It appears that mass and energy are two forms of the same thing. The relation between the mass m lost and the energy E released is: $E = mc^2$, where c is the velocity of light. Since c is about 3×10^8 m/s, it will be realized that a very small loss in mass results in a relatively large release of energy.

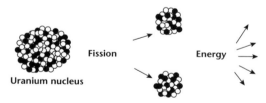

Fission   Energy

Uranium nucleus

Fig. 17.9

For example, the energy released by the complete conversion of 1 kg of any substance from matter into energy would be about $(3 \times 10^8)^2$ J, 90×10^6 GJ, or $2 \cdot 5 \times 10^{10}$ kW h. This is equivalent to the calorific value of 2000 million litres of petrol or 3 million tonnes of coal. However, it is only the nucleus of ^{235}U that undergoes fission readily and the decrease in mass of the uranium nucleus as it splits is only about 0·1 per cent. Hence if 1 kg of ^{235}U were fissioned the energy released would be 0·1 per cent of 90×10^6 GJ, i.e. 90 000 GJ, equivalent to the calorific value of about 3000 tonnes of coal. Only about 0·7 per cent of the mass of natural uranium consists of ^{235}U and therefore the energy theoretically available from 1 kg of natural uranium, the heaviest known element, is equivalent to that obtained by burning about 21 tonnes of coal.

This energy, released under controlled conditions in a *nuclear reactor*, is a considerable source of heat and is the basis of a modern nuclear power station. Similarly, an 'atomic' bomb is the result of a similar, but uncontrolled, release of nuclear energy.

Fusion

It will be recalled that some chemical reactions release energy in the form of heat when the decomposition of molecules takes place, while in other cases the combination of some atoms into compounds also releases heat energy. A similar situation exists with nuclear reactions. Energy can be released when certain 'light' nuclei combine to form a heavier nucleus. This process is termed *fusion*.

For example, consider the fusion of hydrogen nuclei to form a helium nucleus. The normal form of the hydrogen nucleus contains a single proton. Another form, or isotope, of this atom has a nucleus containing both a proton and a neutron. This isotope is called *deuterium* or 'heavy hydrogen' and is available from water. Another isotope is *tritium* which has a proton and *two* neutrons. These two isotopes are the primary fusion fuels. At a temperature of several tens of millions degrees Celsius two heavy hydrogen nuclei may combine to produce a 'light' form of the helium nucleus having two protons and one neutron. This fusion reaction results in the shedding of a single neutron, and in very small loss in total mass which reappears as energy, Fig. 17.10. If it were possible to achieve fusion of all the atoms in 1 kg of heavy hydrogen, the energy released would be equivalent to that obtained by burning 3×10^6 kg of coal. It is believed that the great quantity of energy released by the sun is due to fusion reactions of this type. The hydrogen bomb is also based on a similar fusion reaction.

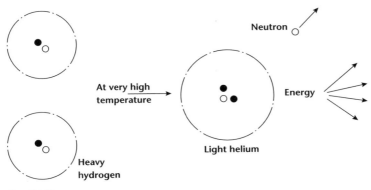

Fig. 17.10

For fusion of hydrogen atoms to take place the nuclei have to be 'contained', while the temperature is raised to the extremely high values required. The containment has to be sustained long enough for the reactions to take place. There are two methods of achieving fusion at high temperature:

1. *Magnetic confinement*, where an intense magnetic field holds a hot plasma in a ring-shaped (toroidal) reactor called a *tokamak*; the plasma is a 'maelstrom' of molecules, atoms and subatomic particles produced by various processes from a combination of the primary fusion fuels, deuterium and tritium isotopes. The Joint European Torus (JET) research station has completed successful experiments in laboratory conditions verifying the principles involved but the difficulties in proceeding to the next stages and industrial production are immense.

2. *Implosion* or *laser fusion*, where a number of powerful laser or electron beams bombard deuterium and tritium isotopes in pellet form so that they collapse inwards, i.e. implode, and thereby produce the temperature conditions for fusion.

For some time experiments have been carried out without success using chemically assisted nuclear reactors in attempts to create *cold fusion* at laboratory temperatures. Cold fusion would meet all future energy needs from a fuel available in water and without producing dangerous waste. One experiment used a 'reactor cell' on the lines of a fuel cell. Another approach involves 'muonic hydrogen' where the electron orbiting the nucleus in the deuterium isotope is replaced by the much heavier relatively short-lived particle called the 'muon' or 'mumeson' (*see* page 390).

Chapter 18

Steam

···

Water and steam are two of the most important working fluids in engineering. The phases in which they can exist are ice, water and wet, dry or superheated steam; the energy content (enthalpy) of unit quantities are considered here for each phase. Examples given include the use of tables of properties of water and steam when applied to simple heat transfer problems.

18.1 Heat flow and enthalpy change

Heat transfer is energy transfer. A heat flow to a body therefore produces a corresponding rise in its content of energy. The energy content of a given mass of body is called the *enthalpy, H*. A precise definition is not attempted here; however, if the pressure remains constant,

heat flow to body (Q) = increase in enthalpy (ΔH)

or

$$Q = \Delta H$$

The *specific enthalpy* is the enthalpy per unit mass.

We are usually concerned with *changes* in enthalpy. Nevertheless it is convenient to have a scale of enthalpy. For water and steam, the zero of the scale of enthalpy is taken at atmospheric pressure and a temperature of $0\,^{\circ}\text{C}$. The basic unit of enthalpy is the unit of energy, i.e. the *joule* (J), but it is more convenient in steam work to use the kilojoule or the larger multiples, mega- and gigajoule.

In a solid or liquid, the change in enthalpy ΔH is measured very closely by the product

specific heat capacity (c) × mass (m) × temp. change (Δt)

i.e. $\Delta H = cm\Delta t$

For unit mass, $m = 1$, and the change in specific enthalpy is $c\Delta t$.

18.2 Enthalpy (sensible heat) of water

The specific heat capacity of water is $4\cdot2\,\text{kJ/kg\,K}$, hence for $1\,\text{kg}$ of water at $t\,^{\circ}\text{C}$ its specific enthalpy (h) is given by

h = specific heat capacity × mass × (temperature above 0 °C or 273 K)

= 4·2 × 1 × t

= 4·2t kJ/kg

For water the enthalpy is also the *total heat* which must be supplied to raise the temperature of the water from 0 °C to t °C. The term *sensible heat* is also used, since the changes in temperature are sensible to the touch or to the sight if a thermometer is used.

18.3 Change of phase

Matter may exist in one of three *phases*: solid, liquid and gas; for example, ice, water and steam. For water, a change of phase occurs at a definite temperature. At the ice point, atmospheric pressure and temperature 0 °C, ice melts and water freezes; at the steam point, atmospheric pressure and 100 °C, water boils and steam condenses. The processes of melting and freezing and of boiling and condensation each take place at a constant temperature corresponding to a particular pressure.

At the ice point, ice and water exist together in equilibrium and at the steam point, liquid water and water vapour exist together. There is also a *triple point* where the pressure (4·6 mm Hg or 610 N/m^2) temperature (+0·01 °C) are such that all three phases, ice, liquid water and water vapour, exist together in equilibrium. This triple point of water is used to define the fixed point of the Kelvin or absolute scale of temperature (*see* section 14.5). There is a triple point for every pure, chemically stable substance.

18.4 Melting and freezing

Both melting and freezing involve a rearrangement of molecules, and energy must be supplied or taken away for this to happen. In melting ice the regularly close-packed molecules of the ice are loosened by a supply of heat at the melting temperature so that they take up the relatively disordered arrangement of molecules in water. Freezing is the reverse of melting and occurs at the same temperature. To freeze water requires that heat energy be extracted at the freezing point. This allows disordered liquid molecules to take up a relatively ordered arrangement in the solid state.

18.5 Latent heat of fusion

To melt ice at atmospheric pressure the heat which must be added to the ice at temperature 0 °C amounts to 335 kJ/kg. The corresponding change of enthalpy is called the *latent heat of fusion of ice*, and 0 °C is the *fusion temperature*. Similarly, in order to freeze water, a quantity of heat equal to 335 kJ/kg must be extracted from the water at 0 °C.

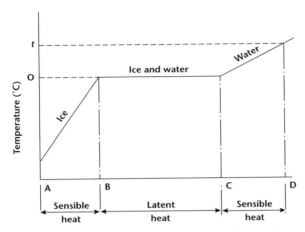

Fig. 18.1   Enthalpy changes during change of state – ice to water.

The latent heat of fusion of a substance is defined as the heat required to change unit mass of the solid into liquid at constant temperature. The heat energy required for a change of state is called *latent* heat since it is not detected by a thermometer.

Figure 18.1 shows the relation between the temperature and heat absorbed when 1 kg of ice below freezing point is converted to warm water. The heat absorbed is made up of three quantities:

1. The enthalpy change (sensible heat) required to raise the temperature of the ice to the melting point, 0 °C. This is shown by AB.
2. The enthalpy change (latent heat) required to melt the ice, shown by BC.
3. The enthalpy change (sensible heat) required to raise the temperature of the water formed to temperature t °C, shown by CD.

Example

Calculate the heat required to convert 8 kg of ice at –10 °C to water at 60 °C. The specific heat capacity of ice is 2·1 kJ/kg K, and its latent heat is 335 kJ/kg.

Solution
Heat to raise temperature of the ice from –10 °C to melting point 0 °C

$$= cm\Delta t$$

$$= 2{\cdot}1 \times 8 \times 10$$

$$= 168\,\text{kJ}$$

Heat to melt the ice at constant temperature

$$= \text{mass} \times \text{latent heat}$$

$$= 8 \times 335$$

$$= 2680\,\text{kJ}$$

Heat to raise temperature of 8 kg of water from 0 °C to 60 °C

$= cm\Delta t$

$= 4{\cdot}2 \times 8 \times 60$

$= 2016 \,\text{kJ}$

Total heat required

$= 168 + 2680 + 2016$

$= 4864 \,\text{kJ}$

This is also the heat which must be extracted from 8 kg of water at 60 °C to form ice at –10 °C.

18.6 Evaporation and condensation

When a liquid changes to the gaseous phase, it is said to *evaporate* or *vaporize*. *Condensation* is the reverse process in which a gas or vapour becomes liquid. Both processes require a flow of heat. Evaporation requires heat to be added to loosen the forces between molecules so that they become separated and move at random as in a gas. Similarly, condensation requires that heat be released or extracted from the gas to become liquid. *Both evaporation and condensation take place at the same constant temperature.*

18.7 Enthalpy (latent heat) of vaporization

The *enthalpy* or *latent heat of vaporization* is the heat required to change the state of unit mass of substance from liquid to gas, or from gas to liquid, at constant temperature. It is denoted by h_{fg} and is the increase in enthalpy of the substance accompanying evaporation.

The latent heat of vaporization of water at atmospheric pressure is 2257 kJ/kg.

18.8 Saturated water vapour

Consider a closed cylinder partly filled with water, Fig. 18.2. Rapidly moving water molecules near the surface tend to escape and fill the space above the water, forming *water vapour*. This change of state from liquid to gas occurring at the surface of the liquid is called *evaporation* and takes place at all temperatures. Also, molecules of water vapour moving rapidly about the space may return to the surface of the water and condense there. Under steady conditions the number of molecules leaving the water surface by evaporation is equal to the number returning to it by condensation. The space above the water then contains a constant number of molecules and is said to be *saturated* with water vapour.

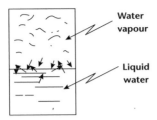

Fig. 18.2

If water is placed in the open air, evaporation again takes place but air currents remove the water vapour formed. The reverse process of condensation on the surface cannot therefore take place and the water evaporates away. This is the process of drying. Of course, drying in the open air does not require a temperature of 100 °C.

The latent heat required for evaporation from the surface of water is provided by the cooling of the remaining water and of the surrounding air. This explains the cooling effect of a lake in hot weather.

18.9 Vapour pressure

Consider again the water in the closed cylinder of Fig. 18.2. The molecules of saturated water vapour in the space above the water are in random motion and therefore exert a pressure on the cylinder walls and on the water. This is called the *vapour pressure*. For example, at 30 °C the vapour pressure of water is about 4·24 kN/m² abs.

If the temperature is raised, a greater number of molecules are continually evaporated and condensed and a larger number are present as water vapour. The mass of water vapour above the water surface is increased and the vapour pressure rises. For every temperature there is found to be a corresponding vapour pressure. Table 18.1 gives a short list of vapour pressures for water at various temperatures.

Table 18.1

Temperature (°C)	Vapour pressure (kN/m² abs.)
10	1·227
50	12·34
100	101·325
200	1555·0

18.10 Boiling

At a temperature of 100 °C, the vapour pressure of water is 101·3 kN/m², i.e. atmospheric pressure. If water is placed in a vessel open to the atmosphere and heated to 100 °C, it is found that at this temperature vapour is formed as bubbles

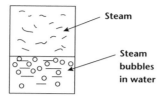

Fig. 18.3

throughout the entire mass of liquid and not merely at the surface, Fig. 18.3. In addition, the rate at which vapour is formed is much greater than the rate of evaporation from the surface. The process of vaporization by rapid formation of bubbles is called *boiling*.

Note the distinction between evaporation and boiling. Evaporation takes place at all temperatures and occurs only at the surface of the liquid; boiling of water occurs at a definite temperature and throughout the volume of the water.

(18.11) Effect of pressure on boiling; saturation temperature

Consider a closed cylinder containing water, Fig. 18.4. The cylinder is connected by a valve to a reservoir as shown. When the valve is open the space above the water is connected to the reservoir, in which the pressure is held constant. If heat is supplied to the water, the temperature at which boiling starts may be found, corresponding to the known pressure. If the pressure in the reservoir is raised experiment shows that the temperature at which boiling starts is also raised.

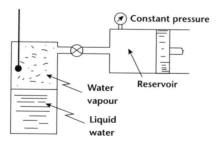

Fig. 18.4

An explanation is as follows: the formation of steam from liquid water involves an expansion. Water expands many times in volume when converted to steam. An increased pressure on the water surface, and hence throughout the volume of water, tends to act against this expansion, i.e. the greater the pressure the more difficult it is for bubbles to form. Therefore more energy per molecule is required to start boiling as the pressure rises, i.e. a higher temperature is required. Similarly if the pressure in the cylinder is lowered the boiling point is lowered. It is found experimentally that the relation between the boiling point and the pressure follows the relation between temperature and vapour pressure (Table 18.1). That is, boiling occurs when the vapour pressure is equal to the pressure in the space above the water.

The temperature at which boiling starts is called the *saturation temperature*, denoted by t_f. Water at the saturation temperature is called *saturated water*. Water at a temperature below the saturation temperature is termed *subcooled* water. The vapour formed by evaporation at the water surface at a temperature below the saturation temperature is water vapour, i.e. water in its gaseous form, but not steam.

For example, at a pressure of 1·55 MN/m² the saturation temperature or boiling point of water is 200 °C (Table 18.1). Water at 1·55 MN/m² and 150 °C is subcooled and does not boil until the temperature reaches 200 °C.

18.12 Flash steam

At constant pressure, boiling is started by raising the temperature to the boiling point. Or, at constant temperature, boiling may be started by lowering the pressure to the vapour pressure corresponding to the water temperature, Fig. 18.5. If the pressure falls suddenly below the vapour pressure, boiling occurs rapidly, steam bubbles being formed explosively throughout the water volume. Steam formed by a sudden drop in pressure is called *flash steam*. Similarly boiling can be stopped by raising the pressure. A rise in pressure causes any steam bubbles in the liquid to collapse to form liquid. The latent heat given up tends to warm the water and bring it up to the saturation temperature corresponding to the new, higher, pressure.

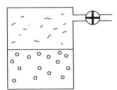

Fig. 18.5   Steam bubbles formed by a drop in pressure

18.13 Dry and wet steam

Dry steam is water completely transformed to vapour, i.e. there are no particles of liquid water in suspension in the vapour formed. Since any further addition of heat to dry steam raises its temperature it is termed *dry saturated steam*. It is said to be saturated since a given volume now contains the largest possible amount of steam at that pressure.

When 1 kg of water at atmospheric pressure and at the boiling point is given a further 2257 kJ, the water is converted to dry saturated steam. If less than this quantity of heat is added to the water, the steam formed will have drops of liquid in suspension and the steam is said to be *wet*.

If dry steam is suddenly raised in pressure some of the vapour condenses out and wet steam is formed. Similarly, if dry steam at a particular pressure loses heat, wet steam results. For example, when steam issues from the spout of a kettle, the initial part of the jet is almost invisible but at a short distance from the spout a fine mist can be seen. The mist is water droplets formed by contact with cold air. The steam issuing from the spout is practically dry steam and wet steam forms because of chilling by cold air.

(18.14) Dryness fraction of wet steam

If the steam is not dry saturated but contains water droplets in suspension, it is *wet steam*. The *dryness fraction* or *quality* x is defined as the fraction by mass of dry steam in the total mass of wet steam, i.e.

$$\text{dryness fraction } x = \frac{\text{mass of dry steam present}}{\text{total mass of steam and water}}$$

If m is the mass of moisture in suspension and m_s the mass of dry steam then the total mass is $m + m_s$. Therefore

$$\text{dryness fraction } x = \frac{m_s}{m + m_s}$$

For example, if 5 kg of wet steam is 98 per cent dry the fraction of moisture by mass is $1 - 0.98 = 0.02$. The mass of moisture present in 5 kg of wet steam is therefore $5 \times 0.02 = 0.1$ kg.

(18.15) Enthalpy of dry saturated steam

The total amount of heat that must be added to 1 kg of water at 0 °C to form 1 kg of dry saturated steam at saturation temperature t_f °C is made up of two parts.

1. The heat h_f required to raise the temperature of the water from 0 °C to t_f °C, i.e. $h_f = ct_f$ kJ/kg where c is the specific heat capacity of water. This is known as the *specific enthalpy of saturated water* (or *sensible heat*).
2. The latent heat h_{fg} required to evaporate 1 kg of water completely.

Therefore, the total heat required is $(h_f + h_{fg})$ kJ/kg. This amount of heat, measured above 0 °C, is called the *specific enthalpy* of the dry steam, denoted by h_g. Thus

$$h_g = h_f + h_{fg}$$

$$= ct_f + h_{fg} \text{ kJ/kg}$$

(18.16) Enthalpy of wet steam

If 1 kg of water at 0 °C is converted to steam of dryness fraction x the total heat required is again equal to the change of enthalpy. Since only x kg of water is converted to dry steam (per kg of wet steam) the latent heat required is only $(x \times h_{fg})$ kJ/kg. The enthalpy h of wet steam is the sum of the enthalpy h_f of water at the saturation temperature t_f °C and the increase of enthalpy due to the change of state from water to steam, i.e. the sum of the sensible and latent heats. Thus

specific enthalpy of wet steam $h = h_f + xh_{fg}$

$$= ct_f + xh_{fg} \text{ kJ/kg}$$

(18.17) Steam tables

The specific enthalpy of water at the saturation temperature t_f °C is given by ct_f. The specific heat capacity of water, c, is 4·1855 kJ/kg at atmospheric pressure between 15 and 16 °C. It varies slightly, however, with temperature and pressure. Similarly the values of latent heat vary greatly at different saturation temperatures. It is convenient therefore to have the exact values of the enthalpy and latent heat of water at various saturation temperatures. These values have been found by experiment and tabulated in the *steam tables*. Table 18.2 is a modified extract from the steam tables, made up as follows:

Table 18.2 Properties of steam

1		2	3	4	5
Pressure (abs.)		Saturation temperature t_f	Enthalpy (sensible heat) of saturated water h_f	Enthalpy of evaporation (latent heat) h_{fg}	Enthalpy (total heat) of dry saturated steam h_g
(kN/m²)	(bar)	(°C)	(kJ/kg)	(kJ/kg)	(kJ/kg)
15	0·15	54·0	226·0	2373·2	2599·2
40	0·4	75·9	317·7	2319·2	2636·9
100	1	99·6	417·5	2257·9	2675·4
101·325	1·01	100·0	419·1	2256·9	2676·0
150	1·5	111·4	467·1	2226·2	2693·4
300	3	133·5	561·4	2163·2	2724·7
400	4	143·6	604·7	2133·0	2737·6
550	5·5	155·5	655·8	2095·9	2751·7
700	7	165·0	697·1	2064·9	2762·0
850	8·5	172·9	732·0	2037·9	2769·9
1000	10	179·9	762·6	2013·6	2776·2
1100	11	184·1	781·1	1998·5	2779·7
1200	12	188·0	798·4	1984·3	2782·7
1400	14	195·0	830·1	1957·7	2787·8
1600	16	201·4	858·6	1933·2	2791·7
1800	18	207·1	884·6	1910·3	2794·8
2000	20	212·4	908·6	1888·6	2797·2
2800	28	230·0	990·5	1811·5	2802·0
3000	30	233·8	1008·4	1793·9	2802·3
4000	40	250·3	1087·4	1712·9	2800·3

- *Column* 1 gives the absolute pressure.
- *Column* 2 gives the saturation temperature corresponding to the pressure.
- *Column* 3 gives the enthalpy h_f of saturated water.
- *Column* 4 gives the latent heat h_{fg}.
- *Column* 5 gives the enthalpy of dry saturated steam, h_g.

Example

A boiler working at a pressure of $700 \, kN/m^2$ generates steam 98 per cent dry. Calculate the enthalpy per kg of steam leaving the boiler. If the water is supplied to the boiler at $21 \, °C$, what is the total heat supplied to each $1 \, kg$ of water?

Solution

At $700 \, kN/m^2$, from Table 18.2,

$\qquad h_f = 697 \cdot 1 \, kJ/kg$

$\qquad h_{fg} = 2064 \cdot 9 \, kJ/kg$

The enthalpy of the wet steam is

$\qquad h = h_f + x h_{fg}$

$\qquad\quad = 697 \cdot 1 + (0 \cdot 98 \times 2064 \cdot 9)$

$\qquad\quad = \mathbf{2721 kJ/kg}$

The enthalpy of water at $21 \, °C$

$\qquad = cm\Delta t$ approximately

$\qquad = 4 \cdot 2 \times 1 \times 21$, measured above $0 \, °C$

$\qquad = 88 \cdot 2 \, kJ/kg$

Therefore, total heat supplied per kg of water

$\qquad = 2721 - 88 \cdot 2$

$\qquad = \mathbf{2633 \, kJ/kg}$

Example

Steam enters an engine at $850 \, kN/m^2$ and is exhausted at atmospheric pressure, $101 \cdot 3 \, kN/m^2$. The steam enters 10 per cent wet and at exhaust is 18 per cent wet. Calculate the 'drop' in enthalpy of the steam through the engine.

Solution

From Table 18.2,

$\qquad$ at $850 \, kN/m^2$, $h_f = 732 \cdot 0 \, kJ/kg$

$\qquad\qquad\qquad\qquad h_{fg} = 2037 \cdot 9 \, kJ/kg$

$\qquad$ at $101 \cdot 3 \, kN/m^2$, $h_f = 419 \cdot 1 \, kJ/kg$

$\qquad\qquad\qquad\qquad h_{fg} = 2256 \cdot 9 \, kJ/kg$

Initial enthalpy of the steam,

$$h_1 = h_f + xh_{fg}$$

$$= 732 \cdot 0 + (0 \cdot 9 \times 2037 \cdot 9)$$

$$= 2566 \, kJ/kg$$

Final enthalpy of the steam,

$$h_2 = 419 \cdot 1 + (0 \cdot 82 \times 2256 \cdot 9)$$

$$= 2270 \, kJ/kg$$

Therefore

$$\text{drop in enthalpy} = h_1 - h_2$$

$$= 2566 - 2270$$

$$= \mathbf{296 \, kJ/kg}$$

(**Example**)

10 t of steam per hour at a pressure of 3 MN/m² and 10 per cent wet is condensed to water at 101·3 kN/m² and temperature 22 °C. What is the total heat extracted from the steam per hour?

Solution
From Table 18.2,

initial enthalpy of steam at 3 MN/m² = $1008 \cdot 4 + (0 \cdot 9 \times 1793 \cdot 9)$

$$= 2622 \cdot 9 \, kJ/kg$$

Enthalpy of water at 22 °C = $cm\Delta t$ approx.

$$= 4 \cdot 2 \times 1 \times 22$$

$$= 92 \cdot 4 \, kJ/kg$$

The heat extracted per kg of steam is therefore 2622·9 – 92·4, i.e. 2530·5 kJ/kg, and since 10 t (= 10 000 kg) of steam is condensed per hour, the rate of heat extraction is

$$10\,000 \times 2530 \cdot 5 = 25 \cdot 3 \times 10^6 \, kJ/h$$

$$= \mathbf{25 \cdot 3 \, GJ/h}$$

(**Problems**)

(Use Table 18.2)

18.1 A cylinder contains 0·5 kg of dry saturated steam at a temperature of 165 °C. What are the steam *gauge* pressure and the enthalpy?

(599 kN/m²; 1380 kJ)

18.2 A boiler generates steam at a pressure of $1.4 \, MN/m^2$. If the steam is dry saturated what is the enthalpy per kg? If the steam has a dryness fraction of 90 per cent what is its enthalpy per kg?   (2788; 2590 kJ/kg)

18.3 Calculate the mass of moisture in 200 kg of steam, 92 per cent dry, at a pressure of $550 \, kN/m^2$. What is its enthalpy?   (16 kg; 517 MJ)

18.4 Steam leaves a boiler 6 per cent wet at a pressure of $400 \, kN/m^2$. How much additional heat per kg of steam is required to form dry saturated steam?

(128 kJ/kg)

18.5 1 kg of steam 60 per cent dry at an initial pressure of $700 \, kN/m^2$ suffers a rapid fall in pressure to $150 \, kN/m^2$. Calculate the final dryness of the steam, assuming that there is no change in enthalpy of the steam.   (66 per cent)

18.6 The steam at the inlet to a turbine is dry saturated at a pressure of $2 \, MN/ m^2$. The exhaust steam has dryness fraction 0.9 and is at atmospheric pressure, 1.01 bar. Calculate the change in enthalpy per kg of steam passing through the turbine.   (347 kJ/kg)

18.7 A gas-heated boiler generates dry saturated steam at $400 \, kN/m^2$ at the rate of 100 kg/h from water supplied at 25 °C. How much heat does the water receive per hour? If the steam is generated at atmospheric pressure, how much heat would be required per hour?   (263 MJ/h; 257 MJ/h)

18.8 A laundry process uses dry steam at $550 \, kN/m^2$ and exhausts the steam at a pressure of $150 \, kN/m^2$. If the process uses 1000 kg of steam per hour and the exhaust steam is 90 per cent dry calculate the heat energy required per hour.   (281 MJ/h)

18.9 Prepare graphs of saturation temperature, latent heat and enthalpy per kg of dry saturated steam plotted against *absolute* pressure. The values are to be taken from Table 18.2. Read off from your graph the temperature, latent heat and enthalpy corresponding to a *gauge* pressure of $278.7 \, kN/m^2$.

(141 °C; 2139 kJ/kg; 2735 kJ/kg)

18.10 In a heating installation where steam is used for heating air the steam enters heating coils at $300 \, kN/m^2$ dry saturated. There is a drop in pressure through the coils and the steam leaves 8 per cent wet at 111.4 °C. What are the pressure at exit and the loss of enthalpy per kg of steam?

($150 \, kN/m^2$; 210 kJ/kg)

18.11 8000 kg of steam per hour at $150 \, kN/m^2$ and 10 per cent wet is condensed to water at $101 \, kN/m^2$ and temperature 27 °C. What is the total heat extracted from the steam per hour?   (18.9 GJ/h)

(**Example**)

Steam at a pressure of $550 \, kN/m^2$ is blown into a tank containing 200 kg of water at 15 °C. The final mass of water in the tank is 210 kg and the final temperature

43 °C. Calculate the initial dryness fraction of the steam. The tank has a mass of 40 kg and its specific heat capacity is 0·5 kJ/kg K.

Solution

This problem is an exercise in the 'method of mixtures' (section 15.8). We equate the heat released by the condensed steam to the heat gained by the water and tank.

Let x be the dryness fraction of the steam blown into the tank; then the enthalpy per kg of steam at 550 kN/m^2 is

$$h_1 = h_f + xh_{fg}$$

$$= 655·8 + (x \times 2095·9) \text{ kJ/kg (Table 18.2)}$$

When each kg of steam is condensed to water at 43 °C, its enthalpy is just the sensible heat, i.e. 4·2 × 43 = 180·6 kJ/kg.

Enthalpy given up by the steam = $h_1 - 180·6$

$$= (655·8 + 2095·9x) - 180·6$$

$$= 475·2 + 2095·9x \text{ kJ/kg}$$

The mass of steam condensed = 210 − 200 = 10 kg

Therefore,

heat given up by 10 kg of steam = 10(475·2 + 2095·9x) kJ

Rise in temperature of water and tank = 43 − 15 = 28°C

Therefore,

heat gained by water and tank = (4·2 × 200 × 28) + (0·5 × 40 × 28)

$$= 24\,080 \text{ kJ}$$

Heat given up by steam = heat gained by water

10(475·2 + 2095·9x) = 24 080

Therefore

$$x = 0·92$$

Problems

18.12 1 kg of steam, 0·9 dry at atmospheric pressure, is blown into an open vessel containing 10 kg of water at 21 °C. Calculate the resulting temperature if all the steam is condensed.   (72 °C)

18.13 Dry saturated steam at a pressure of 550 kN/m^2 is blown into 180 kg of water at 25 °C. The mass of steam condensed is 8 kg. Calculate the final temperature of the water.   (51·9 °C)

18.14 Wet steam at a pressure of 700 kN/m² is blown into 300 kg of water at 20 °C. The mass of steam condensed is 24 kg and the final temperature 60 °C. Calculate the initial quality of the steam.   (0·8)

18.15 3 kg of steam 95 per cent dry is condensed in a tank containing 100 kg of water at 15 °C. The tank has a mass of 60 kg and a specific heat capacity of 0·5 kJ/kg K. Calculate the final temperature of the water. The initial pressure of the steam was 4 bar.   (31·7 °C)

(18.18) Superheated steam

If water in a *closed* cylinder is heated, the vapour pressure in the cylinder increases with temperature. If the space above the water surface contains water vapour only, then boiling will not occur. All that happens is that the rate of evaporation and condensation at the water surface increases. As the temperature increases there is a greater mass of water vapour in the space above the water and this extra vapour produces an increase in pressure. The additional quantity of water vapour is taken from the liquid water and the water level drops slightly.

Now suppose the cylinder of water and water vapour to be connected to a reservoir so that the *pressure may be maintained constant*, Fig. 18.6(a). When the temperature is raised to the saturation temperature t_f, boiling starts and continues as long as heat is added to the water. If sufficient heat is added all, the water is evaporated until only saturated steam is left in the cylinder, Fig. 18.6(b). If, after all the water has disappeared to form steam, further heat is added at constant pressure, the temperature of the steam in the cylinder rises above the saturation temperature, Fig. 18.6(c). The steam is said to be *superheated*.

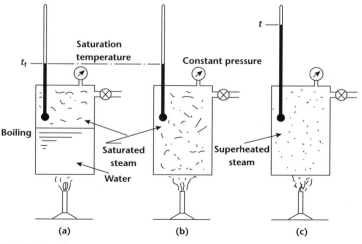

Fig. 18.6

This superheated steam is steam at a temperature above the saturation temperature. The *degree of superheat* is measured by the difference between the temperature of the superheated steam and the saturation temperature at the same pressure.

Thus steam at 180 °C and a pressure of 300 kN/m² has a saturation temperature of 133·5 °C, from Table 18.2. The degree of superheat is therefore 180 − 133·5 = 46·5 °C.

18.19 Gas and vapour: critical temperature

The terms gas and vapour are sometimes used interchangeably in a rather confusing manner. Both 'gas' and 'vapour' describe a fluid in its gaseous form which can be condensed to liquid by increasing the pressure alone, i.e. without lowering the temperature.

A vapour is sometimes thought of as the gaseous phase in contact with its liquid form, e.g. steam in contact with water. A vapour is then always at, or below, the saturation temperature corresponding to its pressure.

Similarly, a gas is the gaseous phase which is not in contact with its liquid, and which may be at a temperature equal to or greater than the saturation temperature (e.g. superheated steam).

A *permanent gas* is a gas at a temperature above a certain *critical temperature* (374 °C for superheated steam). Above this temperature the liquid form cannot exist and the gas cannot be condensed to liquid by pressure change alone.

18.20 Enthalpy of superheated steam

The enthalpy of 1 kg of superheated steam at a temperature t_{sup} °C, measured above 0 °C and 101·3 kN/m², is made up of two parts, Fig. 18.7:

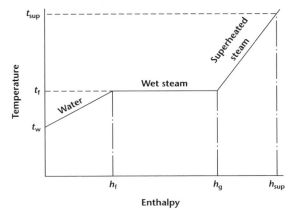

Fig. 18.7

1. The enthalpy h_g of 1 kg of dry saturated steam at the saturation temperature t_f °C, corresponding to the pressure of the superheated steam.
2. The additional heat required to raise the temperature of dry steam from t_f °C to t_{sup} °C, at constant pressure.

The specific heat capacity c of superheated steam varies with temperature and pressure and ranges from 1·98 to 2·7 kJ/kg K for normal industrial purposes. For our purpose we will use an average value of 2 kJ/kg K. The heat required to raise the temperature of 1 kg of the dry steam is therefore

$$c(t_{sup} - t_f)$$

and the enthalpy of superheated steam, denoted by h_{sup}, is

$$h_{sup} = h_g + c(t_{sup} - t_f)$$
$$= h_f + h_{fg} + c(t_{sup} - t_f)$$

where c is the average specific heat capacity for the temperature range $t_{sup} - t_f$.

Example

Calculate the heat which must be added to 1 kg of water at 15 °C to form 1 kg of superheated steam at a pressure of 850 kN/m² and temperature 205 °C. Take the specific heat capacity of superheated steam as 2 kJ/kg K.

Solution

Enthalpy of water at 15 °C = ct

$$= 4·2 \times 15$$
$$= 63 \text{ kJ/kg}$$

From Table 18.2,

enthalpy h_g of dry saturated steam at 850 kN/m² = 2769·9 kJ/kg

Saturation temperature at 850 kN/m² = 172·9 °C

Heat required to superheat the steam = $c(t_{sup} - t_f)$

$$= 2 \times (205 - 172·9)$$
$$= 64·2 \text{ kJ/kg}$$

Enthalpy of superheated steam

$$h_{sup} = 2769·9 + 64·2$$
$$= 2834·1 \text{ kJ/kg}$$

The heat that must be added to the water is the difference between the enthalpy of the superheated steam and the initial enthalpy of the water at 15 °C, i.e.

heat added = 2834·1 − 63

$$= 2771 \text{ kJ/kg}$$

Example

Steam flows into a reducing valve with a pressure of 14 bar and dryness 0·97. It leaves the valve at 1·5 bar and is superheated at this pressure. Assuming that there is no change in the enthalpy of the steam before and after the valve find the degree of superheat of the steam leaving the valve. Specific heat capacity of superheated steam = 2 kJ/kg K.

Solution

At 14 bar and 0·97 dry,

enthalpy $h_1 = h_f + xh_{fg}$

$= 830·1 + (0·97 \times 1957·7)$ (Table 18.2)

$= 2729·1$ kJ/kg

At 1·5 bar, dry saturated, enthalpy = 2693·4 kJ/kg. At 1·5 bar, superheated to temperature t_{sup} °C,

enthalpy $h_2 = 2693·4 + c(t_{sup} - t_f)$

$= 2693·4 + 2(t_{sup} - t_f)$ kJ/kg

and since $h_1 = h_2$,

$2729·1 = 2693·4 + 2(t_{sup} - t_f)$

i.e. $t_{sup} - t_f = 18$ °C approx.

Hence degree of superheat is **18 °C**.

Problems

(Take average value of the specific heat capacity of superheated steam as 2 kJ/kg K).

18.16 Calculate the heat required to form 600 kg of dry saturated steam at 1·2 MN/m² from water at 15 °C. How much additional heat is required to form superheated steam at a temperature of 260 °C?   (1·63 GJ; 86·4 MJ)

18.17 Calculate the heat required in MJ/kg to form superheated steam at a pressure of 1·4 MN/m² and temperature 250 °C, from water at 30 °C.
(2·77 MJ/kg)

18.18 Steam is generated 0·985 dry in the boilers of a steam plant at 12 bar and passes to a superheater where its temperature is raised to 285 °C. How much heat is supplied per kg of steam at the superheater?   (224 kJ/kg)

18.19 Steam is admitted to a turbine at 3 MN/m² and 285 °C. What are the degree of superheat and the enthalpy per kg of steam?   (51·2 °C; 2905 kJ/kg)

18.20 The boilers of a steam turbine power plant generate 250 t of steam per hour at 1 MN/m² and 70 °C superheat. What is the enthalpy per hour available?
(730 GJ/h or 0·73 TJ/h (1 TJ = 10^{12} J))

18.21 1 kg of steam at 700 kN/m² and temperature 245 °C has its pressure raised to 2 MN/m², the temperature remaining constant. What is the change in its enthalpy?   (−59·6 kJ/kg)

18.22 A turbine is supplied with steam at 2 MN/m², superheated 160 °C above the saturation temperature at this pressure. Steam tables give the enthalpy of steam at this condition as 3181·2 kJ/kg. Estimate the average value of the specific heat of the superheated steam for the temperature range considered.   (2·4 kJ/kg K)

18.23 In a heating installation where steam is used for heating air, the steam is generated in the boiler at 850 kN/m², 0·98 dry. Before entering the circulating pipes, the steam passes through a control valve which reduces its pressure to 40 kN/m² and the steam leaves the valve superheated. Assuming the enthalpy of the steam before and after the valve to be unchanged, find the degree of superheat.   (46·2 °C)

18.24 In passing to an engine, steam at entry to the steam pipe is at 1·2 MN/m² and 250 °C. Find its enthalpy per kg. If at exit the pressure is reduced to 1·1 MN/m² and there is assumed to be no loss of enthalpy, calculate the degree of superheat at exit from the pipeline.   (2907 kJ/kg; 63·5 °C)

18.25 Steam at 700 kN/m² with 50 °C superheat is desuperheated by contact with a spray of water at 27 °C. Find the quantity of water at 27 °C needed per kg of steam so that the mixture formed may be dry saturated steam at 700 kN/m².   (0·038 kg)

18.26 10 kg of ice at −10 °C is melted at atmospheric pressure and the water formed is converted into superheated steam at atmospheric pressure. The temperature of the steam is 122 °C. Calculate the total heat supplied. Latent heat of fusion of ice = 335 kJ/kg. Specific heat capacity of ice = 2·1 kJ/kg K.   (30·8 MJ)

18.27 At atmospheric pressure, 1 bar, the saturation temperature of steam is 99·6 °C and its enthalpy is 2675·4 kJ/kg. From steam tables the enthalpy of superheated steam at a pressure of 1 bar is given as 2875 kJ/kg at temperature 200 °C, 3074 at 300 °C, 3278 at 400 °C, 3488 at 500 °C. Find the average values of the specific heat capacity of superheated steam at this pressure for the ranges of temperature 99·6–200 °C, 200–300 °C, 300–400 °C, 400–500 °C, and also for the overall range 99·6–500 °C.

(1·99, 1·99, 2·04, 2·1, 2·03 kJ/kg K)

Chapter 19

Energy, heat and work

Consideration of the equivalence of heat and work together with further notes on energy and power lead to the relationship between a practical heat engine cycle and the ideal (Carnot) cycle.

19.1 No perpetual motion

It is not possible for a heat engine to do work continuously without an external supply of energy. That is, there is no perpetual motion. To understand why this should be so we need to know more about the relation between energy, work and the flow of heat.

19.2 Equivalence of heat and work; Joule's experiment

The performance of work and the flow of heat may each produce the same effect. This was first demonstrated experimentally by Joule (1840). Figure 19.1 is a sketch of the apparatus. It consists essentially of a calorimeter containing water, carefully lagged to prevent flow of heat to or from the surroundings. A paddle wheel P is immersed in the water. The blades of the paddle are caused to rotate by the fall of a load W, the load being connected to the shaft of the paddle wheel by the pulley system shown.

The load is allowed to fall through a known distance and the work done by the load is measured by the product of its weight and the distance. As a consequence of the work done in churning the paddles through the water, *the temperature of the water rises*. By definition, heat flows because of a temperature difference. The

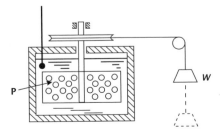

Fig. 19.1

calorimeter is lagged to prevent heat flow and yet the temperature of the water has increased. In fact the same effect – a rise in temperature – is produced by doing work as by a heat flow into the calorimeter. We say that the *internal energy content* of the water (and calorimeter) has been increased.

The 'water equivalent' or 'thermal capacity' of the calorimeter and its contents is known; hence the heat flow required to produce the *same* rise in temperature as in the paddle wheel experiment may be calculated. If proper allowance is made for pulley friction, heat loss through the lagging, thermometer corrections, etc., it is found that the ratio

$$\frac{\text{work done at the paddle}}{\text{heat flow required}}$$

is always a constant for the *same* temperature rise. If the load is W newtons and it falls a height h metres then the work done is Wh J. In fact every $4200\,\text{N m}$ of work done is found to raise the temperature of $1\,\text{kg}$ of water by $1\,\text{K}$ (or $1\,°\text{C}$), i.e. the specific heat capacity of water is $4200\,\text{J/kg K}$ (*see* section 15.4). Joule's experiment showed clearly the relationship between heat and mechanical energy, and where different units are used for heat and work, a factor known as *Joule's equivalent* or the *mechanical equivalent of heat* must be introduced. In the international system of units (SI), of course, this factor is unity.

19.3 Units of energy and power

Energy does not involve time. The SI unit of energy is the *joule* (**J**) and the larger multiples are kJ, MJ, GJ and TJ ($10^{12}\,\text{J}$).

Power does involve time. Power is defined as the rate of doing work or the rate of expenditure of energy. The SI unit of power is the *watt* (**W**) and one watt represents a rate of working of *one joule per second* i.e. **1 J/s**. The larger multiples of the watt are kW, MW, GW, TW. Note that

1 kW = 1000 W = 1000 J/s = 1 kJ/s

Two other units of energy are used which include the units of power. The *watt hour* (**W h**) is the energy equivalent of $1\,\text{W}$ developed for one hour. The *kilowatt hour* (**kW h**) is the energy equivalent of $1\,\text{kW}$ developed for one hour. The relationship between the kilowatt hour and the joule is

$$1\,\text{kW h} = 1000\,(\text{W}) \times 1\,(\text{h})$$

$$= 1000\,(\text{J/s}) \times 3600\,(\text{s})$$

$$= 3.6 \times 10^6\,\text{J}$$

$$= 3600\,\text{kJ} \quad \text{or} \quad 3.6\,\text{MJ}$$

Thus:

One kilowatt hour is equivalent to 3·6 MJ

Note that the watt hour and the kilowatt hour are specific quantities of energy.

Example

A pump lifts 20 000 litres of water to a height of 3 km every hour. Calculate the power output of the pump.

Solution

Mass of 20 000 litres of water = 20 000 kg

Weight of 20 000 kg of water = 20 000 × 9·8 N

Weight of water pumped per second = $\dfrac{20\,000 \times 9\cdot8}{3600}$

$$= 54\cdot4 \text{ N/s}$$

The work done per second is equal to the increase in potential energy of the water per second; therefore

power output = work done per second

$$= \text{weight of water per second} \times \text{height raised}$$

$$= 54\cdot4 \times 3000 \text{ W}$$

$$= \mathbf{163\,kW}$$

Example

A train of mass 612 t descends a slope of 1 in 60 at a uniform speed of 45 km/h with its engine shut off. Assuming the heat generated by the braking to be absorbed entirely by the wheels, of mass 40 t, estimate the rise in temperature per minute of the wheels. Specific heat capacity of the material of the wheel = 0·5 kJ/kg K.

Solution

To maintain a uniform speed the braking force must be equal to the component of the weight of train down the slope, i.e.

braking force = $\dfrac{1}{60} \times (612 \times 10^3 \times 9\cdot8)$

$$= 10^5 \text{ N}$$

Distance travelled in one minute = $\dfrac{45 \times 10^3}{60} = 750 \text{ m}$

Heat absorbed by wheels per minute = work done by braking force per minute:

i.e. $cm\Delta t = 10^5 \times 750$

i.e. $(0\cdot5 \times 10^3) \times (40 \times 10^3) \times \Delta t = 10^5 \times 750$

so that

$$\Delta t = \mathbf{3\cdot75\,K/min} \quad \text{or} \quad \mathbf{3\cdot75\,^{\circ}C/min}$$

Example

Calculate the energy theoretically available in kilowatt hours (a) from the burning of 1 tonne of fuel of calorific value 36 MJ/kg, and (b) from 1 tonne of steam, dry saturated, at 1·2 MN/m², measured above water at 15 °C.

Solution

$$1 \text{ kW h} = 3600 \text{ kJ} = 3·6 \text{ MJ}.$$

(a) Energy in 1 tonne of fuel $= 1 \times 10^3 \times 36 \text{ MJ}$

$$= \frac{1 \times 10^3 \times 36}{3·6}$$

$$= 10\,000 \text{ kW h per tonne}$$

(b) From Table 18.2,

enthalpy of dry saturated steam at 1·2 MN/m² = 2782·7 kJ/kg

and

enthalpy of water at 15 °C $= 4·2 \times 15$

$$= 63 \text{ kJ/kg}$$

Therefore

energy available from the steam $= 2782·7 - 63$

$$= 2719·7 \text{ kJ/kg}$$

$$= 2719·7 \times 10^3 \text{ kJ/tonne}$$

$$= \frac{2719·7 \times 10^3}{3600}$$

$$= 756 \text{ kW h per tonne}$$

Problems

19.1 What is the energy equivalent of 50 000 litres of water falling through a height of 20 m in 1·5 s? Density of water = 1 Mg/m³.   (6·54 MW)

19.2 A diesel locomotive burns 1 kg of oil per minute. What is the power equivalent if the calorific value of the oil is 45 MJ/kg?   (750 kW)

19.3 A dam holds 10 million tonnes of water. The level falls 0·3 m in 24 hours. Calculate the *average* power equivalent to this loss of potential energy.
            (170 kW)

19.4 The load on the cutting tool of a planing machine is 270 N at a cutting speed of 20 m/min. Estimate the rate of heat generation at the tool.
            (90 W)

19.5 The calorimeter described in section 19.2, Fig. 19.1, is of copper, having a water equivalent of 0·2 kg. The calorimeter contains 1 kg of water. The falling load is of weight 135 N. When the load falls 2·6 m the rise in temperature of the water (and the calorimeter) is 0·07 °C. Neglecting friction and heat losses, estimate the specific heat capacity of water.

(4·18 kJ/kg K)

19.6 Electrical generators at a power station 'send out' 800 MW from their terminals. What is the energy output in a period of 24 hours in kW h and MJ?

(19·2 × 10⁶ kW h; 69·1 × 10⁶ MJ)

19.7 A turbine shaft rotates at 40 rev/s. If the shaft torque is 160 kN m what is the output for a period of three hours in kilowatt hours and in megajoules?

(121 × 10³ kW h; 434 × 10³ MJ)

19.8 The brakes for the Bluebird racing car were designed to dissipate 75 million joules of kinetic energy in 60 s. To what average power is this equivalent?

(1·25 MW)

19.9 A train is brought to rest from a speed of 48 km/h. The wheels and brakes have a mass of 40 t and specific heat capacity 0·5 kJ/kg K. If the total mass of the train is 400 t estimate the average temperature rise of wheels and brakes. Neglect heat losses.

(1·8 °C)

19.10 In a machining process, cooling compound is passed over the tool at the rate of 2 litres/min and its rise in temperature is 2·5 °C. If the specific heat capacity of the compound is 3·6 kJ/kg K and the density of the compound is 1·1 Mg/m³, estimate the power equivalent of the heat energy generated at the tool. Neglect losses.

(0·33 kW)

19.11 Find the mass of cooling water required per minute to dissipate the heat equivalent of 3 MW from an electrical transformer. The inlet and outlet temperatures of the water are 15 °C and 55 °C respectively.   (1070 kg/min)

19.12 The electrical input to a lathe motor is 1·5 kW and a power distribution chart shows that the losses in the motor, spindle transmission and feed are respectively 11, 18 and 1 per cent, of the input power. Find (a) the torque on the motor shaft if the motor speed is 1650 rev/min, (b) the efficiency of the spindle transmission and (c) the power available at the tool. Show that the product of the efficiencies of the motor, transmission and feed is equal to that obtained by summing the power losses.

(7·7 N m; 79·8 per cent; 1·05 kW)

(19.4) ## Conservation of energy

As a result of Joule's paddle wheel experiment, we may conclude that the law of conservation of energy applies to heat flow and changes in internal energy content, as well as mechanical energy. Thus in the paddle wheel experiment potential energy lost by the falling load reappears as an increase in the internal energy of the calorimeter and water. If, subsequently, the lagging were removed from around the calorimeter then heat would flow to the surroundings if these were at a lower temperature.

It is often convenient to think of the work done on the water being converted into heat. For example, the mechanical energy lost in bearing friction is often stated as transformed into heat.

19.5 Efficiency of a heat engine

The purpose of a heat engine is to convert to mechanical energy, or useful work, as large a portion as possible of the heat energy liberated by the combustion of a fuel. In the paddle wheel experiment, it was possible to transform nearly all the paddle wheel work into internal energy – 'heat' if you like. On the other hand, *it is not possible to rotate the paddles of Joule's apparatus by merely heating the water.* Similarly, it is not possible to convert all the heat energy liberated by combustion into useful work. The *efficiency* of a heat engine is a measure of its ability to utilize the heat liberated from the fuel. The efficiency η is defined as the ratio

$$\frac{\text{work output } (W)}{\text{energy input } (Q)}$$

or

$$\eta = \frac{W}{Q}$$

Note that both W and Q must be measured in the same units.

Now, even in the most efficient of heat engines the efficiency is always less than 100 per cent. To see why this should be so we must study the cycle of operations of such an engine.

19.6 Heat engine cycle

All heat engines require a *working fluid*. This might be air, burnt gas or steam. For *continuous* operation of an engine the working fluid goes through a cycle of events, continuously repeated. At the end of each cycle, the fluid must return to precisely the same *state* as at the start of the cycle; that is the condition of the fluid as regards pressure, temperature, specific volume, etc., must be unchanged. For example, in a petrol engine cylinder the working fluid is initially a petrol–air mixture which is compressed, ignited, allowed to expand and do work, and finally exhausted from the engine cylinder as hot products of combustion. A fresh fuel–air supply is required at the end of each cycle.

Again, in the steam plant, Fig. 19.2, steam or water flows continuously in a closed cycle round the plant. The water is converted to steam in the boiler. The steam does work in a turbine or engine cylinder and is converted back to its original state (water) in a condenser.

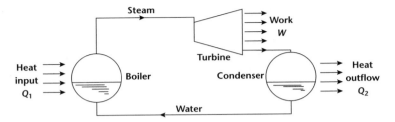

Fig. 19.2

More detailed explanations of the principles and operations of internal combustion engines, steam and turbine plants are given in the following chapters.

19.7 Efficiency of a perfect heat engine

Consider the steam plant of Fig. 19.2. In a complete cycle let the heat supplied be Q_1, work done W and the heat rejected to the condenser Q_2. Q_1 is the heat available from complete combustion of the fuel and W is the net work output from the turbine or engine.

In a perfect plant, there are no 'losses' whatsoever and since the internal energy of the water is the same at beginning and end of the cycle, we have

heat supplied = work done + heat rejected

or

$$Q_1 = W + Q_2$$

Hence

work done = heat supplied – heat rejected

or

$$W = Q_1 - Q_2$$

If we draw an imaginary boundary round any heat plant then, assuming perfection, we can reduce the quantities involved to W, Q_1 and Q_2. In the steam plant Q_2, the heat rejected, is the heat taken away by the condenser cooling water.

The efficiency of a heat engine is

$$\eta = \frac{\text{work output}}{\text{energy input}}$$

$$= \frac{W}{Q_1}$$

and for a *perfect* engine or plant, since $W = Q_1 - Q_2$,

$$\eta = \frac{Q_1 - Q_2}{Q_1}$$

$$= 1 - \frac{Q_2}{Q_1}$$

The efficiency of a heat engine therefore is 100 per cent only when the heat rejected, Q_2, is zero. Now all engines must work continuously in a cycle, the working fluid being returned to its original state at the end of the cycle. Some heat must always be rejected (or unavailable) therefore, and hence even for our perfect engine, the efficiency is necessarily always less than 100 per cent. A more rigorous proof of this statement is necessary, but it cannot be given here.

In practice the efficiency of a heat engine is always less than that given by the expression $(Q_1 - Q_2)/Q_1$ because of various heat losses and mechanical friction (section 21.13). Nevertheless, the efficiency of *any* heat engine is *always* the ratio W/Q_1, and the efficiency may be found from the expression $(Q_1 - Q_2)/Q_1$ if in the heat rejected, Q_2, we include all losses such as friction, radiation, etc.

(**Example**)
_____

In a steam plant the heat received at the boiler per kg of steam is 1140 kJ. The energy flow from the steam turbine in the form of work is 300 MJ/t of steam. Neglecting losses, calculate (a) the efficiency of the plant and (b) the energy rejected to the condenser.

Solution

(a) The work output, $W = 300$ MJ/t

$$= 300 \text{ kJ/kg}$$

Heat supplied $Q_1 = 1140$ kJ/kg

$$\text{Efficiency} = \frac{\text{work done per kg of steam}}{\text{heat supplied per kg of steam}}$$

$$= \frac{W}{Q_1}$$

$$= \frac{300}{1140}$$

$$= 0{\cdot}263 \quad \text{or} \quad \textbf{26·3 per cent}$$

(b) If there are no losses:

heat rejected to the condenser $Q_2 = Q_1 - W$

$$= 1140 - 300$$

$$= \textbf{840 kJ/kg}$$

_____

(**19.8**) **Carnot efficiency**

In 1824, Sadi Carnot, a French engineer, devised a hypothetical ideal engine based on a closed, reversible cycle, in which a working substance takes in heat from a hot source at temperature T_1 and rejects heat to a cold sink at temperature T_2 after converting some of the heat intake to mechanical energy. The working substance is finally brought back to its initial state at the end of the cycle, each process in the cycle being thermodynamically reversible. He showed that a heat engine cannot convert all the heat energy supplied to it into useful work since some of it must be rejected. He stated the principle that a reversible heat engine has the highest possible theoretical thermal efficiency between two given limits of temperature. The efficiency of the Carnot cycle is $1 - T_1/T_2$, solely a function of the source and sink temperatures. The principles he laid down, his concept of the ideal engine and his investigations into the law of conservation of energy, were all fundamental to the development of the laws of thermodynamics. His work, together with that of Clapeyron and Joule, led to Lord Kelvin's absolute scale of temperature.

The conditions that have to be met to achieve the Carnot efficiency are:

- all the heat supplied must be at the highest temperature of the cycle;
- all the heat rejected must be at the lowest temperature of the cycle;
- all the processes must be reversible.

The Carnot cycle is not a practical working cycle.

Example

A petrol engine, 28 per cent efficient, develops 80 kW. If the petrol used has a calorific value of 37 000 kJ/kg and a relative density of 0·78, calculate the rate in litres per hour at which fuel is burnt.

Solution

Rate of doing work = 80 kW

$$= 80 \, kJ/s$$

If the engine is 28 per cent efficient, then

$$0.28 = \frac{\text{work done per second}}{\text{heat supplied per second}}$$

Therefore

$$\text{heat supplied} = \frac{80}{0.28} = 286 \, kJ/s$$

The heat supplied per kg of fuel is 37 000 kJ, and therefore the rate at which fuel is supplied is

$$\frac{286}{37\,000} = 0.00773 \, kg/s$$

The relative density of the petrol is 0·78, i.e. its density is 0·78 Mg/m^3, or 0·78 kg/litre; therefore

$$\text{fuel consumption} = \frac{0.00773}{0.78} \, litres/s$$

$$= \frac{0.00773}{0.78} \times 3600 \, litres/h$$

$$= 35.7 \, \textbf{litres/h}$$

Example

In a gas turbine plant raising steam for heating from its exhaust gases, the net useful power and heat output is 35 MW. The gas fuel is burned at the rate of 3 kg/s
and the calorific value of the gas at the conditions of entry to the combustion
chamber is 42 MJ/kg. The upper and lower limits of temperature of the working
cycle are 400 °C and 25 °C. What is the ratio of the actual thermal efficiency to
that of the Carnot cycle efficiency?

Solution

Energy input $= 42 \times 3 = 126$ MJ/s   or   MW

$T_1 = 273 + 400 = 673$ K; $T_2 = 273 + 25 = 298$ K.

$$\text{Thermal efficiency} = \frac{\text{output}}{\text{input}} = \frac{35}{126} = 0\cdot28$$

$$\text{Carnot efficiency} = \frac{T_1 - T_2}{T_1}$$

$$= \frac{673 - 298}{673}$$

$$= 0\cdot56$$

$$\frac{\text{Thermal efficiency}}{\text{Carnot efficiency}} = \frac{0\cdot28}{0\cdot56} = \textbf{0\cdot5}$$

Example

A steam turbine develops 2 kW h for every $1\cdot2$ kg of oil burned in the boiler
furnace. Find the thermal efficiency of the plant. The calorific value of the fuel is
48 MJ/kg. If the lowest temperature of the working fluid cannot fall below 70 °C,
what must be its upper temperature for the Carnot cycle efficiency to be 40 per
cent?

Solution

2 kW h $= 2 \times 10^3$ (J/s) $\times 3600$ (s)

$$= 7\cdot2 \times 10^6 \text{J}$$

(Note conversion factor, 1 kW h $= 3\cdot6 \times 10^6$ J.)

$$\text{Efficiency} = \frac{\text{output}}{\text{input}}$$

$$= \frac{7\cdot2 \times 10^6}{1\cdot2 \times 48 \times 10^6}$$

$$= 0\cdot125 \quad \text{or} \quad \textbf{12\cdot5 per cent}$$

$T_2 = 70 + 273 = 343$ K

$$\text{Carnot efficiency} = \frac{T_1 - T_2}{T_1}$$

therefore

$$0.4 = \frac{T_1 - 343}{T_1}$$

i.e.   upper temperature, $T_1 = 572\,\mathrm{K}$ or   **299 °C**

Problems

19.13 A steam plant, 30 per cent efficient, burns coal of calorific value $34\,\mathrm{MJ/kg}$ in the boiler furnace. Calculate the energy available for useful work per kg of coal, in megajoules and kilowatt hours.   (10·2 MJ; 2·83 KW h)

19.14 A motor cycle engine develops $8\,\mathrm{kW}$ and consumes petrol at the rate of $2.4\,\mathrm{kg/h}$. If the calorific value of the petrol is $40\,\mathrm{MJ/kg}$, calculate the efficiency of the engine and estimate the rate at which heat is rejected to the exhaust. Neglect all other losses.   (30 per cent; 67·2 MJ/h)

19.15 A nuclear power plant produces $250\,\mathrm{MW}$ of electrical energy at the generators. The efficiency of plant is 27·1 per cent. Calculate the energy in megawatts actually supplied by the nuclear reactor. How many tonnes of coal of calorific value $30\,\mathrm{MJ/kg}$ would have to be burned per hour to supply the same amount of energy?   (922 MW; 111 tonnes)

19.16 A pumping set driven by a diesel engine pumps $90\,000$ litres of water per hour to a tank $20\,\mathrm{m}$ vertically above the pump. The overall efficiency of the pump and engine is 27·6 per cent. Calculate the quantity of fuel used in litres per hour. The fuel has a calorific value of $42\,\mathrm{MJ/kg}$ and a relative density of 0·75.   (2·03 litres/h)

19.17 A steam plant consumes $3.46\,\mathrm{t/h}$ of coal having calorific value $36\,\mathrm{MJ/kg}$. The output of the turbine is $9\,\mathrm{MW}$. Calculate the overall efficiency of the plant.

Assuming that 15 per cent of the available heat in the fuel is lost by inefficiency in the boiler furnace, estimate the rate at which heat is rejected to the condenser. Neglect all other losses. If the upper and lower temperatures of the working fluid are $450\,°\mathrm{C}$ and $30\,°\mathrm{C}$, what is the Carnot efficiency?   (26 per cent; 73·4 GJ/h; 58 per cent)

19.18 A power station consumes 36.4 tonnes of coal per hour of calorific value $37\,\mathrm{MJ/kg}$. If the overall efficiency of the boilers and turbines is 32 per cent what is the power output of the turbines? If the electrical output of the station is $39.6\,\mathrm{MW}$, what is the overall efficiency of the station? The upper limit of temperature of the steam in the turbines is limited to $400\,°\mathrm{C}$. What is the lowest temperature of the condensate required for a possible Carnot efficiency of 55 per cent?   (120 MW; 10·6 per cent; 30 °C)

19.19 The *drawbar power* of a locomotive is defined as the power developed at the coupling when pulling the load.

At a given instant a diesel-electric locomotive exerts a drawbar power of 1·2 MW. The power required to drive the locomotive itself is 150 kW. The overall efficiency of the diesel engine, generator and motors is 27 per cent. If the engine burns fuel of calorific value 45 MJ/kg, calculate the fuel consumption in kg/h.   (400 kg/h)

19.20 A water turbine receives 50 t of water per second from a head of 30 m and delivers 12 MW to the generators. Find the efficiency of the turbines.

(81·7 per cent)

19.21 The net force on the piston of a steam-operated hammer is 21 kN and the stroke of the piston is 400 mm. Find the work done per stroke and the power developed if there are 20 working strokes per minute.

If 75 per cent of the heat in the steam is wasted in the conversion of heat energy to mechanical energy, find the total heat supplied in the steam per minute.   (8·4 kN m; 2·8 kW; 672 kJ)

19.22 A cast-iron block of mass 20 kg is machined at a cutting speed of 0·7 m/s and the cutting force at the tool tip is 1 kN. Assuming that 80 per cent of all the work done in cutting is converted to heat and absorbed by the block, calculate the heat generated and the rise in temperature of the block. Specific heat capacity of cast-iron = 0·51 kJ/kg K.   (560 J/s; 0·055 °C/s)

19.23 An aluminium piston of mass 0·8 kg is machined to bring it to a specified diameter of 60·000 ± 0·025 mm at 20 °C. Owing to the heat generated during machining, the temperature of the piston rose to 50 °C and at this temperature the diameter was measured as 60·000 mm. State whether the work should be accepted or rejected and find the heat absorbed by the piston. Specific heat capacity of aluminium = 0·92 kJ/kg K. Coefficient of linear expansion of aluminium = 24×10^{-6}/K.   (rejected; 22·1 kJ)

Chapter 20

Heat exchangers; power generation; emissions

The work covered in previous chapters on the properties of water, steam and gases, together with cycle efficiencies and heat energy, is now put to use in a study of power generation plants. Heat exchangers, key elements in such plants, are described with typical calculations on mass flow rates of the working fluids. The main elements and processes involved in fossil-fuelled steam turbine plants and the modern combined-cycle gas turbine plants are indicated and their performances compared. The increasingly important subject of emissions of pollutants (and their climatic effects) issuing from power stations, vehicles and other sources is dealt with in some detail, together with notes on nuclear power generation and renewable sources of energy.

20.1 The working fluids

Steam is still the most convenient and economical carrier of energy. Other substances have been employed, such as mercury and diphenyl, but water, being in plentiful supply, gives steam an immense economic advantage. Steam is used for producing power in a turbine (a *prime mover*), with water in heat exchangers and for many processes, e.g. paper making.

20.2 Heat exchangers

A *heat exchanger* is a device for the transference of heat from one fluid to another without mixing or contact between the fluids. It usually consists of one or more tubes within a shell, Fig. 20.1. One fluid flows through the tubes, the other through the shell and over the outer surface of the tubes. If the fluids are at different temperatures heat flows from the hotter to the colder and an exchange of heat is said to occur.

Let Δh_1 be the *decrease* in enthalpy per kg of the hotter fluid and $\dot{m}_1$ the mass flow of fluid per unit time. Then the rate of heat flow $\dot{Q}_1$ from the hot to the cold fluid is

$$\dot{Q}_1 = \dot{m}_1 \Delta h_1$$

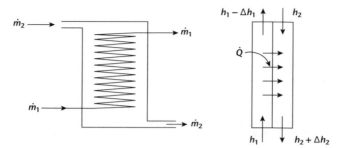

Fig. 20.1

Similarly, if Δh_2 is the *increase* in enthalpy of the colder fluid and $\dot{m}_2$ its flow rate then the heat flow rate $\dot{Q}_2$ into the cooler fluid is

$$\dot{Q}_2 = \dot{m}_2 \Delta h_2$$

If there are no heat losses, as in a perfectly lagged heat exchanger, then the heat flow rate from the hot fluid must equal the rate at which heat flows into the cooler fluid. Thus

$$\dot{Q}_1 = \dot{Q}_2$$

or

$$\dot{m}_1 \Delta h_1 = \dot{m}_2 \Delta h_2$$

The performance of a heat exchanger in terms of the rate of heat transfer and shift in temperature achieved is affected by a number of factors which impinge on each other. The most important are the choice of metal, design and configuration of the tubing, loss of pressure through the system and the rate of flow, turbulence and difference in temperatures of the working fluids. In seeking to improve the performance of a heat exchanger, a balance has to be struck between these and other demands.

There are many different designs of heat exchangers in use throughout industry, employing a variety of working fluids, but the following examples relate particularly to water and steam units in power stations.

- *Boilers* (or *evaporators*) bring water to its saturation temperature corresponding to the boiler pressure either by direct firing or by heat from furnace gases.
- *Economizers* exchange heat between water at below saturation temperature and hot gases or steam. The economizer pre-heats water going into a boiler or evaporator.
- *Superheaters* exchange heat between wet or dry steam and hot gases in order to raise the temperature of the steam above the dry saturation temperature corresponding to the pressure in the tubes carrying the steam. A superheater may be heated directly, in which case it is simply an extension of the boiler heating surface.
- *Heat-recovery steam generators* (HRSGs) are used in conjunction with gas turbines. The turbine's exhaust gases flow past tubing which comprises an economizer,

evaporator and superheater, in a single unit in which the circulation may be natural convection or assisted. The feedwater entering the economizer is usually below atmospheric pressure. In a dual-pressure HRSG, steam is taken off at two pressure levels to drive a steam turbine or for some other purpose; triple-pressure units are becoming common.

● *Water-cooled tube condensers* consist of a bank of tubes carrying steam over which cooling water passes or vice versa. The steam condenses out by giving up its latent heat to the cooling water. The cooling water from the condenser is passed to a *cooling tower* and drops down from the top against an up-coming draught of cold air induced by fans. An alternative type of condenser, not a true heat exchanger, is the *spray* or *jet* condenser which relies on the complete mixing of the steam and cold water.

● *Air-cooled condensers* eliminate the need for a water-cooled condenser with cooling tower and reduce the demand on the water supply (*see* page 434). 'Make-up' water for cooling systems is now often severely restricted, perhaps to a few hours in any 12-hour tide interval cycle.

Example

One million kilograms of carbon dioxide per hour enters the economizer section of a nuclear reactor heat exchanger shell at 315 °C, and leaves at 205 °C. The rate of flow of water in the tubes is 150 000 kg/h. If the water enters at 16 °C, what is its temperature on leaving the heat exchanger? The specific heat capacity of carbon dioxide is 0·85 kJ/kg K.

Solution

Decrease in enthalpy per kg of gas

= mass × specific heat capacity × temperature drop

$= c \times \Delta t$

Rate of heat flow from the gas is

$\dot{Q} = \dot{m} c \Delta t$

$= 1\,000\,000 \times 0{\cdot}85 \times (315 - 205)\,\text{kJ/h}$

$= 93{\cdot}5 \times 10^6\,\text{kJ/h}$

Similarly for the water,

$\dot{Q} = 150\,000 \times 4{\cdot}2 \times (t - 16)\,\text{kJ/h}$

where t is the outlet temperature of the water.

Hence, equating the heat flow rates,

$150\,000 \times 4{\cdot}2(t - 16) = 93{\cdot}5 \times 10^6$

so that

$$t = 164\,°\text{C}$$

Example

A tube condenser condenses steam initially 85 per cent dry at a pressure of $40\,kN/m^2$ to form water at the corresponding saturation temperature. The cooling water enters at $19\,°C$ and leaves at $43·8\,°C$. Calculate the mass of cooling water required per kg of steam condensed.

Solution

At $40\,kN/m^2$, latent heat $= 2319·2\,kJ/kg$.

Since the steam is condensed to saturated water (being at the saturation temperature), only the latent heat has been extracted; therefore

heat lost per kg of steam $= xh_{fg}$

$$= 0·85 \times 2319·2$$

$$= 1970\,kJ$$

Heat gained by m kg of cooling water

$$= 4·2 \times m \times (43·8 - 19)\,kJ$$

$$= 104\,m\,kJ$$

Heat gained by m kg of water = heat lost by 1 kg of steam; therefore

$$104\,m = 1970$$

and

$$m = 18·9\,kg$$

Hence for each kg of steam condensed **18·9 kg** of cooling water are required.

Example

A heat exchanger is to dissipate 4 GJ of heat energy per hour when the coolant (water) is restricted to a rise in temperature of $40\,°C$. Find the rate of flow of cooling water required in litres per hour. Specific heat capacity of water $= 4·2\,kJ/kg\,K$.

Solution

Heat taken up by cooling water = Heat energy dissipated per hour

$$c\,\dot{m}t = 4 \times 10^6\,kJ$$

$$4·2 \times \dot{m} \times 40 = 4 \times 10^6$$

$$\dot{m} = 23\,800\,kg/h$$

Since 1 litre of water has a mass of 1 kg, the rate of flow of cooling water required is **23 800 litres per hour.**

Example

The feedwater to an HRSG is at pressure $1·01\,bar$ and temperature $80\,°C$. Its rate of flow is $\dot{m}$ kg/s. Dry saturated steam is taken off from the evaporator at 12 bar at the rate of 0·3 $\dot{m}$ kg/s and the remaining steam, superheated, is taken from the exit at 40 bar, 350 °C. Hot gas flowing over the tubes carrying the water and steam has a shift in temperature from 520 °C to 95 °C.

The flow rate of the gas is 23 t/h and its mean specific heat capacity is 1·15 kJ/kg. Find the value of $\dot{m}$, neglecting thermal and pressure losses. Take the mean specific heat capacity of superheated steam as 2 kJ/kg K.

Solution

Refer to Table 18.2 (steam table).

Enthalpy of water at 1·01 bar 80 °C h_f = 4·2 × 80 = 336 kJ/kg

Enthalpy of dry saturated steam at 12 bar, h_1 = 2782·7 kJ/kg

Enthalpy of dry saturated steam at 40 bar = 2800·3 kJ/kg

Saturation temperature of steam at 40 bar = 250·3 °C

Enthalpy of the superheated steam at 40 bar, h_2 = 2800·3 + 2(350 − 250·3)

$$= 3000 \text{ kJ/kg}$$

Heat *flow* to feedwater/steam in the HRSG = $(0{\cdot}3\,\dot{m}h_1 + 0{\cdot}7\,\dot{m}h_2) - \dot{m}h_f$

$$= \dot{m}[0.3 \times 2782.7 + 0.7 \times 3000 - 336] = 2600\,\dot{m}\,\text{kJ/s}$$

Heat *flow* from hot gases

= flow rate of gas × sp. heat capacity of gases × temperature drop

$$= \frac{23 \times 10^3}{3600} \times 1{\cdot}15 \times (520 - 95)$$

$$= 3123 \text{ kJ/s}$$

Equating the heat flow from the hot gases to the heat flow to the steam gives

$$2600\,\dot{m} = 3123$$

i.e. $\dot{m} = 1{\cdot}2$ kg/s   or   **4·2 t/h**

Problems

(For properties of steam *see* Table 18.2, page 403. Take specific heat capacity of superheated steam as 2 kJ/kg K and that of water as 4·2 kJ/kg K.)

20.1 Lubricating oil enters the tubes of a heat exchanger at 74 °C and leaves at 32 °C: the cooling water flows into the exchanger at 15 °C and leaves at 20·5 °C. The specific heat capacity of the oil is 2·28 kJ/kg K. Find the mass of oil flowing per kg of coolant.   (0·24 kg/kg)

20.2 The exhaust gas flow rate from an engine is 5300 kg/h and its temperature 460 °C. The gas flows through a heat exchanger where heat is extracted for use in a factory process. The heat demand is 1·8 GJ/h. If the specific heat capacity of the exhaust gas is 1·03 kJ/kg K calculate the final temperature of the gas.   (130 °C)

20.3 Water flowing through the jacket of a 450 kW diesel engine installation is required to dissipate 3·12 GJ/h. The engine cooling water is itself cooled by a second stream of water in the tubes of a heat exchanger. This water enters the heat exchanger at 21 °C and is pumped at the rate of 20 000 litres per hour. Calculate the exit temperature of this water.   (58 °C)

20.4 20 kg of water per kg of steam are injected into a spray condenser at a temperature of 18 °C. The dryness of the steam at entry is 0·85 and the steam pressure 40 kN/m². Calculate the final temperature of the *mixture* of condensed steam and injected water leaving the condenser.   (43 °C)

20.5 Calculate the mass of cooling water required per kg of steam circulated through a tube condenser if the steam enters 0·82 dry. The steam pressure is 15 kN/m². The cooling water has a temperature of 20 °C at inlet and 44 °C at outlet. The temperature of the condensed steam leaving the condenser is 38 °C.   (20 kg)

20.6 A pressurized water reactor, such as is used in a nuclear submarine, produces hot *subcooled* water at 315 °C. This water passes through the tubes of a heat exchanger where it is used to produce steam at a pressure of 20 bar from feedwater at 120 °C. If the steam is dry saturated and the water flow rate is 7 kg per kg of steam, calculate the temperature of the water returning to the reactor. (Subcooled water – *see* section 18.11.)   (237 °C)

20.7 Exhaust gases from the combustion chamber of a closed-cycle gas turbine pass through the shell of a tubular heat exchanger where they exchange heat with the air flow to the turbine. The exhaust gas temperature at entry is 870 °C and at exit 510 °C. The corresponding air temperatures are 70 °C and 480 °C respectively. Calculate the mass of exhaust gases per kg of air. Specific heat capacity of air = 1 kJ/kg K, specific heat capacity of exhaust gases = 0·97 kJ/kg K.   (1·18 kg/kg)

20.8 The exhaust gases from an oil engine are passed through a cooler. The gas temperatures entering and leaving the cooler are 315 °C and 55 °C respectively. The cooling water flow rate is 20 kg/min. The oil fuel used by the engine is 0·09 kg/min and the air supplied is 2·7 kg/min. If the rise in temperature of the cooling water is 8 °C, estimate the mean specific heat capacity of the exhaust gases.   (0·93 kJ/kg K)

20.9 A cooler for compressed air deals with 340 m³/min of air at NTP (1·01 bar and 16 °C). The air enters at 94 °C and leaves at 25 °C. The cooling water enters at 15 °C and leaves at 45 °C. Calculate (a) the mass of air per minute delivered by the compressor and (b) the water flow rate required. Specific heat capacity of air = 1 kJ/kg K. Gas constant R for air = 287 J/kg K.
                                                        (413 kg/min; 227 kg/min)

20.10 A refrigerating plant includes a heat exchanger into which liquid carbon dioxide (CO_2) flows at 27 °C and leaves at 16 °C. The coolant used is carbon dioxide vapour, which enters the exchanger with enthalpy 280 kJ/kg and leaves with enthalpy 338 kJ/kg. If 25 kg of liquid CO_2 has to be cooled per minute, what is the flow rate of the coolant vapour? Specific heat capacity of liquid $CO_2 = 4.12$ kJ/kg K.   (19·6 kg/min)

20.11 Feedwater is pumped into the economizer of a dual-pressure heat-recovery steam generator (HRSG) at temperature 76 °C and pressure 0·4 bar. Steam is taken off slightly superheated at the rate of 16 kg/s from a point where the pressure is 5·5 bar and temperature 170 °C. Steam from the HRSG's super-heater is taken off at the rate of 75 kg/s, pressure 40 bar, temperature 410 °C. Find the total heat received in MW by the feedwater. If the flow of hot gas into the HRSG is at the rate of 490 kg/s and the temperature of the gas is 540 °C estimate, neglecting losses, the temperature of the gas leaving the HRSG. The mean specific heat capacity of the gas is 1·15 kJ/kg.

                                                          (250 MW; 96 °C)

20.12 Feedwater at 54 °C and 0·15 bar enters the economizer of a HRSG flowing at the rate of 63 kg/s. Steam leaves the superheater at 425 °C, 30 bar. The hot gas raising the steam has a shift in temperature from 505 °C to t °C. The gas flows at the rate of 400 kg/s and its mean specific heat capacity is 1·15 kJ/k. Neglecting losses, estimate the value of t.   (100 °C)

20.13 In a HRSG the hot gas flows in at 530 °C and leaves at 90 °C with a rate of flow of 367 kg/s. Feedwater (enthalpy 334 kJ/kg) is pumped into the econo-mizer at the rate of 60 kg/s and steam is taken off only at the superheater exit at conditions 80 bar, 400 °C. Show that there is a 5 per cent loss of energy in the exhange of heat between the working fluids. The enthalpy of the superheated steam may be taken as 3274 kJ/kg.

20.14 A triple-pressure HRSG has steam taken off at three levels – low-, interme-diate and high-pressure points. The conditions of pressure, temperature and rate of flow of steam are respectively 7 bar, 192 °C, 9 kg/s; 28 bar, 520 °C; 46 kg/s; 80 bar, 520 °C, 42 kg/s. The feedwater enters at 1·01 bar, 85 °C. Confirm that the enthalpy of the feedwater and steam at the low and inter-mediate pressure levels are respectively 357, 2816 and 3382 kJ/kg. The enthalpy of the high-pressure steam can be taken as 3447 kJ/kg. What is the rate at which heat is received by the steam in passing through the HRSG? If the gas flow rate is 640 kg/s and thermal losses in the HRSG are 2·5 per cent, estimate the shift in temperature of the gas. The mean specific heat capacity of the gas is 1·15 kJ/kg.   (291 MJ/s; 406 °C)

20.3 Power generation

Most of the electricity generated throughout the world is steam-turbine powered and burning coal, oil or gas. However, an increasing part is being taken by another prime mover – the gas turbine, burning natural gas. Whatever the type of power station, the power produced by the turbines is used to drive a generator. The

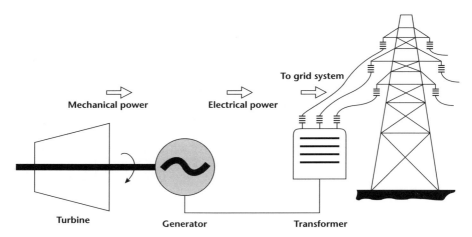

Fig. 20.2

turbine shaft is directly coupled to the generator shaft (Fig. 20.2) which carries a large electromagnet rotating in a series of stationary copper coils thereby inducing a high three-phase alternating current (a.c.) at a fairly low voltage, up to 25 kV. The a.c. voltage is raised by a step-up transformer to the low current, high voltage of 400 kV necessary for supplying current over long distances with minimum losses and in order to reduce the cross-sectional area of the current-carrying cables. Alternators and transmission lines can be rated at various voltages and the generators have to be synchronized in relation to the grid system as to correct phase and speed.

Some stations produce electrical power only but others on industrial sites such as paper mills or refineries are used for *co-generation*, the simultaneous production of heat and power on the same site.

(20.4) Power-producing steam plant

A typical steam plant fired by coal or oil consists of the following major items, Fig. 20.3:

- The steam-raising unit including furnace, boiler, economizer and superheater.
- The steam turbine.
- The condenser; cooling tower.
- The boiler feed-pump; condensate extraction pump; de-aerator.
- The water supply, continuous or 'make-up'; feedwater storage tank (hotwell, sump), purge and settling ponds.
- Auxiliary plant.
- The emission control plant; flue stack.

The **economizer** raises the temperature of the water fed to the boiler, i.e. the feedwater, to within roughly 22 °C of the saturation temperature corresponding to

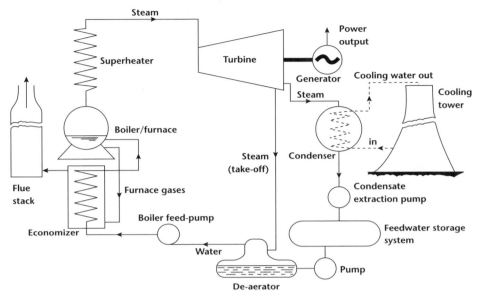

Fig. 20.3   Steam plant

the boiler pressure. The feedwater passes through the economizer tubes placed in the path of the flue gases from the boiler furnace. A considerable portion of the heat made available by burning of the fuel is lost to the chimney stack with the flue gases. The economizer serves to recover some of this heat.

The **boiler** acts mainly as an evaporator: the heat required to bring the water to the saturation temperature and evaporate it is obtained directly from the furnace gases.

The **superheater** raises the temperature of the steam above that of the saturation temperature corresponding to the boiler pressure.

The **steam turbine** receives the steam from the boiler at slightly below boiler pressure and usually superheated. In passing through the turbine the steam expands to atmospheric pressure or below. The steam does work on the turbine and in consequence its 'energy content' is reduced. The effect on the steam is to reduce its superheat. In today's turbines very high pressures are employed, dropping to the low condenser pressure through stages. As the steam expands, even when initially greatly superheated, it rapidly becomes wet and it is necessary to extract it at intermediate points for *reheating* at constant pressure. After reheating it is returned to the turbine and fully expanded to the condenser pressure.

The **water-cooled condenser** converts the wet exhaust steam to water at the lowest possible pressure and temperature. The cooling water passes to a cooling tower or pond and the condensate is pumped to a feedwater storage tank from where it is returned to the boiler by the feed-pump. The feed-pump forces the water into the boiler against the boiler pressure but does not, of course, create the boiler pressure which is maintained by the evaporation of water in the boiler to form steam. Some plants may operate without a condenser, the turbine exhaust being simply discharged to the atmosphere; this requires a continuous and ample supply of water.

Where the water in the system is continually recycled only 'make-up' water is required to be supplied, but for a large power station this still represents a considerable amount. In general, water-cooled condensers with their cooling towers rely on vast quantities of water from rivers and ponds and this is becoming more difficult to obtain.

An **air-cooled condenser** considerably reduces the demand on water resources but has the disadvantage of requiring a very large area to house it. Rye House, Hertfordshire, operates the largest such condenser in the UK at present. Recently constructed, it has a 700 MW output and a condensate flow of 720 t/h. The condenser is the largest item of plant on the site covering some 8000 m². The wet steam from the turbine is passed downwards through 600 radiators in a 'roof-shape' arrangement while fans operating in a vacuum send air upwards. It is designed as 100 modules in a 10 × 10 formation with each module consisting of a 6 m diameter fan and radiators. The fans, rotating at 68 rev/min, direct air over some 160 km of oval-finned tubing carrying the steam.

20.5 Boilers

Basically, boilers for power stations fall into one of two classes: the *fire-tube* or the *water-tube*. In the fire-tube boiler the furnace gases flow through one or more tubes; steam is generated from water on the outside of the tubes and within the boiler shell. In the water-tube boiler the tubes (a large number) contain water and the furnace gases pass over the outside of the tubes.

20.6 The fire-tube boiler; fluidized-bed combustor

Figure 20.4 shows the essential features of a simple fire-tube boiler having a single large tube. The boiler shell A contains water and steam under pressure. The water level is at B, the space above B being filled with wet steam. The fire-tube C contains the grate D, extending from the fire-door E to the brick fire-bridge F. Air for combustion of the fuel on the grate passes through the firing door openings and both over and underneath the grate. Air flowing beneath the grate flows

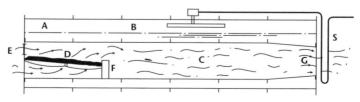

Fig. 20.4   Fire-tube boiler

upward through the fire-bars. Burnt flue gases and any excess air then pass along the fire-tube to the exit G. The superheater tubes S are usually placed at this exit.

The boiler shell is surrounded by a brick structure so that hot gases leaving the fire-tube pass back along the outside of the shell and finally flow back underneath the shell to the chimney stack. The economizer tubes are generally placed at the exit leading to the chimney.

A *fluidized-bed combustor* reduces emissions of pollutants by improving the mixing of air and fuel and the rate of heat transfer. A typical unit is a bed of hot fine sand, with additives for sulphur removal, having air blown evenly through it as fuel particles are fed in. The beds may be classified as 'atmospheric pressure circulating' or 'pressurized'. A bed behaves like a fluid and steam pipes may be placed within or above it. Such combustors are capable of burning all types of coal as well as peat, coke, etc. They are also used to gasify fine-grained coal, converting it into carbon dioxide and hydrogen, using oxygen or air and steam as gasifying agents.

20.7 The water-tube boiler

The water-tube boiler, Fig. 20.5 consists essentially of a bundle of parallel tubes containing water and connected to headers at each end. The headers are in turn connected to the boiler drum which contains a further supply of water. The feed-water supply is to the drum and the steam generated leaves through a stop-valve at the top of the drum.

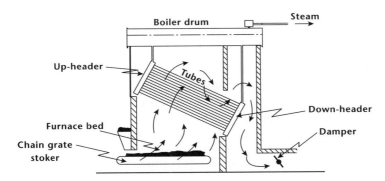

Fig. 20.5 Water-tube boiler

Within the boiler, water flows by natural circulation to the down-header along the tubes, where evaporation takes place, and returns to the drum via the up-header. Natural circulation does not take place due to a temperature difference as might be supposed. The boiler tubes contain steam and water at saturation temperature and this temperature is nearly uniform along the tube. As steam is formed by evaporation the space in a tube is taken up by vapour bubbles which tend to rise. A natural circulation is then set up since the water in the down-header is more dense than the mixture of steam and water in the up-header, Fig. 20.6.

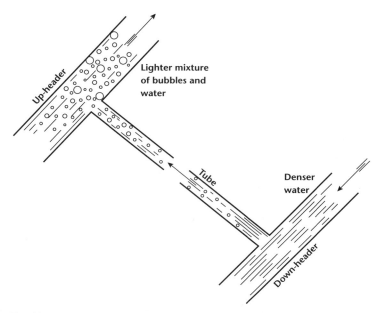

Fig. 20.6

There are other designs of boilers with different arrangements and types of tubes. For example, the tubes may be vertical and internally ribbed (rifled).

20.8 Auxiliary plant

For small boilers, the air for combustion is supplied by *natural circulation*. The hot flue gases in the chimney stack are less dense than the cold air outside the chimney. A natural circulation is therefore set up and maintains a continuous supply of fresh air to balance the outflow of hot burnt gases. The flow of air increases with increasing chimney height, but for large boilers the supply of air is augmented by various types of fans.

The *forced-draught fan* blows the large quantities of air needed for combustion into the furnace and is therefore located before the furnace.

The *induced-draught fan* directly aids the natural chimney draught and is placed between the chimney and the flue gas exit from the furnace. The fan draws the flue gases from the furnace thus encouraging the induction of air into the furnace.

Another auxiliary plant in connection with the air supply is the *air pre-heater*. This is a device for recovering waste heat from the flue gases and utilizing the heat to raise the temperature of cold air going into the furnace. The pre-heater is usually placed between the forced-draught fan and the furnace.

The *de-aerator* is used with the feedwater to remove air and incondensable gases such as oxygen and carbon dioxide from the condensate; steam bled from the turbine is used as the heating agent.

20.9 Boiler efficiency

The *output* of a boiler may be stated as the number of kg of steam produced per hour. To calculate the *thermal efficiency* of a boiler we take the output as the increase in enthalpy of the water supplied to the boiler when converted into steam. The input is the heat available in the fuel fired.

Let enthalpy of steam produced = h_1 (kJ/kg)

enthalpy of water supplied = h_2 (kJ/kg)

calorific value of the fuel = C (kJ/kg)

steam generated = m_s (kg/h)

fuel fired = m_f (kg/h)

The *output* per hour from the boiler is the increase in enthalpy of the water/steam passing through the boiler per hour, i.e.

output = $(h_1 - h_2) \times m_s$ (kJ/h)

The *input* to the boiler is the heat available in the fuel fired per hour, i.e.

input = $C \times m_f$ (kJ/h)

The thermal efficiency is then given by

$$\text{thermal efficiency} = \frac{\text{output}}{\text{input}}$$

$$= \frac{(h_1 - h_2)m_s}{m_f \times C}$$

The steam produced may be wet, dry or superheated. The enthalpy h_1 of the steam is found from the steam tables, as in Chapter 18.

If the temperature of the water entering the boiler is $t\,°C$ then its enthalpy h_2 is *approximately*

$4.2t$ kJ/kg

where 4.2 kJ/kg K is the specific heat capacity of water.

An alternative method of arriving at the thermal efficiency is to estimate the total heat losses Q_2, and by subtraction from the heat Q_1 supplied from the fuel, to estimate the enthalpy gain of the steam produced. Thus

net gain in enthalpy of the steam = $Q_1 - Q_2$

Hence

$$\text{efficiency} = \frac{Q_1 - Q_2}{Q_1}$$

The major source of heat loss arises from the heat content of the hot flue gases which pass up the chimney. Other losses are due to radiation, incomplete

combustion, excess air which chills the fuel, cold air entering flues through defect-
ive brickwork and loss due to combustible matter in the ash.

(20.10) Overall efficiency of a power station

The input and output are measured over a given period. The input is the energy
value of the fuel consumed (gross, as fired). The output is the electrical power *sent
out* to the National Grid, plus in some cases, the energy in the useful heat produced. The ratio of output to input gives the *overall efficiency* which covers losses
in all individual items of the plant.

Example

In a boiler test, steam at a pressure of $1.4 \, \text{MN/m}^2$, having a dryness fraction 0.95,
is generated at the rate of 8 kg per kg of coal burnt. The calorific value of the coal
as fired is 35 000 kJ/kg and the temperature of the feedwater is 47 °C. Calculate the
thermal efficiency of the boiler.

Solution
Enthalpy of the steam generated,

$h_1 = h_f + xh_{fg}$

$= 830.1 + (0.95 \times 1957.7)$

$= 2690 \, \text{kJ/kg}$

Enthalpy of the feedwater,

$h_2 = 4.2 \times 47 = 197.4 \, \text{kJ/kg}$

Output per kg of coal burnt

= increase in enthalpy of water/steam per kg of coal

$= (2690 - 197.4) \times 8$

$= 19\,940 \, \text{kJ}$

Input per kg of coal burnt = heat value of coal as fired

$= 35\,000 \, \text{kJ/kg}$

$\text{Efficiency} = \dfrac{\text{output}}{\text{input}}$

$= \dfrac{19\,940}{35\,000}$

$= 0.57 \quad \text{or} \quad \textbf{57 per cent}$

Example

A generating station in a 3-hour test gave the following results:

$$\text{Power 'sent out'} = 80\,000\,\text{kW h}$$

$$\text{coal used} = 40\,\text{tonnes}$$

$$\text{calorific value of coal} = 33\,\text{MJ/kg}$$

Calculate the overall thermal efficiency of the station.

Solution

Output from the station in three hours

$$= 80\,000\,\text{kWh}$$

$$= 80\,000 \times 3\text{·}6\,\text{MJ since } 1\,\text{kW h} = 3\text{·}6\,\text{MJ}$$

$$= 288\,000\,\text{MJ} \quad \text{or} \quad 288\,\text{GJ}$$

Input to the station in three hours

$$= \text{heat in fuel burnt}$$

$$= \text{mass of coal used} \times \text{calorific value of coal}$$

$$= 40 \times 10^3 \times 33$$

$$= 1\,320\,000\,\text{MJ} \quad \text{or} \quad 1320\,\text{GJ}$$

$$\text{Efficiency} = \frac{\text{output}}{\text{input}} = \frac{288}{1320}$$

$$= 0\text{·}218 \quad \text{or} \quad \textbf{21·8 per cent}$$

Problems

(For properties of steam *see* Table 18.2, p. 403. For calorific values of carbon and hydrogen *see* section 17.16. Take specific heat capacity of superheated steam as 2 kJ/kg K and specific heat capacity of water as 4·2 kJ/kg K.)

20.15 Find the heat required to produce 1 kg of steam at 1·1 MN/m² pressure, 95 per cent dry, from feedwater at 48 °C. This steam is passed through a super-heater and its temperature at exit is 210 °C. Calculate the heat supplied in the superheater.   (2480 kJ/kg; 51·8 kJ/kg)

20.16 Steam leaves a boiler at a pressure of 2 MN/m² and dryness 0·94. If the heat supplied to the feedwater to produce this steam is 2620 kJ/kg, calculate the temperature of the feedwater.

The steam temperature is raised to 260 °C in a superheater. Calculate the heat supplied in the superheater.   (15·2 °C; 95·2 kJ/kg)

20.17 A boiler burns coal of calorific value 32·5 MJ/kg and consumes 2·8 t/h. The steam is produced at a pressure of 1·4 MN/m², 96 per cent dry, from feedwater at 28 °C. Calculate the efficiency of the boiler if the steam is generated at the rate of 30 t/h.   (86 per cent)

20.18 In a test of a steam plant, the water supplied to the boilers was 370 t at a temperature of 155 °C. The steam was generated at a pressure of 2·8 MN/m² and superheated to 337 °C. 36 tonnes of coal were consumed during the test. The calorific value of the coal was 32 MJ/kg. Estimate the boiler efficiency.

(76 per cent)

20.19 A boiler is supplied with feedwater at 32 °C and steam is taken off 93 per cent dry and at a temperature of 195 °C. The coal used has a calorific value of 33 MJ/kg. If 8·5 kg of feedwater is evaporated per kg of coal, estimate the boiler efficiency.   (65 per cent)

20.20 Coal burnt in a boiler contains 88 per cent carbon and 10 per cent hydrogen. The feedwater temperature is 32 °C and the boiler pressure 16 bar. The boiler efficiency is 80 per cent. Calculate the theoretical calorific value of the coal and hence estimate the mass of dry saturated steam produced per kg of coal.   (44 MJ/kg; 13 kg)

20.21 A Lancashire boiler working at a pressure of 700 kN/m² burns 1 kg of coal for every 8 kg of steam produced. The calorific value of the coal is 35 MJ/kg. The temperature of the steam is 184 °C and of the feedwater 16 °C. Calculate (a) the heat given to each kg of water and (b) the efficiency of the boiler.

(2733 kJ/kg; 62·5 per cent)

20.22 In a three-hour boiler test the coal consumed was 3·4 t and its calorific value was 32·6 MJ/kg. Dry saturated steam at 1·4 MN/m² was generated at the rate of 11 t/h. The feedwater temperature was 49 °C. After leaving the boiler the steam enters a separate oil-fired superheater which it leaves at 1·4 MN/m² and superheated to a temperature of 288 °C. The mass of oil used in the test was 180 kg and its calorific value 44·2 MJ/kg. Estimate (a) the boiler efficiency and (b) the efficiency of the superheater.

(77 per cent; 77 per cent)

20.23 A boiler uses oil with a calorific value of 45 MJ/kg. The analysis of the oil is 86 per cent carbon, 14 per cent hydrogen. The air supplied is double the theoretical mass required for complete combustion of the fuel. Find the mass of exhaust gases per kg of oil burnt.

It the exhaust gas temperature is 430 °C and the air temperature 18 °C, find the percentage heat lost to the exhaust gases per kg of oil burnt, measured above air temperature. Mean specific heat capacity of the exhaust gases = 1·04 kJ/kg K.   (30·7 kg; 29 per cent)

20.24 A water-tube boiler generated steam at 18 bar dry saturated, from feedwater at 88 °C. The calorific value of the coal used was 32·5 MJ/kg. Three tests were carried out for those conditions but at different furnace draughts. The figures obtained were:

Test no.	1	2	3
Water evaporated per kg of coal (kg)	10·5	10·25	8·8

Compare the boiler efficiencies for these tests. (78·4, 76·4, 65·7 per cent)

20.25 A power station consumes 0·45 kg of coal per kilowatt hour. The calorific value of the coal is 30 MJ/kg. Estimate the overall efficiency of the station.
(26·7 per cent)

(20.11) Combined-cycle gas turbine (CCGT) plant

The generation of electricity was a fairly static field of technology until the impact of natural gas and advanced gas turbine designs, together with further reductions in permitted levels of discharge of pollutants, brought about the introduction of the *combined-cycle gas turbine* (CCGT) plant. The conventional furnace and boiler are eliminated and *two* prime movers, a gas turbine and a steam turbine, produce power. The gas turbine's exhaust gases are used to raise steam in a *heat-recovery steam generator* (HRSG) which incorporates an economizer, evaporator and super-heater in one unit (for description *see* page 426).

Typical CCGT plant module

Figure 20.7 shows a schematic diagram of a typical module using the *single-shaft concept* where the two turbines and generator are on the same shaft. An alternative arrangement has each turbine serving its own generator and there are arguments for and against the different arrangements as to their merits and reliability. The sequence of operations is as follows.

Gas turbine cycle (open circuit):

- Air is continuously supplied to the compressor at atmospheric conditions.
- The compressed (heated) air, mixed with gas fuel, is injected through nozzles into a combustion chamber.
- Burnt gases at very high temperature, upwards of 1000 °C, pass to the gas turbine and expand through the blades on the rotor attached to the shaft; work is done on the blades and the turbine supplies mechanical power via the shaft to the air-cooled electric generator.
- The turbine's exhaust gases, at a temperature upwards of 500 °C, enter an unfired HRSG where they pass, in turn, over the economizer, evaporator and superheater tubes.
- The exhaust gases vent to the atmosphere via a flue stack at as low a temperature as possible.

Steam turbine cycle (closed circuit):

- Feedwater is pumped into the economizer of the HRSG, usually at below atmospheric pressure, having been pre-heated directly or by the heat remaining in the gas turbine's gases before they leave the flue stack.

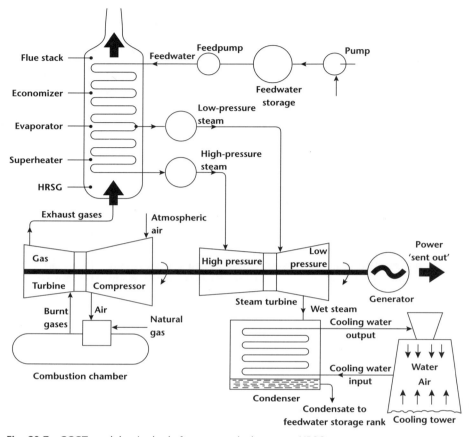

Fig. 20.7   CCGT module: single-shaft concept, dual pressure HRSG

- A dual- or triple-pressure assisted circulation HRSG provides superheated steam for the different pressure stages of the steam turbine. Figures of 110 bar, 540 °C are typical conditions of steam entering the high-pressure stage.
- The steam expands through the turbine blades doing work on the shaft and finally power is supplied to the hydrogen-cooled generator.
- As for a coal-fired station the steam is condensed, de-aerated and returned to the HRSG. The circuit is then complete.

Arrangement of modules in a CCGT plant

There are many CCGT plants in service or construction across the world with outputs in the range 50–1500 MW. A station is made up of one or more modules as described above, depending on the output required. A large station, for example, rated at 1450 MW could have four modules, whereas a 50 MW plant may supply a peak demand using only a single module with both the gas and steam turbines serving their own generators. A typical medium 650 MW station has two gas turbines, one HRSG, one steam turbine and three generators.

The use of other fairly clean fuels besides natural gas is possible although 'dirty' fuels are normally precluded, not only because of pollution, but also because of the possibility of turbine blade corrosion. The introduction of air-blown gasification and air-separation units has produced an integrated plant (ICCGT) utilizing a wider range of fuels.

Thermal efficiencies

The overall efficiency of a CCGT plant is the ratio of the electrical power 'sent out' at the generator terminals to the energy input from the gas fuel. The input is the product of the mass or volume flow rate of fuel burned and its calorific value at the conditions of entry to the combustion chamber. This input is also that of the gas turbine section. The input to the steam section, however, is the heat energy in the steam leaving the HRSG and, neglecting losses, this may be taken as the heat transferred from the exhaust gases of the gas turbine plus the heat energy already in the feedwater. The ability now to support very high temperatures in gas turbines over long periods enables a greater *temperature difference* to be achieved between that of the gas entering the turbine (the *source*) and the temperature of the condensate in the steam cycle (the *sink*). This permits a higher theoretical Carnot efficiency for the cycle. This efficiency can never be attained since all the heat supplied and rejected must be at the highest and lowest temperatures respectively and all the processes must be thermodynamically reversible.

(20.12) Comparison of CCGT and coal-fired plants

Besides the use of gas turbines in CCGT plants as a primary producer of power most power stations have gas turbines operating for auxiliary power and to meet peak load demands. Novel cycles of operation are now being employed in CCGT stations and in plants that are being retrofitted for the use of gas as an alternative fuel.

Compared with an equivalent output coal-fired station at present, a CCGT plant is less expensive per unit of power 'sent out', its emissions of pollutants and cooling water requirements are much less, the effect of low-profile cooling towers, chimneys and buildings is less visually intrusive on the landscape and the delivery of fuel is below ground, resulting in a clean site.

Coal, in general, will remain plentiful for many years and there have been some advances in 'clean-coal' technology and improving efficiencies of coal-fired stations. CCGT plants at present rely on relatively inexpensive clean gas but there is a finite limit to this resource; reaching this limit may be deferred through advances in the technologies of obtaining gas from other fuels.

Over-riding all of the above technical and environmental points is the question of the long-term reserves of different fuels and access to them.

Example

A CCGT plant has three compressor–gas turbine-HRSG-generator sets serving a single steam turbine which has its own generator. The output from each gas turbine generator is 140 MW; total gas consumption, 33 m³/s (calorific value 38 MJ/m³; mean specific heat capacity of exhaust gases from turbine, 1·15 kJ/kg K; mass flow rate of exhaust gases entering *each* of the three HRSGs, 473 kg/s; turbine exhaust gas temperatures entering and leaving the HRSGs, 545 °C and 95 °C respectively; efficiency of steam turbine, 34 per cent. If the enthalpy of the feedwater entering each HRSG is 160 kJ/kg and its mass flow rate is 88 kg/s, estimate the output of the steam turbine and the overall efficiency of the station. What is the mean value of the enthalpy of the steam leaving each HRSG?

Solution
Diagrams should be drawn showing the input and output to each element in the plant.

The heat H given up by the exhaust gases in three HRSGs is the product of the mass flow rate of gas, the mean specific heat capacity of the gas and the temperature difference in passing through the HRSGs. Thus

$$H = (3 \times 473) \times 1·15 \times (545 - 95)$$

$$= 734 \times 10^3 \text{ kJ/s}$$

$$= 734 \text{ MJ/s or MW}$$

This is the heat gained by the steam leaving the HRSGs.

Total enthalpy of feedwater entering HRSGs = $3 \times 88 \times 160 \text{ kJ/s}$

$$= 42 \times 10^3 \text{ kJ/s}$$

$$= 42 \text{ MJ/s}$$

Therefore total input energy to steam turbine = 734 + 42 = 776 MJ/s

$$\text{Steam turbine efficiency} = 0·34 = \frac{\text{output}}{\text{input}}$$

Therefore

$$\text{steam turbine output} = 0·34 \times 776 \text{ MW}$$

$$= \textbf{264 MW}$$

$$\text{total gas turbine output} = 3 \times 140 = 420 \text{ MW}$$

$$\text{output from station} = 420 + 264 = 684 \text{ MW}$$

$$\text{input to station} = \text{flow rate of gas fuel} \times \text{calorific value of gas}$$

$$= 33 \times 38$$

$$= 1254 \text{ MJ/s or MW}$$

Therefore

$$\text{overall efficiency} = \frac{684}{1254} \times 100 = \textbf{55 per cent}$$

The heat given up by the exhaust gases flowing through the HRSGs is 734 MW and this is equal to the change in enthalpy of the feedwater being raised to steam, i.e. for three HRSGs,

$$3 \times (H_s - 160) \dot{m} = 734 \times 10^3 \, \text{kJ/s}$$

where H_s = mean value of the enthalpy of the steam leaving each HRSG. Hence

$$H_s = \textbf{2940 kJ/kg}$$

Problems

(For properties of steam, *see* Table 18.2, page 403. Take mean specific heat capacity of superheated steam as 2 kJ/kg K, water 4·2 kJ/kg K and turbine exhaust gases, 1·15 kJ/kg K.)

20.26 A CCGT plant burns natural gas at the rate of 6 kg/s. Its calorific value is 38 MJ/m³ and density 1·24 kg/m³ at standard conditions on entering the combustion chamber. The output from the gas turbine generator is 81 MW and from the steam turbine 49 MW. Find the overall thermal efficiency of the plant.

If the plant operates between a maximum temperature in the gas cycle of 1000 °C and a minimum temperature of 36 °C in the steam cycle, what is the theoretical maximum possible efficiency that can be achieved?

(46 per cent; 76 per cent)

20.27 The gas and steam turbines of a CCGT station send out 220 MW from their generators to the grid system. The overall efficiency of the plant is 49 per cent and the calorific value of the gas fuel is 38 MJ/m³. What is the rate of consumption of fuel? The gas turbine output is 135 MW and the steam turbine is supplied with steam at the rate of 83 kg/s from the HRSG. If the heat energy of the steam at entry to the turbine is 3200 kJ/kg, estimate the gas and steam turbine thermal efficiencies.

(11·8 m³/s; 30 per cent; 32 per cent)

20.28 The HRSG of a CCGT module receives feedwater into its economizer and delivers steam to the low-pressure stage of a single steam turbine at the rate of 10 kg/s. The remaining steam from the superheater is delivered to the high-pressure stage at 50 kg/s. The enthalpy values are: feedwater, 230 kJ/kg; low-pressure steam 2400 kJ/kg; high-pressure steam 3340 kJ/kg. The efficiency of the steam turbine plant is 34 per cent. Find the heat energy *transferred* from the exhaust gases to the feedwater in passing through the HRSG and estimate the steam turbine output.   (177 MW; 64·9 MW)

20.29 The steam turbine generator set of a CCGT plant has an output of 155 MW. There are three gas turbines, each with an output at its generator of 120 MW. A single HRSG supplies highly-superheated steam to the turbine at 80 bar (enthalpy, 3447 kJ/kg), at the rate of 94 kg/s for the high-pressure stage and to the low-pressure stage at 7 bar, 192 °C and 113 kg/s. The heat flow rate into each gas turbine is 390 MJ/s. (i) What is the enthalpy of the low-pressure steam? (ii) What is the total energy in MJ/s of the steam entering the turbine? Neglecting losses, estimate the efficiency of the steam cycle part of the plant and the overall efficiency.

(2816 kJ/kg; 642 MJ/s; 24 per cent; 44 per cent)

20.30 A CCGT plant has four modules each consisting of a combustion chamber, gas turbine, HRSG and a generator. There are two steam turbine generator sets with each set being supplied with steam from two HRSGs. The feed-water enters each HRSG at 1·01 bar, 100 °C at the rate of 23 kg/s. Gas fuel (calorific value 38 MJ/m^3) is supplied to each of the combustion chambers at the rate of 7 m^3/s. The output from each gas turbine is 90 MW and from each steam turbine, 100 MW. Assuming that for each module 60 per cent of the input energy to the combustion chamber is transferred to the feed-water to produce steam for the HRSG, estimate (i) the total heat energy in MW of the steam leaving one HRSG, (ii) the steam turbine thermal efficiency and (iii) the overall efficiency of the plant.

(169 MW; 30 per cent; 53 per cent)

20.31 A module of a CCGT station consists of a gas turbine, steam turbine and generator carried on a single shaft. The gas turbine supplies exhaust gases to raise steam in one triple-pressure HRSG. The steam turbine thermal efficiency is 30 per cent; the steam leaving the HRSG has enthalpy (mean value) 3140 kJ/kg; the feedwater (enthalpy 350 kJ/kg) enters the HRSG at the rate of 97 kg/s. The gas turbine thermal efficiency is 29 per cent, measured from the input energy in the gas fuel intake to the combustion chamber. The flow rate of exhaust gas is 610 kg/s and the temperature of the gases entering and leaving the HRSG are t °C and 105 °C respectively. At standard conditions, the fuel has a calorific value of 38 MJ/m^3 and flow rate 19 m^3/s. Neglecting pressure and thermal losses and auxiliary power needs, find (i) the output power from the gas turbine, (ii) the output power from the steam turbine, (iii) the temperature of the exhaust gases entering the HRSG and (iv) the overall efficiency of the plant.

(209 MW; 91 MW; 491 °C; 42 per cent)

(20.13) Emissions from power stations

The chemical and engineering problems associated with controlling emissions from power stations and conserving fuel stocks are complex and expensive to solve. As soon as combustion begins, numerous chemical reactions are initiated and chains of reactions take place until the final products of combustion form. When these gases enter the atmosphere, further reactions occur. Besides hydrogen, oxygen and carbon, both fuels and air contain other elements. Most do not take part in the combustion process but may produce so-called 'greenhouse' gases.

Pollution to some extent depends on prevailing atmospheric conditions since the gases are affected by the movement of air, 'washing-out' of particles by rainstorms, etc. International legislation has set new lower limits on greenhouse gas emissions into the atmosphere based on a reduction of 12·5 per cent overall by the year 2010 on the 1990 levels and the UK for some gases such as CO_2 has set even stricter limits. Pollution of the air is measured in parts per million (ppm) by mass and there is a complex methodology as to particle size, measurements at source or end user, type of fuel, etc. Cutting pollution using present methods and criteria could result in more micron harmful particles in the flue gases.

Some of the factors affecting emissions from a power station are:

- condition of the plant;
- composition of the fuel;
- temperature of combustion;
- efficiency of combustion;
- atmospheric conditions.

Comparison of fuels

Fuels free of substances causing pollution are said to be 'clean'. Natural gas is referred to as a 'clean-burning' fuel but from some sources it may be 'sour' in that it contains sulphur, nitrogen and water. Most natural gas is almost free of sulphur and nitrogen; methane is its main constituent followed by ethane, hydrocarbons and other gases. North Sea (Bacton) gas is particularly clean, consisting of 95 per cent methane, a mix of hydrocarbons and some nitrogen. Coal contains sulphur and the primary pollutants produced by its combustion are oxides of sulphur and nitrogen, dust and ash. Of the fuels burned in power stations, per unit of power output, coal produces the largest amount of sulphur dioxide, followed by petroleum, gases in general and then natural gas. When pre-heated and burned efficiently, natural gas combustion products are free of sulphur dioxide and contain appreciably lower amounts of carbon dioxide and nitrogen dioxide than are produced by coal.

Greenhouse gases

The sun's energy in the form of solar radiation is reflected or absorbed by the earth. The amount of the sun's energy reaching the earth is called the *solar constant*, although it varies slightly over long periods. The heat reflected back into the atmosphere becomes involved in 'global warming' when it is held in the lower layers of the atmosphere because of the action of carbon dioxide, water vapour and other greenhouse gases. The high-temperature, low-wavelength radiation (light and ultra-violet) reaching the greenhouse gas layers from the sun, pass without difficulty through to the earth's surface, but the greenhouse gases partially absorb and reflect downwards the rising infra-red, long-wave radiation to produce a warming effect. In a greenhouse, the glass permits the short-wave radiation through to the plants inside but it is opaque to the resulting outgoing long-wave radiation. There must, of course, be a measure of 'greenhouse effect' for the earth to survive.

The most important greenhouse gas is carbon dioxide followed by water vapour. Ignoring water vapour, carbon dioxide occupies only 0·03 per cent of the lower atmosphere and the remaining gases take up less than 0·06 per cent; the other greenhouse gases, hardly more than traces, in order of magnitude, are methane, nitrous oxide (N_2O), chlorofluorocarbons (CFCs) and perhaps ozone. Carbon monoxide (CO) has an indirect effect through forming carbon dioxide (CO_2). Nitrogen oxides occur naturally, arising from the earth's soil but also from the use of fertilizers, manufacturing processes and, unfortunately, from the catalytic process used in cars. Methane also arises naturally as well as from land-fill sites, gas distribution pipes, etc. Water vapour's presence in the atmosphere depends almost entirely on the temperature and hence it varies greatly across the globe and at different layers of the atmosphere. The trace gases have a very low presence but a high shielding effect, particularly against infra-red radiation, much more so than carbon dioxide. Some polluting gases persist in the atmosphere for many years and in the case of CFCs, the longest lasting, it is thought to be hundreds of years; the production and use of CFCs are being phased out as rapidly as possible.

Carbon dioxide is abundant, persists in the atmosphere for a long time and an increasing amount is caused by industrial activity and vehicle emissions. The gas plays a large part in nature through plants, algae on the surface of the oceans, animals and humans and there is an enormous amount of global movement of this gas between the earth and its atmosphere. Trees, wood, vegetables and the oceans are all reservoirs and sinks of enormous, incalculable quantities of carbon dioxide. Trees are a part of the natural cycle regulating the atmosphere since they absorb the carbon dioxide as they grow and emit roughly the same amount when rotting down or burnt.

Oxides of sulphur

The sulphur in a fuel is the only source of the pungent gas, sulphur dioxide, in the products of combustion. After sulphur dioxide, further reactions lead finally with atmospheric rain to dilute sulphuric acid, i.e. acid rain. Coal-fired stations take measures to reduce pollutants, particularly the corrosive sulphurous gases, by treating both fuel and flue gases. The removal of dust and ash requires filters, scrubbers, settling chambers or precipitators. As much as possible of the sulphur in the fuel is removed using a material to absorb it in conjunction with a fluidized-bed combustor. For flue gases one method employs sea-water scrubbers but a more usual method is the wet limestone plant, retrofitted to the 4000 MW Drax power station aimed at removing 90 per cent of the sulphur dioxide produced; the station's flue-gas flows at 21 000 t/h, necessitating ducts 10 m in diameter and absorber towers 20 m in diameter, 40 m high, all specially lined to resist corrosion. The process requires a great deal of back-up work quarrying and delivering lime-stone. In all power stations the flue stack temperature has to be maintained above the condensation temperature of sulphur dioxide.

Oxides of nitrogen

Nitrogen, being an inert gas, does not burn with oxygen but may form oxides of nitrogen at high temperatures when combined with oxygen and there can be a chain of complex reactions concerning these oxides resulting in nitric acid. Ultraviolet radiation and collisions of molecules may split nitrogen dioxide into nitric oxide and oxygen. Ozone may react with nitric oxide to produce nitrogen dioxide. Such are the complexities of the interactions concerning oxides in the atmosphere.

Flame size, structure and peak flame temperature during combustion are important and for gas fuels 'low NO_x' burners have design features including axial and swirl streams of air–gas mixture to achieve optimum conditions for minimum discharge of the pollutants.

For solid fuels, one solution is conversion to clean gas using a combined air-gasification and air-separation plant; the separation unit produces oxygen which is mixed with steam as the gasification agent.

Ozone

Ozone (O_3), a rare, highly toxic gas, in general appears as slightly more than a trace in the lower atmosphere, even less at higher levels and its presence varies greatly across the globe. It occurs when oxides of nitrogen or hydrogen and unburned carbons react to sunlight or lightning. The gas is a three-atom molecule of oxygen resulting from the separation of the common two-atom oxygen molecule. It is unstable and the molecule is easily split, so that there is a constant interchange between the common oxygen and ozone molecules. The so-called 'ozone layer' is at the 20–25 km level and since it absorbs ultra-violet radiation it provides a shield for the earth's surface. Artificial CFCs are stable and inert but release chlorine and fluorine atoms which deplete the ozone in this layer. Another chemical, halon-1202, of unknown source, has also been found to attack ozone.

(20.14) Emissions from vehicles and aircraft

Emissions from vehicles, the cause of 'photochemical smog', are a main source of pollutants; smog results from the chemical combinations of oxides of nitrogen and unburned hydrocarbons at ground level under sunlight. Vehicles are now subject to increasing regulations on lead in fuel and on permissible levels of CO_2, CO, NO_x and unburned hydrocarbons. These regulations have positive and negative effects on fuel consumption, costs involved in fuel processing, thermal efficiency, etc.

Apart from less polluting alternative transport systems and legislation to reduce the number of vehicles on the roads, the solutions lie with various technologies and technical improvements being developed to reduce the output of polluting gases and conserve fuel. These include:

- all-electric vehicles for certain purposes;
- 'hybrid' vehicles, e.g. where a petrol, oil or turbine engine motoring at constant speed drives a generator to power an electric motor or recharge batteries;
- dual-fuel engines;
- the use of compressed natural gas, liquefied petroleum gas, hydrogen or compressed air as fuel;
- the hydrogen–oxygen fuel cell;
- light, efficient, more rapidly rechargeable batteries such as nickel–metal hydrides (compounds containing hydrogen and some other element) and lithium-ion types;
- advanced design of the common lead–acid traction battery based on boosting the charge to give a high-power, rapid recharging battery with a large number of recharge–discharge cycles and an increased range of journey at low cost;
- lean-burn technology with computer control of air–fuel ratio and ignition systems;
- direct injection systems;
- catalytic converters with exhaust traps for gases emitted during the warm-up periods when the cold converters are ineffective;
- chemical treatment of exhaust gases.

A wide variety of hybrid vehicles are being developed combining the internal combustion engine, fuel cell, batteries, electric motors and different fuels. The problems to be solved include reducing the weight incurred in dual systems, the provision of charging stations and producing such vehicles economically to have a reasonable range of journey per unit of fuel.

The presence of nitrogen oxides and unburned carbon in their exhaust gases is greater for spark-ignition engines than for those working on compression-ignition but the latter engines have the problem of emitting 'soot'. The products of combustion from internal combustion engines are affected by a number of factors such as ignition-timing and air–fuel ratios.

Aircraft engine emissions (which are not yet covered by international agreement) are substantial and increasing rapidly. Local concentrations of pollutants occur at take-off and landing but also, when aircraft cruise close to the stratosphere, at around 10 km altitude, they exhaust pollutants including NO_x gases which create ozone under sunlight; at this level the ozone acts as a greenhouse gas but the NO_x gases may at the same time eliminate any methane, another greenhouse gas, present in the atmosphere. Resolving the pollution problems of aircraft requires similar solutions as for vehicles but the question of additional weight is more critical with further costs for complex equipment and maintenance.

(20.15) Climate; natural forces; global warming

The earth's climate is governed primarily by the sun's radiation through the *difference* in the amount of heat received and reflected by the earth. The average global temperature has risen several degrees over the past century giving rise for concern, although it is considered by many scientists to be within the limits of normal climate variability; this is a most complex subject, however, about which to draw conclusions. The sun's slow cycle of variation of output, the occurrence of sunspots, the interaction of the solar winds with the magnetic fields of the sun and earth and the earth's tilt and wobble all have effects in varying degrees on the global temperature. The relevance of cloud formation and the location of the recording instruments when measuring these temperatures are important. Then there are the natural processes and events related to the earth's land–sea–atmosphere system, some of which are listed here:

- Earthquakes, tornadoes, volcanic eruptions: for example, there are hundreds of live volcanoes and a single major volcanic eruption throws up enormous quantities of pollutants producing clouds of dust that may circulate in the stratosphere for several years.
- Variations in the extent of the earth's polar ice-caps and sea-levels.
- The El Niño–Southern Oscillation, La Niña and North Atlantic Oscillation phenomena associated with the oceans, winds and air pressures (*see* below).

In terms of the natural forces listed above affecting climate, emissions of pollutants would appear to be only one of many contributors but unfortunately one of constant and ever-increasing significance in relation to greenhouse gases.

From the use of satellites and sophisticated, computerized monitoring instruments natural phenomena of all kinds are becoming better understood. For example, satellite images of the movements of ocean currents and stretches of warm and cold waters are now produced using information obtained from buoys moored in the oceans; these buoys transmit data on currents and sea-water temperatures at various depths, to their monitoring satellites. This particular information relates to the following brief notes on the natural climatic processes at work in the region of the Pacific and Atlantic Oceans. These notes serve to illustrate the important current arguments on global warming.

El Niño–Southern Oscillation (Enso)

El Niño

The Pacific Ocean is bounded by Australia and Indonesia on the west and stretches to the east 10 000 km across the vast, open waters before reaching the coasts of the South American countries. In so-called normal times there is usually a 'pool' of warm water in the western equatorial region, north of Australia, and at the eastern end flows the cold, deep, streams of the *Peru Current* (formerly the *Humboldt*); this great current travels northwards from the Antarctic along the coast of South America, passing Peru and then turning westwards when it reaches Ecuador, just below the equator. The cold waters of the current rise to the surface off the coasts of Peru, Ecuador and Chile and since they are rich in nutrients, plankton and oxygen the region benefits from thriving sea-vegetation, fisheries and bird colonies. In these circumstances there is usually a fairly regular pattern of climatic events, with a high atmospheric pressure over the western region and the prevailing north-easterly trade winds confining the warm water pool to the west, resulting in the sea level there being maintained some 20 cm or more above the level of the eastern equatorial waters. Occasionally the coastal waters off Peru become slightly warm around the Christmas period.

So far we have described the usual state of the climate in normal years but at intervals of several years the warming of the coastal waters off Peru becomes excessive, rising by as much as 7 °C. The name *El Niño de Navidad* for this event connected with Christmas derives from the Spanish for 'the boy child'. The interval between the *strong* El Niño years appears to be shortening and is now on average, about four to seven years. The El Niño varies in intensity on each occasion and usually lasts up to 18 months, but may be prolonged before the system corrects itself: the 1990–91 occurrence continued for nearly five years.

In a strong El Niño year the prevailing east to west trade winds weaken or die down altogether, allowing the warm water pool in the west to spread across the Pacific to the coasts of Peru, Ecuador and Chile; this warm water suppresses the cold waters of the Peru Current and prevents their *upwelling* so that the surface waters are deprived of enrichment by the nutrients. The most immediate effect in exceptional years, such as 1982–3 and more recently 1997–8, is on the land and coastal seas of Peru and Indonesia; the wider effects are experienced around the world in the form of devastating losses in marine and bird life, decimation of coral reefs, forest fires, crop failures, changes in desert areas and in the pattern of monsoons, floods, etc. Other countries, of course, such as East Africa, benefit in a number of ways from the El Niño event.

The recurring El Niño phenomenon primarily refers to the 'ocean element', i.e. the creation of an expanse of warm water predominantly off the coast centering on Peru.

Southern-Oscillation

There is an atmospheric component to the El Niño since it appears to be initiated by the loss or reversal of the trade winds and these are determined by the air pressures over the Pacific. The El Niño, therefore, is closely linked with the 'Southern-Oscillation' which refers to the seesawing of pressure over the south-east and central to western regions. When a high pressure exists over the south-east it is lower over the west and vice versa; these pressures 'oscillate', maintaining a kind of equilibrium and the strength of this oscillation is measured by an index registered by the difference in pressure over two places, Darwin and Tahiti. Such is the close relationship between El Niño and the Southern-Oscillation that they are now referred to as *Enso*.

La Niña

This is the reverse phenomenon to that of El Niño, when the prevailing east to west trade winds become exceptionally strong and extremely cold water is driven northwards to the eastern equatorial waters and as far as the central Pacific zone.

Again, these abnormal conditions are thought to be the cause of climatic extremes, particularly those affecting America.

North Atlantic Oscillation (NAO)

The atmosphere and the pattern of winds, currents and air pressures over the North Atlantic are generally chaotic but a similar pressure oscillation is observed to exist as for the Pacific Ocean. Its strength is measured by the difference in pressure over Iceland and the Azores, off Portugal. A low index corresponds to weak westerly winds with a high-pressure area over Iceland and is observed to bring one type of weather, whereas a high index with low pressure over Iceland and strong westerly winds brings another. This pressure oscillation is closely linked to large stretches of water at various depths found to have irregular (anomalous) temperatures, a degree or two higher or lower than normal and which are observed to 'move about' as separate bodies of water. The movements are in various directions but generally from west to east, on a timescale of years; these large, warm patches of water directly affect marine life. Another aspect of the North Atlantic is the enormous flow of water taking place from the depths of the ocean, southwards towards and around the South Pole region, then onwards to become the surface waters of the Pacific. As distinct from the El Niño and Southern-Oscillation association, the ocean component of the NAO is not yet defined but one school of thought suggests the oceans act as a memory-bank storing up the actions and effects of the climatic events that take place.

Global warming

The phenomena described above appear to involve fundamental long-term cycles of climate changes and events as well as dramatic short-term incidents. The importance of the subject is now recognized world-wide but how extreme the climatic events were over past centuries, how long they lasted and their relationship to global warming and ozone depletion are not yet fully known by any means. Scientists conflict with each other over the interpretation of events and their causes.

The arguments in support of the greenhouse effect being an increasingly vital factor in climate change refer to human interference in the biosphere in respect of world population increase, waste products and carbon dioxide emissions pouring into and from the earth, etc.

From recent tree-ring, ice-cores and past records of climate studies it is claimed that the climate in this century has been the most unusual in regard to global warming for several hundred years and that the evidence pointing to the cause (or the additional critical factor) being greenhouse gases rather than entirely natural processes has now been reinforced. However, there are many uncertainties and other climatologists interpret information from ice layers near the north polar cap as indicating that similar climatic events to those currently being recorded were common occurrences in the past.

Computer models are now simulating climate patterns of the past and present in attempts to predict future events. The usefulness of such models depends on the quality of the data supplied. Other approaches to aid prediction of climate are being pursued, including the study of changes in the condition of algae in the Pacific.

(20.16) Renewable sources of energy

Fossil-fuelled power stations across the world deliver the bulk of the power required for electricity generation with the remainder being supplied mainly by nuclear energy which has its attendant problems and the pollution-free large-scale hydro-electric power stations. Nuclear technology may advance in the future with safer designs of fission or fusion reactors. Hydro-electric schemes are limited by lack of suitable water power and have associated environmental problems.

Renewable sources of energy that are pollution-free include wind, fuel cells, solar cells, and wave and tidal power. Other sources which result in emissions are geothermal, burning of 'energy crops' (biomass) and gas from landfill sites. If all the sources of energy mentioned could be brought into action globally in a substantial manner there would no doubt be a fall in the demand made on fossil-fuelled stations; the demand for power, however, is constantly increasing and large numbers of pollution-free units would be required to replace even one mainstream power station. The prospects for wave power, in particular, are improving with the introduction of underwater power stations using new designs of 'windmill' type turbines anchored on the sea-bed to harness off-shore currents. The following sections cover briefly the prospects and state of the other main alternative power sources, i.e. fuel cells, solar cells and panels, wind power and nuclear energy.

Government legislation is now setting targets for the use of renewable energy sources. Safety, reliability, costs per kWh and the environmental costs are the elements of debate regarding any power source. The difficulties lie in the calculations of the true cost of the various power sources as instanced in the use of nuclear power when account is taken of the enormous costs arising from land use, radiation leaks and decommissioning of plant over perhaps 100 years.

(20.17) Fuel cell

The *fuel cell* is increasing in importance as an alternative power source to small power plants and large storage batteries. The cell converts chemical energy *continuously* to electrical energy using a controlled reaction between oxygen or air and a fuel, typically pure hydrogen. Catalysts such as nickel plates are needed to speed up the reactions and a connecting acid medium, commonly phosphoric acid, permits the passage of *ions* to and from the plates.

The principle of operation has been known since 1839 from the demonstration cell shown by Sir W. Grove but the first practical applications were in the *Gemini*

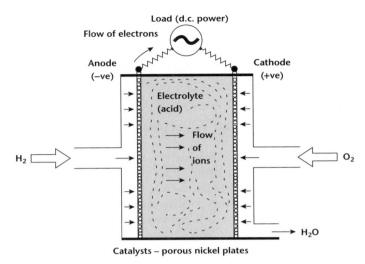

Fig. 20.8   Fuel cell

and *Apollo* spacecraft, culminating in the Space Shuttle's use of cells providing over 7000 hours of useful life with an output of about 4·5 kW.

A typical H_2–acid–O_2 cell has two *porous* plates, the positive and negative *electrodes*, separated by acid, the *electrolyte*, Fig. 20.8. Hydrogen bubbles through one plate and oxygen through the other so that the gases are in contact with the plates and acid at the same time. Connecting the electrodes by an external circuit causes a different reaction at each nickel/acid interface. Charged hydrogen ions flow through the acid electrolyte to the positive side of the cell and electrons through the external circuit. A simple description is that the external result is direct current (d.c.) in the external circuit and the internal result is the formation of water (a useful by-product) at the negative side, plus a small amount of heat. The chemical equation for the *overall* reaction is

$$2H_2 + O_2 \rightarrow 2H_2O + d.c. \text{ power}$$

The oxygen is available in the surrounding air but the storage of hydrogen fuel presents problems. It may be stored in steel cylinders supplied through 'fuel reformers' which convert natural gas to methane and then hydrogen or it may be stored in metal hydrides that act like a sponge; these hydrides are compounds containing hydrogen and an element such as palladium that absorbs hydrogen. Another example, where a cell is used in a vehicle, employs an aluminium foam holding a metallic nickel powder which releases hydrogen when heated slowly.

The cells are relatively heavy, pollution-free, inherently reliable and silent in operation. The thermodynamic limitation (Carnot) on efficiency does not apply and efficiencies of 60–70 per cent can be achieved on full or part load. However, power is limited so far to 2 MW and a large number of cells in series are required to obtain enough power per unit volume of fuel to be economical. Step-up transformers cannot operate with d.c. output so that other devices have to be used.

The design of a cell takes into account the situation in which it has to operate, e.g. in an isolated region, hospital, car, submarine or small heat-and-power plant. A variety of plate materials and acids are in use and in some designs the acid is replaced by a 'plastic resin membrane' or a 'proton exchange membrane' in order to prevent the gases from igniting.

20.18 Solar (photovoltaic) cells: solar collectors

These semiconductor cells operate by absorbing the sun's radiant energy. Light is converted to electricity in a cell using small pieces of light-sensitive silicone crystals sandwiched between sheets of glass; each cell contains two different types of silicon, the upper layer being the positive-charged layer and the lower the negative-charged. Light striking the cell frees electrons from the atoms of the silicon, electrons flow and a current is generated. Large arrays of cells are combined to form panels. Constant advances are being made in the technology and materials involved and there is now the non-crystalline amorphous type mounted in flexible steel sheeting; this type of cell is lighter, less expensive, easily processed and more suitable for roof panels. A great deal of energy is expended in constructing such cells, particularly the crystalline type, so that it is the net energy gained that is important. The panels depend on their location and orientation as to sunlight levels, the power output is intermittent and the cost per unit of power produced is relatively high. Apart from their use in isolated regions and such places as satellites there are some large-scale projects including a facade of $400\,m^2$ yielding over $30\,MW\,h$ per year, a large solar roof supplying $340\,kW$ of supplementary power to the grid system and a commercial 'solar' building with a $900\,m^2$ facade having $530\,m^2$ of active cells generating $55\,MW\,h$ of electricity per year.

Solar flat-plate collector panels and parabolic mirrors which concentrate the sun's rays for water and space heating face the problems of cost, intermittent supply, suitable positioning and net energy gain.

20.19 Wind power

In certain geographical situations, rural 'wind farms' are being created throughout the world. The environmental impact if vast numbers of these farms were to be constructed would be considerable. There are also large off-shore developments and some off the coast where the turbines are mounted high on rocks.

Wind is intermittent and diffuse; therefore the first requirement for success with wind turbines is to have them in a position to take advantage of high average wind speeds, around 30–$40\,km/h$. A second requirement is that the blades or 'sails' should sweep as large an area as possible. The power output is directly proportional to the *square* of the diameter of the circle swept out and to the *cube* of the wind speed.

A turbine generates power on average for less than three-quarters of its useful life and the maximum power output from a turbine is about $3\,MW$ (in Orkney);

UK output from wind farms is expected to reach 2000 MW in due course. For example, National Power's turbine sites currently exceed 120 MW output of commercial electricity. These figures have to be judged alongside 1500 MW from one fossil-fuel fired station.

The modern turbines, developed from water and corn-grinding windmills, now have paired aerodynamically contoured blades 50–60 m in diameter carried on towers up to 80 m high. The rotor shaft usually rotates about a horizontal axis at a relatively low speed of 1 rev/s and is directly coupled to an electrical generator; since the generator operates at a high speed a gear box is required. Electricity cannot be stored in bulk so preferably output should match demand and storage batteries are necessary unless the excess electricity can be taken by the grid.

A new technology incorporates an old idea – smaller diameter (about 7 m) shrouded blades – but using modern, light and extremely strong materials for the shrouds. The design results in a region of low-pressure air behind the blades so that air is drawn through the blades at a greatly increased speed and since the power varies with the cube of the speed, the output is increased several times over. The small diameter blades reduce the required height of the supporting towers to about 40 m. The action of one turbine affects others adjacent to it so that with an array of turbines in a field, the spacing, layout and orientation have to be considered carefully.

There are objections to wind farms apart from the cost per unit of power output. These are intermittent supply, take-up of land, dealing with excess power, noise and visual aspects. One source of noise arises from the gearbox, caused by the harmonics of the gear tooth contact frequencies. A reduction in noise can be achieved through gearbox design, isolating the box from the structure, accoustic enclosures and also by employing auxiliary mass-damping techniques using mass-spring resonators or damping plate devices.

(20.20) Nuclear power plant

Figure 20.9 illustrates the layout of a gas-cooled nuclear power plant of the Calder Hall type. The heat source is a nuclear reactor contained in a steel pressure vessel and cooled by carbon dioxide gas under a pressure of about 2 MN/m². The gas coolant is circulated through the reactor and four boilers by an electrically driven gas circulator. Heat is given up by hot gas to the water and steam in the tubes of the boiler. The latter consists of an economizer, an evaporator and a superheater. Hot water from the economizer passes into a steam drum and from there into the evaporator. A mixture of steam and water passes from the evaporator back into the steam drum where the steam is separated and passed on into the superheater. The water is recirculated continuously through the drum and the evaporator. Dry steam from the superheater is supplied to the turbine where its heat energy is utilized to provide the mechanical energy required to drive the electrical alternator. Steam from the turbine is condensed in the condenser and returned by the boiler feedpump to the economizer.

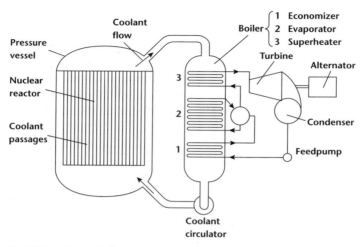

Fig. 20.9   Gas-cooled nuclear power plant

Another example is the Sizewell B station which has a pressurized-water reactor moderated and cooled by water with two steam-driven turbine generators, producing 3000 MW of heat, resulting in 600 MW of electricity. In this most common type of reactor, water under pressure in a sealed vessel is superheated in the reactor core and its heat is then transferred in a heat exchanger to raise steam for the turbines.

The construction of nuclear power stations is in abeyance because of rising costs of construction relative to other power stations but mainly because of the difficulties of radiation, disposal of waste, the possibility of accidents and the enormous costs of decommissioning of plants of present design.

Chapter 21

The prime mover

*This chapter is intended as an introduction to the technology of **prime movers**, devices for the conversion of energy to mechanical work. Two examples, steam and gas turbines, were considered in the previous chapter as elements in power-generation plants; in vehicles the prime movers are mainly petrol and oil engines. Only the elementary principles concerning prime movers and the methods of measuring their performance can be dealt with here as each of them requires specialist textbooks to be properly covered. The problems of emissions of pollutants from internal combustion engines has been discussed in section 20.14 with reference to possible solutions in regard to innovations, design changes and new 'hybrid' forms of engines.*

21.1 The steam turbine

The shaft of a turbine is rotated by the action of flowing steam on a ring of blades. Steam turbines are generally classified according to the design or arrangement of the blading. Figure 21.1(a) shows a section through the blading of an *impulse turbine*. Steam expands in a suitably shaped stationary nozzle. The potential energy of the steam, possessed by virtue of its pressure, is thus converted into kinetic energy of motion. The steam leaves the nozzle with a high velocity and enters a ring of moving blades. The flow passages in the moving blades are parallel and do not form a nozzle. There is no further drop in pressure through the blades and therefore no further expansion of the steam. The blades are curved, however, in order that some of the kinetic energy of the steam is transferred to the moving blades by the impact or *impulse* of the steam on them.

To extract the maximum amount of energy from the steam using only one set of blades, however, would require too high a speed. The energy conversion is therefore divided into a number of *stages*, each stage having alternative fixed and moving blades. The fixed blades merely serve to redirect the steam on to the next stage of moving blades.

Figure 21.1(b) shows the blades of a *reaction turbine*. The nozzles in this case are replaced by a ring of fixed blades. The steam expands in the fixed blades and leaves with a high velocity. The moving blades are also shaped to serve as nozzles. In passing through the moving blades the steam expands. An increase in velocity due to expansion produces, by Newton's third law of motion, a reaction force on the moving blades, whereby they gain kinetic energy.

In practice, pure reaction or impulse blading is not often used. The fixed blades between the moving rings are not simply guide blades but are also shaped to provide expansion of the steam and act as nozzles. The action is then a combination of impulse and reaction.

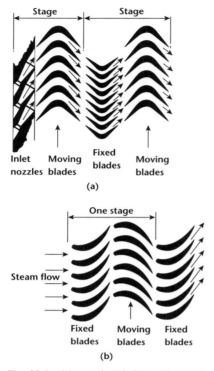

Stage   Stage

Inlet   Moving   Fixed   Moving
nozzles blades   blades   blades

(a)

One stage

Steam flow

Fixed   Moving   Fixed
blades   blades   blades

(b)

Fig. 21.1   (a) Impulse blading; (b) reaction blading.

The *overall* thermal efficiency of a steam turbine was explained in Chapter 20 as part of a power-generation plant. The energy input to the turbine was the enthalpy of the steam supplied by the heat-recovery steam generator in a CCGT plant or by the boiler in a fossil-fuelled plant and the output was the power delivered to the electrical generator. The thermal efficiency of a steam turbine is low unless various devices are employed; these include the use of low-, intermediate and high-pressure stages as explained above and also a reheat cycle where steam is taken off at some intermediate point and heat added to it before re-entering the turbine to complete its expansion. As for all thermodynamic cycles the efficiency can be improved by increasing the difference between the maximum and minimum temperatures of the working fluids. There are a number of other efficiencies to be considered in steam turbine design such as the efficiency of the blades and of the separate stages.

21.2 Cycle of operation in a reciprocating engine

The reciprocating engine, Fig. 21.2, usually consists of a cylinder A, piston B, connecting rod and crank. For each revolution of the crank the piston makes two

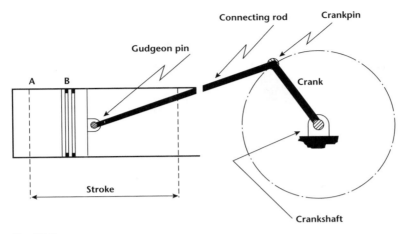

Fig. 21.2

strokes, a stroke being a single movement of the piston from one end of its travel to the other. The length of the stroke is, of course, *twice* the length of the crank. The positions of the piston, or crank, at the ends of each stroke are known as the *outer* or *bottom dead centre* (*o.d.c.* or *b.d.c.*) and *inner* or *top dead centre* (*i.d.c.* or *t.d.c.*) respectively. The terms inner and outer usually apply to horizontal engines and the terms top and bottom to vertical engines. These points are the points of least and greatest cylinder volumes respectively.

An engine *cycle* is defined as the sequence of operations required to perform mechanical work. Each cycle usually requires either two or four strokes of the piston, and hence one or two revolutions of the crank.

When the cycle of operation takes place on one side of the piston, the engine is said to be *single-acting*; in a *double-acting* engine a cycle of operations takes place on both sides of the piston.

(21.3) The internal combustion engine

In the *internal combustion* (*IC*) engine the working fluid is air and burnt gases. The fuel is admitted to, and burnt inside, the cylinder. IC engines are classified as four-stroke or two-stroke engines, according to the cycle of operations used. They may also be classified as gas, petrol or oil engines according to the fuel used. A further classification depends on the method of ignition; in petrol and gas engines a fuel–air mixture is ignited by a spark and they are known as *spark-ignition* engines; in oil or diesel engines air alone is compressed to a high temperature and fuel is ignited by injection into the hot air, hence the term *compression-ignition* is used for such engines.

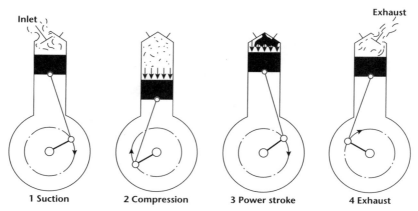

Fig. 21.3 Four-stroke cycle

(21.4) **Petrol engine: four-stroke cycle**

The sequence of operations in a four-stroke petrol engine, Fig. 21.3, is as follows:

1. *Suction or induction stroke*. Induction of a petrol–air mixture into the cylinder.
2. *Compression stroke*. Compression of the mixture to a high temperature. (At the end of compression the mixture is ignited by means of a spark and the pressure in the cylinder rises rapidly.)
3. *Expansion, working or power stroke*. The burnt gases expand doing work on the piston.
4. *Exhaust stroke*. The burnt gases are exhausted from the cylinder.

Each of the four operations takes place during one stroke of the piston. The operation of the cylinder valves is as follows:

suction stroke	inlet valve open, exhaust valve closed
compression stroke	both valves closed
expansion stroke	both valves closed
exhaust stroke	inlet valve closed, exhaust valve open

The *timing* of the opening and closing of the valves is most important for the correct running of the engine.

The cycle of operations for a four-stroke engine requires two revolutions of the crank. Therefore, if an engine crank rotates at n rev/s the number of cycles completed per second is $\frac{1}{2}n$.

Note that there is only one power or working stroke in every cycle, the remaining three strokes simply being necessary to prepare the engine for the power stroke.

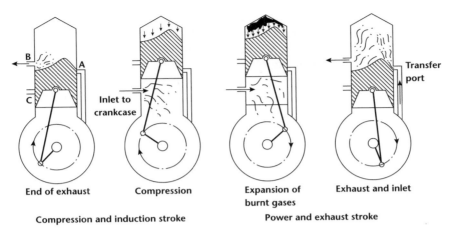

End of exhaust   **Compression**   **Expansion of**   **Exhaust and inlet**
                                                 burnt gases

Inlet to
crankcase

Transfer
port

Compression and induction stroke   **Power and exhaust stroke**

Fig. 21.4   Two-stroke cycle

21.5 Petrol engine: two-stroke cycle

Figure 21.4 illustrates the operation of a two-stroke petrol engine, the valves being replaced by ports which are opened and closed by the movement of the piston. The same cycle of operations is carried out as for the four-stroke engine but it is completed in *two* strokes of the piston or *one* revolution of the crank. The number of cycles of operations per second, i.e. the number of working strokes per second, is therefore equal to the number of revolutions per second of the crank. The strokes are as follows:

1. *Compression and induction stroke.* The inlet or transfer port A and the exhaust port B are closed in succession and the petrol–air mixture in the cylinder is compressed as the piston moves upward. At or near the top dead centre position the mixture is ignited and the pressure rises rapidly. Also, as the piston moves upwards, a fresh supply of air and petrol is drawn into the crankcase below the piston from the inlet port C.

2. *Power and exhaust stroke.* On the top side of the piston the high-pressure burnt gases expand doing work on the piston. Near the end of the power stroke, the exhaust port B is uncovered by the piston and the burnt gases exhaust to the atmosphere. The inlet port A is then uncovered in turn and a fresh mixture, which has been slightly compressed in the crankcase, is forced into the cylinder, thus displacing or scavenging the remaining exhaust gases. At bottom dead centre position, the cycle is completed.

21.6 The oil engine

The *oil* or *diesel* engine operates on a very similar cycle (two-stroke or four-stroke) to that of the petrol engine. The major difference is that air only is admitted to the cylinder during the suction stroke. The fuel is injected into the cylinder at the end of the compression stroke and ignition occurs spontaneously due to the high temperature of the compressed air. The fuel injector therefore replaces the spark plug of the petrol engine; the fuel is oil in place of petrol.

The construction of the oil engine is very similar to that of a petrol engine but is more robust to allow for the greater pressures involved in the cylinder. Oil engines cover a wide range in size and purpose varying from single-cylinder, four-stroke, normally aspirated engines to very efficient two-stroke, single-acting, turbo-charged multi-cylinder engines. An example is a 12-cylinder in-line engine running at below 200 rev/min with output power up to 40 MW. Cylinders may be arranged in several forms, e.g. *vee* formation, or *opposed pistons* where the air charge is compressed between opposed pistons.

(21.7) The gas engine

Petrol and oil engines are in use throughout industry and transport but the reciprocating gas engine has a limited role in particular situations, for power generation and combined heat and power plant. The gas engine has the advantage of low-level emissions and is similar in design and construction to petrol and oil engines. Although many are specially designed, some are simply modified diesel engines. On off-shore oil platforms and well-heads in general, spark-ignition, four-stroke engines utilize by-product natural gas employing a compression ratio of about 9·5:1. Also in use are dual-fuel compression-ignition engines where both oil and gas are available. When using gas, the gas fuel is mixed with a small quantity of 'pilot' oil to enable ignition to take place by compression of the charge. Dual-fuel engines are often used where standby power generation is required, such as in hospitals. Typical modern examples of gas engines range from a six-cylinder in-line engine, water or air-cooled, running at 800 rev/min with a brake output of 1 MW, up to a very large engine having eighteen cylinders, in-line or 'vee' formation, turbo-charged, water-cooled, giving 4 MW at 1500 rev/min. Some engines are designed to take the gas from large landfill and sewage-treatment sites and from other sources such as coal-mining operations where drained methane is available. In many cases the waste heat is utilized, for example, in assisting digesters in water-treatment works to break down material.

Gas for use in an engine must have a combustible component of at least 95 per cent methane by volume and a calorific value greater than about 19 MJ/m^3 at NTP. Natural gas from North Sea fields has a calorific value 38·5 MJ/m^3, from other fields the range is 33 to 44 MJ/m^3 and for sludge and landfill gas the value may be as low as 20 MJ/m^3.

(21.8) Ideal air standard cycles for IC engines

There are a number of theoretical cycles on which engine design and analysis of performance can be based; in all cases air is assumed to be the working fluid and to act as a perfect gas. For example, the *Otto* cycle is the *ideal air standard cycle*

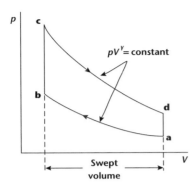

Fig. 21.5

for spark-ignition, gas and high-speed compression-ignition engines. It consists of four thermodynamically reversible operations on the air going through the cycle: (i) adiabatic compression, (ii) heating at *constant volume*, (iii) adiabatic expansion and (iv) cooling at *constant volume*. These operations are represented in a pressure–volume graph, Fig. 21.5, by **ab**, **bc**, **cd** and **da** respectively. The pressure–volume diagram for a practical engine is described in section 21.18 and shown in Fig. 21.10. Although similar to that of the ideal Otto (constant volume) cycle it differs for a number of reasons: (i) the working fluid is not a perfect gas, (ii) combustion does not take place instantaneously at constant volume and (iii) engine valves take time to open and close.

There are other ideal cycles such as the *diesel* and *dual-combustion* cycles for IC engines.

21.9   The rotary IC engine

The *rotary* engine is an attempt to eliminate the crank, connecting rod and reciprocating piston. Many attempts have been made to develop the principle of the rotary engine to try to realize its several advantages, but so far without real commercial success. Figure 21.6 shows the basic elements of the early German NSU Wankel engine.

The specially shaped triangular rotor A turns eccentrically in the casing B. The air inlet port is at C, and the exhaust at D. Clockwise rotation of the rotor admits air–fuel mixture at C until this port is eventually closed. The shape of the rotor then ensures that the air–fuel mixture is compressed as the rotor turns further beyond the point at which the inlet port is closed.

When the mixture is fully compressed the spark plug is uncovered, as shown, and fired. The expanding gases cause rotation of the shaped rotor 'piston'. Finally, the spent gases escape through the port D.

Although the mechanical operation is quite different from that of the reciprocating engine, the fundamental sequence of operations on the working fluid is the same.

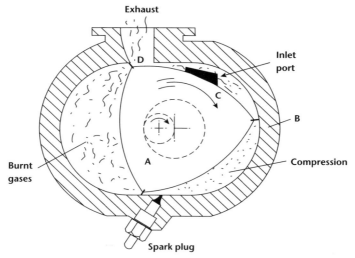

Fig. 21.6   Rotary IC engine.

The rotary engine is light weight, having one, two or three rotors, has a good power-to-weight ratio and is relatively maintenance-free now that sealing problems have been mostly overcome. Many designs have been produced to try to reduce the exhaust emissions and improve fuel consumption. Applications of rotary engines are to be found in light-weight, low- to medium-power situations such as motor-cycles, air compressors and portable generating sets. One application was the Norton 'Commander' motor-cycle, which was equipped with a liquid-cooled, twin-chamber engine, giving 63·4 kW at 9000 rev/min.

21.10 The gas turbine

In the reciprocating IC engine, all the operations involved on the working fluid take place intermittently in the cylinder of the engine. Thus the working fluid is drawn in, compressed, ignited, expanded and exhausted within the cylinder. The gas turbine is an internal combustion engine but the various operations of compression, ignition and expansion are carried out in separate components. The method of operation of a gas turbine is similar to that of a steam turbine, the steam being replaced by burnt gases from a combustion chamber. Often, as distinct from a steam turbine, only a single stage is used, the gas turbine thereby having a very high speed. Figure 21.7 shows diagrammatically the layout of a gas turbine for aircraft propulsion. The flow of air into the engine and the exhaust of burnt gases from the engine are continuous. The air is first compressed in a rotary compressor before passing to a combustion chamber where the fuel is injected and ignited. The hot burnt gases expand through the blades of a rotating turbine thus doing work before being exhausted. References to gas turbine operations and reaction propulsion will be found in Chapters 8, 9, 10 and 20.

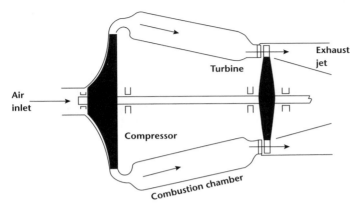

Fig. 21.7   Gas turbine layout

The propelling force (thrust) on the aircraft is due to the rate of change of momentum produced in the air flowing through the engine. Such an engine is used in a *turbo-jet* aircraft where all the air flows through the combustion chamber and is ejected as a high-speed jet. In a *turbo-fan* or *bypass* engine there is a large ducted fan on the shaft ahead of the compressor, both fan and compressor being driven by the turbine. Some of the intake air flows through the fan without combustion and meets up with the burnt gases in the exhaust jet. The thrust is therefore made up partly from the jet force and partly from the force produced by the fan blades. The Rolls-Royce turbo-fan engine RB211, for example, supplies a take-off thrust of nearly 300 kN.

There is no output shaft on an aircraft gas turbine so that measurement of output power cannot be carried out. Thrust, therefore, is used instead of power to assess performance. A gas turbine is not self-starting but must be run up to the speed required by a starter motor. It has a low thermal efficiency but is versatile and of simple construction. The efficiency depends on the maximum temperature of the working fluid and also on the ratio of the maximum to minimum temperatures in the cycle of operations. The efficiency can be improved in industrial power generation situations where the heat energy of the exhaust gases can be utilized. For an industrial turbine the output for external use is the net output at the turbine shaft or generator since the turbine shaft drives the compressor and auxiliaries.

(21.11) The reciprocating engine: definitions

The **bore** of a cylinder is the internal diameter of the cylinder.

The **stroke** of an engine piston is the distance between the inner and outer dead centre positions, and is equal to twice the length of the crank.

The **swept volume** V_s is the volume swept out by the piston during each stroke. Thus, if A is the area of cross-section of the piston and l the length of the stroke, then

$$V_s = A \times l$$

The **clearance volume** V_c is the volume of the space between piston and cylinder head at the inner dead centre position.

The **cylinder volume** V is the sum of the swept volume and the clearance volume. It is the volume of the cylinder when the piston is at the outer dead centre position, i.e.

$$V = V_s + V_c$$

The **compression ratio** r is the ratio of the volume of gas trapped in the cylinder at the instant the inlet valve closes, to the clearance volume. For simplicity, we may assume the inlet valve to close at the outer dead centre position. Therefore

$$r = \frac{\text{cylinder volume}}{\text{clearance volume}}$$

$$= \frac{V}{V_c}$$

The compression ratio used in a petrol engine is usually less than 9:1 and is determined mainly by the composition of the fuel and by the fact that too high a compression ratio causes some mixtures to ignite spontaneously and violently. For oil engines, in which air alone is compressed, much higher compression ratios may be used, up to about 20:1.

(21.12) Work done on an engine piston: indicated power

Let p (N/m²) be the pressure on an engine piston, Fig. 21.8, and A (m²) the area of cross-section of the piston. Then the total force on the piston is

$$P = p \times A \text{ (N)}$$

If the pressure is constant while the piston moves over a distance equal to the length of the stroke l (m) the work done is force × distance, i.e. $P \times l$, or

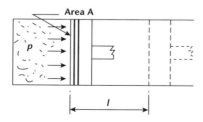

Fig. 21.8

work done $= P \times l$

$= p \times A \times l$

$= plA$ (N m)

If the piston makes n_f *working strokes* per second then the work done per second is

$plAn_f$ (J/s or W)

i.e.   power developed $= \dfrac{plAn_f}{1000}$ kW

In practice, the pressure in an engine cylinder is not constant, but varies throughout the cycle. However, a *mean effective pressure p* may be defined so that the formula $plAn_f/1000$ gives the power in kilowatts developed in the engine cylinder. The power calculated in this way is that developed in the cylinder and is called the *indicated power* (IP). The indicated power is calculated from an indicator diagram. Thus

$$\textbf{indicated power} = \dfrac{plAn_f}{1000} \textbf{ kW}$$

where p is the mean effective pressure of the cycle.

Note that n_f is the number of *firing* strokes per second, i.e. the number of cycles of operation completed per second. In a two-stroke engine, there is one firing stroke per revolution of the crank; in a four-stroke engine there is one firing stroke in two revolutions of the crank; in double-acting engines one cycle of operations is completed on *both* sides of the piston in one revolution of the crank. Thus at n rev/s,

for a single-acting 2-stroke engine $n_f = n$
for a single-acting 4-stroke engine $n_f = \frac{1}{2}n$
for a double-acting 2-stroke (oil) engine $n_f = 2n$

Most engines are single-acting but some oil engines are two-stroke, double-acting. Unless otherwise stated, assume single-acting. Four-stroke engines are the most popular and in general use, whereas two-stroke designs, noted for lack of economy in the use of fuel, are at present used either for low-power petrol and oil engines or for very large power slow-speed marine diesels. The expression 'slow-speed' applies up to about 200 rev/min, 'medium-speed' from 200 to 1000 rev/min and 'high-speed' upwards of 1000 rev/min.

(21.13) Brake power; friction power

The indicated power is the power available at the piston and is measured by an indicator diagram. Because of mechanical friction, the power available at the crankshaft is less than this by the power expended against friction at the bearings and sliding parts. The power output of the engine at the crankshaft is called the *brake power* (BP); so called since it is measured by a brake at the crankshaft (section 21.19). It is also referred to as the *shaft power*.

The *friction power* (FP) is the power taken up in overcoming friction and is the difference between the IP and BP; thus

FP = IP − BP

(21.14) Mechanical efficiency

The *mechanical efficiency* η_m of an engine is defined as the ratio of the power output at the crank to the power available at the piston, i.e. the ratio of the brake to the indicated power; thus

$$\eta_m = \frac{BP}{IP}$$

(21.15) Brake mean effective pressure

The indicated power is calculated from the formula

$$\text{indicated power} = \frac{plAn_f}{1000} \text{ kW}$$

The brake power may be calculated from a similar expression

$$\text{brake power} = \frac{p_m lAn_f}{1000} \text{ kW}$$

where p_m is called the *brake mean effective pressure* (BMEP). Since the ratio of BP to IP is the mechanical efficiency,

$$\textbf{mechanical efficiency} = \frac{\textbf{BMEP}}{\textbf{IMEP}}$$

The BMEP is a fictitious pressure, which is less than the actual indicated mean effective pressure in the cylinder since the mechanical efficiency is always less than 100 per cent. The BMEP may be considered as that portion of the IMEP that produces useful work. The BMEP is often used when comparing the performance of one engine with that of another.

Example

A four-cylinder, four-stroke oil engine develops a brake power of 60 kW at 2000 rev/min. The cylinder bore is 75 mm, stroke 100 mm, compression ratio 12:1. Calculate the clearance volume of the cylinder and the brake mean effective pressure. If the IMEP in each cylinder is 2·4 MN/m², find the mechanical efficiency and the indicated power.

Solution

Swept volume $V_s = l \times A$

$$= 100 \times \frac{\pi}{4} \times 75^2$$

$$= 442\,000\,\text{mm}^3 = 0\cdot000\,442\,\text{m}^3$$

Let V_c be the clearance volume (mm^3); then

total volume $V = V_s + V_c = 442\,000 + V_c$

Compression ratio $r = 12 = \dfrac{V}{V_c}$

or

$$12 = \frac{442\,000 + V_c}{V_c}$$

Hence

$$V_c = 40\,200\,\text{mm}^3 = 40\cdot2\,\text{cm}^3 = \mathbf{0\cdot000\,04\,m^3}$$

For each cylinder,

$$\text{BP} = \frac{p_m l A n_f}{1000}$$

The BP *per cylinder is* 60/4, i.e. 15 kW, and n_f, the number of working strokes per second, is $\frac{1}{2} \times 2000/60$, i.e. 16·67, since the engine works on the four-stroke cycle. Therefore

brake mean effective pressure $p_m = \dfrac{1000 \times \text{BP}}{lAn_f}$

$$= \frac{1000 \times 15}{0\cdot000\,442 \times 16\cdot67}$$

$$= 2\cdot04 \times 10^6\,\text{N/m}^2$$

$$= \mathbf{2\cdot04\,MN/m^2} \quad \text{or} \quad \mathbf{20\cdot4\,bar}$$

Mechanical efficiency $= \dfrac{\text{BP}}{\text{IP}} = \dfrac{\text{BMEP}}{\text{IMEP}} = \dfrac{2\cdot04}{2\cdot4} \times 100$

$$= \mathbf{85 \ per \ cent}$$

$$\text{IP} = \frac{\text{BP}}{\text{mech. eff.}} = \frac{60}{0\cdot85} = \mathbf{71\,kW}$$

Example

A compression-ignition engine has a bore of 100 mm, stroke 120 mm. The indicated mean effective pressure is 5 bar. Calculate the indicated power at a crank speed of 60 rev/s. If the mechanical efficiency at this load and speed is 84 per cent, calculate the brake power output and the power lost in friction. The engine operates on the two-stroke cycle.

Solution

$$\text{IMEP} = 5\,\text{bar} = 5 \times 10^5\,\text{N/m}^2$$

$$\text{Stroke} = 120\,\text{mm} = 0\cdot12\,\text{m}$$

$$\text{Piston area } A = \frac{\pi}{4} \times 100^2 \times 10^{-6} = 0\cdot007\,85\,\text{m}^2$$

The number of working strokes per second, n_f, is equal to the number of revolutions per second.

$$\text{IP} = \frac{plAn_f}{1000}\,\text{kW}$$

$$= \frac{5 \times 10^5 \times 0\cdot12 \times 0\cdot007\,85 \times 60}{1000}$$

$$= \mathbf{28\cdot3\,kW}$$

$$\text{BP} = \text{IP} \times \text{mechanical efficiency}$$

$$= 28\cdot3 \times 0\cdot84$$

$$= \mathbf{23\cdot8\,kW}$$

$$\text{FP} = \text{IP} - \text{BP}$$

$$= 28\cdot3 - 23\cdot8$$

$$= \mathbf{4\cdot5\,kW}$$

Problems

21.1 The engine of a motor car has as bore of 82·55 mm and stroke 79·5 mm. The compression ratio is 7·8:1. Calculate the swept volume and clearance volume for each cylinder in cm³.   (426 cm³; 62·6 cm³)

21.2 The engine of a motor car has a bore of 82·5 mm and stroke 76 mm. The engine has four cylinders and works on the four-stroke cycle. At 4000 rev/min the brake power developed is 59 kW. Calculate the BMEP.

(1090 kN/m²)

21.3 A six-cylinder oil engine, double-acting, two-stroke, has an output of 4·08 MW at 196 rev/min. The cylinder bore is 380 mm and stroke 1100 mm. Find the mechanical efficiency and average IMEP if the friction power loss is 690 kW.   (85·5 per cent; 9·75 bar)

21.4 Calculate the indicated power of a 12-cylinder diesel engine working on the four-stroke cycle at 400 rev/min. For each cylinder the bore is 570 mm, stroke 620 mm and IMEP 27 bar.   (17·1 MW)

21.5 A four-cylinder, four-stroke internal combustion engine has a bore of 153 mm, stroke 203 mm. The mean effective pressure during a working stroke is 420 kN/m² at a crank speed of 20 rev/s. Calculate the indicated and brake powers assuming a mechanical efficiency of 80 per cent.

(62·9, 50·3 kW)

21.6 A 12-cylinder diesel engine, single-acting, two-stroke, has a brake output of 29·5 MW at 78 rev/min. The friction loss is 3·5 MW. Find the indicated power, mechanical efficiency and average IMEP. The cylinder bore is 840 mm and stroke 2900 mm.

(33 MW; 89 per cent; 13·2 bar)

21.7 A two-stroke engine for a light-weight cycle develops 1 kW (IP) at 4000 rev/min. The bore is 38 mm and the stroke 42 mm. Calculate the indicated mean effective pressure at this speed. If the mechanical efficiency is 85 per cent, what is the BMEP?

(315 kN/m²; 267 kN/m²)

21.8 During a test on a four-cylinder, two-stroke oil engine the power developed at 50 rev/s was 80 kW (BP). The stroke is 150 mm and the piston diameter 76 mm. Calculate the brake mean effective pressure. If the mechanical efficiency is 86 per cent, what is the indicated mean effective pressure?

(588 kN/m²; 683 kN/m²)

21.9 An 18-cylinder engine running on natural gas, four-stroke cycle, at 1500 rev/min, has an average IMEP of 16 bar. The cylinder bore is 220 mm and stroke 300 mm. Find the indicated power. If the output measured at the brake shaft is 3·6 MW, what is the mechanical efficiency and the power lost in friction?

(4·1 MW; 88 per cent; 500 kW)

(21.16) ## Thermal efficiency

The *thermal efficiency* of an engine may be based on either indicated or brake power. In both cases it is the ratio of energy output to energy input. The output is the indicated or brake power. The input is the heat available in the fuel supplied to the engine cylinder. Thus

$$\text{thermal efficiency} = \frac{\text{energy output}}{\text{energy input}} = \frac{\text{IP or BP (kW)}}{\text{energy in fuel (kW)}}$$

(Comparative tests on engines must conform to BS 5514 which includes, for example, *Standard Reference Conditions* (SRC) specifying intake air conditions as pressure 1 bar, temperature 27 °C and relative humidity 60 per cent. Performance results must be corrected to SRC conditions.)

The efficiency calculated from the indicated power is called the *indicated thermal efficiency*; the efficiency calculated from the brake power is called the *brake thermal efficiency*.

The energy available in the fuel per second is the product of the mass of fuel used, $\dot{m}$ kg/s, and the calorific value of the fuel C kJ/kg. Thus

$$\text{indicated thermal efficiency} = \frac{\text{indicated power (kW)}}{\dot{m} \times C(\text{kJ/s} = \text{kW})}$$

$$\text{brake thermal efficiency} = \frac{\text{brake power (kW)}}{\dot{m} \times C}$$

The ratio of the brake to the indicated thermal efficiency is equal to the ratio of the brake to the indicated power, i.e. to the mechanical efficiency, thus

brake thermal efficiency

= indicated thermal efficiency × mechanical efficiency

Example

The power developed by a petrol engine at full load is 85 kW (BP) and 105 kW (IP). The fuel used per minute at this load is 0·38 kg. Calculate (a) the fuel consumption in kg/brake kW h, and (b) the thermal efficiency on a brake and on an indicated power basis. Calorific value (CV) of the fuel is 43·8 MJ/kg.

The cooling water flows through the cylinder jackets at the rate of 37 litres per minute. If 45 per cent of the total energy supplied in the fuel is accounted for by friction, exhaust gases and losses to the surroundings, estimate the rise in temperature of the cooling water.

Solution

(a) Fuel used per hour = 0·38 × 60 kg

$$\text{Fuel used per brake kW h} = \frac{0\cdot38 \times 60}{85}$$

$$= 0\cdot268 \text{ kg}$$

Hence the fuel consumption is **0·27 kg/brake kW h**

(b) On a BP basis,

$$\text{thermal efficiency} = \frac{\text{energy to BP}}{\text{energy in fuel}}$$

$$= \frac{\text{BP in kW (kJ/s)}}{\text{kg/s of fuel} \times \text{CV of fuel (kJ/kg)}}$$

$$= \frac{85}{\frac{0\cdot38}{60} \times 43\cdot8 \times 1000}$$

$$= 0\cdot306 \quad \text{or} \quad \textbf{30\cdot6 per cent}$$

On an IP basis,

$$\text{thermal efficiency} = \frac{\text{energy to IP}}{\text{energy in fuel}}$$

$$= \frac{105}{\frac{0\cdot38}{60} \times 43\cdot8 \times 1000}$$

$$= 0\cdot38 \quad \text{or} \quad \textbf{38 per cent}$$

Energy supplied in fuel $= 43\cdot8 \times 0\cdot38\,MJ/min.$

$$= \frac{43\cdot8 \times 0\cdot38 \times 10^3}{60}\,kJ/s$$

$$= 277\cdot4\,kJ/s$$

Of this energy 45 per cent is lost to friction, exhaust gases and surroundings and the remainder is accounted for by energy to brake power (85 kW or kJ/s) and energy to cooling water. Thus working in kJ/s,

energy supplied in fuel = energy to friction, etc. + energy to BP
$$+ \text{ energy to cooling water}$$

i.e. $277\cdot4 = 0\cdot45 \times 277\cdot4 + 85 + \text{energy to cooling water}$

Therefore

energy to cooling water $= 67\cdot6\,kJ/s$

Since 1 litre of water has a mass of 1 kg, the mass flow rate of water is

$\dot{m} = 37\,kg/min = 0\cdot62\,kg/s$

The specific heat capacity of water is $c = 4\cdot2\,kJ/kg\,K$, and if Δt is the rise in temperature of the water through the jackets, then

energy to cooling water $= \dot{m}\,c\Delta t$

i.e. $67\cdot6 = 0\cdot62 \times 4\cdot2 \times \Delta t$

therefore

$$\Delta t = 26\,K \quad \text{or} \quad \textbf{26\,°C}$$

Problems

21.10 A gas engine uses $9\cdot15\,m^3$ of gas per hour. The power developed is $9\cdot8\,kW$ (BP). If the calorific value of the gas supplied is $20\,MJ/m^3$ calculate the thermal efficiency of the engine on a BP basis.   (19·3 per cent)

21.11 An engine using $20\cdot1\,kg$ of petrol per hour, of calorific value $42\,MJ/kg$, develops a brake power of $39\cdot6\,kW$. Calculate the thermal efficiency on a brake power basis.   (16·9 per cent)

21.12 A six-cylinder marine diesel engine uses oil of calorific value $43\,MJ/kg$ at the rate of $0\cdot27\,kg/s$. The brake power output is $700\,kW$ per cylinder. Find the brake thermal efficiency and the fuel consumption in kg/brake kW h.   (36·2 per cent; 0·23 kg/brake kW h)

21.13 A test on an oil engine showed that at a brake load of $20\,kW$ the oil consumption was $7\cdot16\,litres/h$. The mechanical efficiency was 82 per cent. The calorific value of the oil fuel was $43\,500\,kJ/kg$. Calculate (a) the fuel consumption in kg/brake kW h and (b) the indicated thermal efficiency. The density of the oil was 0·8 Mg/m³.   (0·29 kg/kW h; 35·2 per cent)

21.14 An oil engine is supplied with fuel oil at the rate of 85·3 kg/h and the air–fuel ratio is 26:1, by mass. The brake thermal efficiency of the engine is 28 per cent and the calorific value of the oil 43 500 kJ/kg. Find the brake power developed and the mass of exhaust gases per kilowatt hour.

(289 kW; 7·97 kg/brake kW h)

21.15 A four-cylinder, four-stroke cycle petrol engine has a bore 90 mm and stroke 90 mm. The clearance volume of each cylinder is 86 cm³. Find the compression ratio. If the BMEP is 530 kN/m² at 80 rev/s and the engine uses 15·7 litres of fuel per hour, find the brake power and the brake thermal efficiency. The calorific value of the petrol is 42 MJ/kg and its density is 0·8 Mg/m³.   (7·66:1; 48·6 kW; 33·2 per cent)

21.16 A small gas turbine installation has a net output of 152 kW. The working fluid is air which is circulated at the rate of 179 kg/min. The energy supplied is 400 kJ/kg of air. Find the thermal efficiency of the plant.

(12·7 per cent)

21.17 In a gas turbine plant the net work done in the turbine is 41·45 kJ/kg of burnt gas. If the air–fuel ratio by mass is 120 and the calorific value of the fuel 42 MJ/kg, what is the thermal efficiency?   (11·9 per cent)

21.18 The exhaust from an oil engine is to be passed through a waste-heat boiler. The engine consumes 0·27 kg of fuel per brake kW h when developing 550 kW (BP). The calorific value of the oil is 45·6 MJ/kg. Find (a) the brake thermal efficiency of the engine and (b) the heat passing to the boiler in the exhaust gases per hour, assuming 40 per cent of the heat supplied is lost other than in heat to the exhaust gases.   (29·3 per cent; 2·08 GJ/h)

21.19 During a test of a motor vehicle engine, it was found that 30 per cent of the heat supplied in the fuel was converted to work at the coupling and 25 per cent was carried away by the cooling water surrounding the cylinders. If the engine develops 60 kW at the coupling, what is the heat carried away by the cooling water in megajoules per hour?   (180 MJ/h)

21.20 A 12-cylinder gas engine running at 1500 rev/min on the four-stroke cycle has an average IMEP of 10·5 bar. The cylinder bore is 160 mm and stroke 190 mm. The mechanical efficiency is 90 per cent and the engine uses North Sea gas of calorific value 38·5 MJ/m³ at the rate of 144 m³/h. Find the indicated and brake powers and the brake thermal efficiency. If 20 per cent of energy supplied in the fuel is lost to friction and the surroundings, estimate the waste heat available for other use in the exhaust gases and cooling water in GJ/h.   (602 kW; 542 kW; 35·2 per cent; 2·48 GJ/h)

21.17 Engine performance tests

An engine performance test may be designed to suit any one of a number of purposes but certain types of comparative tests have to be carried out under regulated conditions (*see* page 473). A comprehensive test includes information primarily on pressure–crank angle (p–θ) and pressure–volume (p–V or indicator) diagrams, power output, fuel and air flow rates, etc. Pulsating flow and very high temperatures and pressures in modern engines present difficulties. For indicated power it is necessary to have equipment to obtain pressure readings at piston displacement and crank-angle positions for each cylinder of an engine; for brake power a *brake* or *dynamometer* is used to measure the output at the engine.

21.18 Indicator diagrams

The pressure in an engine cylinder is not constant but varies continuously during the engine cycle. The variation in pressure with the position of the piston was recorded in the past for slow-speed, single-cylinder engines by a mechanical indicator consisting of a drum, a small piston located in the cylinder head with a spring control and a connecting linkage. Such an indicator directly gave a p–V diagram but because of the effects of friction, vibration and inertia of moving parts it was not suitable for high-speed (or multi-cylinder) engines. Mechanical indicators have been superseded by modern testing equipment described below.

The shape of an indicator diagram depends on whether the engine operates on spark-ignition or compression-ignition and on the speed of the engine. Figure 21.9(a) shows a typical indicator diagram for a *four-stroke petrol engine*. AB is the induction or suction stroke in which the fuel–air mixture is drawn into the cylinder at approximately constant pressure. BC is the compression stroke. Just before the piston reaches the inner dead centre position the mixture is ignited. The pressure rises rapidly before the piston has moved appreciably from the dead centre position; this is shown by the almost vertical line CD. Line DE represents expansion of the burnt gases as work is done on the piston. At E, just before the end of the stroke, the exhaust valve opens and the pressure drops rapidly to that of the atmosphere. EA represents the exhaust of the remaining burnt gases at approximately constant pressure.

Figure 21.9(b) shows a typical indicator diagram for a *high-speed compression-ignition engine*. Air is drawn into the cylinder at approximately constant pressure, shown by AB. The air is compressed to the ignition temperature corresponding to the point C. The pressure then rises rapidly, shown by CD. The fuel is sprayed into the cylinder at a steady rate so that as the piston moves back on its stroke combustion continues at approximately constant pressure, shown by DE. EF represents expansion, and FA exhaust, of the burnt gases.

For a slow-speed compression-ignition engine the diagram is modified slightly from Fig. 21.9(b). The fuel is injected at the end of the compression stroke; combustion takes place at constant pressure, and due to the slow speed of the piston, the completion of combustion is delayed while the piston moves through a part of the stroke. The modified diagram is shown in Fig. 21.9(c).

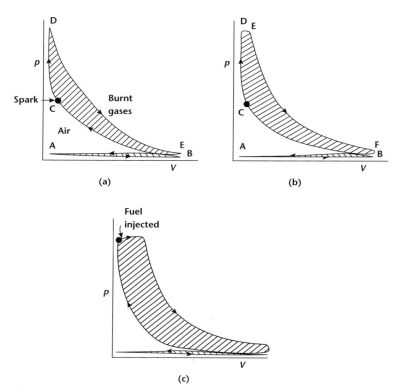

Fig. 21.9 Indicator diagrams for internal-combustion engines: (a) spark-ignition engine; (b) high-speed compression-ignition engine; (c) slow-speed compression-ignition engine.

Calculation of work done from indicator diagram

Consider the work done on a piston by a *constant* pressure p_m over a stroke length l. The force on the piston is $p_m \times A$, where A is the area of cross-section of the piston. The work done by this force is given by the product *force × distance moved*, i.e. $p_m \times A \times l$. But $l \times A$ is the swept volume V_s of the cylinder; hence

work done $= p_m \times V_s$

Figure 21.10(a) is the indicator diagram for this process. Since the horizontal scale is shown as a volume it is also a p–V diagram for the gas in the cylinder. The height of the diagram is the pressure p_m and the length is the swept volume V_s. Hence the product $p_m \times V_s$ is the shaded area of the diagram, A'B'C'D'. Thus

work done $= p_m \times V_s$

$\qquad = \text{area A'B'C'D'}$

$\qquad = \text{area under p–V diagram}$

Now consider a practical indicator diagram as in Fig. 21.10(b). AB represents expansion of the working fluid and CD compression. Suction and exhaust may be

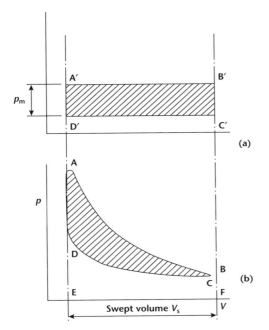

p_m

A′   B′

D′   C′

(a)

p

A

D

E   B   C   F

(b)

Swept volume V_s

V

Fig. 21.10

neglected. During expansion, work is done by the gas *on* the piston. During compression, work is done *by* the piston on the gas. The area ABFE represents work done by the gas and the area DCFE represents work done on the gas. The *net work done* during the cycle is represented by the *difference* between these two areas, i.e. the net work done is the shaded area *enclosed* by the curve ABCD.

Pumping losses

A small correction regarding 'pumping losses' has to be made in some cases when calculating the net work done from the indicator diagram. More importance is now being attached to these losses which are due to viscous friction in the fluid flows during the suction and exhaust strokes. In Fig. 21.9(a), which is the *p–V* diagram for a four-stroke petrol engine, it can be seen that the arrows enclosing the main area are clockwise and there is a small shaded area where they are anticlockwise: this area AEB is the difference in work done between the exhaust and suction strokes and represents work input. It is called the *pumping loop* and the main area is called the *power loop*. Thus it is this *net* area of the indicator diagram that represents the indicated work done. (This does not apply to the two-stroke petrol engine.)

Definition of mean effective pressure

The work done by a constant pressure p_m acting over a stroke length is $p_m V_s$ and is represented by the area of rectangle A′B′C′D′, Fig. 21.10(a). The work done during an engine cycle is given by the net area of the p–V diagram, Fig 21.10(b). If the two areas are equal then the pressure in the cylinder may effectively be taken as constant and equal to p_m for the purpose of calculating the work done in the cycle. The *mean effective pressure* is therefore defined as that pressure which multiplied by the swept volume gives the work done; or as the height of a rectangle which has an area equal to the net area of the p–V diagram and which has the same length (or swept volume). Thus

$$\text{MEP} = \frac{\textbf{net area of } \boldsymbol{p}\textbf{–}\boldsymbol{V} \textbf{ diagram}}{\textbf{length of diagram}}$$

This formula gives the MEP in terms of the mean height of the diagram in mm or cm and a conversion or calibration factor is needed to obtain the actual pressure. (In the mechanical indicator the calibration factor was the spring *constant* or *rate*.) For example, if the net area of the power loop (taking account of any pumping loop) is $20\,\text{cm}^2$, the length of the diagram, $7.5\,\text{cm}$, and the calibration factor, $3\,\text{bar/cm}$, then the MEP is

$$\frac{24}{7.5} \times 3 = 9.6\,\text{bar} \quad \text{or} \quad 960\,\text{kN/m}^2$$

The modern 'test cell', with its follow-up computer programs, delivers all the information and values required from the p–V diagram but otherwise, to find the MEP, it is necessary to use a planimeter or other method to obtain the net area of the diagram; the length of the diagram is known and the mean height can then be found. Alternatively, the mid-ordinate rule can be used to find the mean height, as shown in the example below.

Pressure–crank angle diagram

At the same time as the mechanical indicator was in use, a trace of pressure plotted on a time (crank angle) base would be obtained on the screen of a cathode-ray oscilloscope from the electrical impulses sent out by a pressure-sensor mounted in the cylinder head. The full line in Fig. 21.11 shows the p–θ shape for

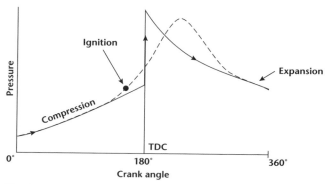

Fig. 21.11

the theoretical Otto (constant volume) cycle where the combustion is assumed to be instantaneous; the dotted line shows the shape for a practical four-stroke petrol engine when combustion actually takes place before and after the top dead centre (TDC) position. This type of diagram is the main product now of a modern test unit and has the advantage of showing in more detail the small region where ignition starts and combustion takes place. Today's testing equipment is electronic, computerized and controlled by microprocessors: it is capable of producing the necessary diagrams, acquiring and analyzing data and monitoring the exhaust gases, etc. over a period of time. For example, by taking a sequence of p–θ diagrams a 'heat release analysis' can be carried out to show the rate at which fuel is burned at different stages of the cycle. The time base can be converted to a piston displacement and volume base assuming the crank rotates at constant speed. The measurement of TDC position is particularly important and is carried out in dynamic 'hot motoring conditions' and as with all such equipment extreme care is necessary as to calibration.

(**Example**)

A six-cylinder, four-stroke petrol engine running at 70 rev/s gave an indicator diagram providing the following mid-ordinates: 3·56, 3·62. 3·1, 2·36, 1·06, 0·53, 0·204, 0·178, 0·13, 0·104 cm. The calibration factor was 40 kN/m² per mm of height on the diagram. If the cylinder bore and stroke are 75 and 100 mm respectively, calculate (a) the mean effective pressure and (b) the indicated power.

Solution
Using the mid-ordinate rule the diagram is divided by a number of *equally spaced* vertical lines. Midway between each vertical the mid-ordinates are measured and the rule states that the mean height is the sum of the mid-ordinates divided by their number.

(a)　Sum of mid-ordinates = 3·56 + 3·62 + 3·1 + . . .

$$= 14·85\,cm$$

No. of mid-ordinates = 10

Mean height of diagram = $\dfrac{14·85}{10}$ = 1·485 cm = 14·85 mm

Indicated mean effective pressure = 14·85 × calibration factor

$$= 14·85 \times 40$$

$$= \mathbf{594\,kN/m^2}$$

(b) Since the engine works on the four-stroke cycle, the number of cycles per second is equal to half the number of revolutions per second, i.e. $\frac{1}{2}$ × 70 or 35. Then indicated power per cylinder = $\dfrac{plAn_f}{1000}$ kW

$$= \frac{594 \times 1000 \times 0 \cdot 1 \times \frac{\pi}{4} \times 0 \cdot 075^2 \times 35}{1000}$$

$$= 9 \cdot 18 \,\text{kW}$$

For six cylinders, the total power developed is therefore **55·1 kW**.

(21.19) Measurement of brake power

The brake or shaft power is the power output developed at the engine crankshaft. In the past rope brakes were used as well as the Prony brake which employed friction pads on the flywheel rim. These are now generally replaced by various types of dynamometers producing drag and acting on the torque-reaction principle. Two examples are the fluid coupling (hydraulic control, *see* page 248) and the electrical (field strength control) dynamometer where an electric motor acts as a generator to absorb the engine output power. To illustrate the principles involved, the simple rope brake, Fig. 21.12 is described here.

The brake drum is keyed to the crankshaft and rotates at the engine speed. The power developed by the engine at the shaft is absorbed in friction work between the drum and a rope wrapped around it, as shown. The rope is attached at one end to a spring balance and to the other end is hung a dead load. Let

R = radius of drum, m

S = spring balance reading, N

W = dead load, N

n = speed of engine in rev/s

T = brake torque (torque at engine coupling)

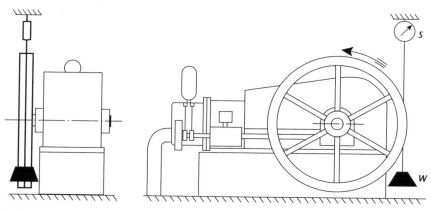

Fig. 21.12

then the friction torque T at the brake drum is

$$T = (W - S) \times R \quad \text{N m}$$

To be correct, R should be the radius of the centre line of the rope.

The power absorbed at the brake drum at a speed n rev/s is

$$\text{brake power} = \frac{\text{work done by brake torque}}{1000}$$

$$= \frac{T \times 2\pi n}{1000} \text{ kW}$$

or

$$\textbf{brake power} = \frac{2\pi n T}{1000} \textbf{ kW}$$

(**Example**)

A small petrol engine fitted with a rope brake gave the following results on test: speed, 10 rev/s; deadweight load, 200 N; spring balance reading, 40 N. The effective diameter of the brake drum was 1·2 m. Calculate the shaft power developed by the engine.

Solution

Net torque on drum, $T = (200 - 40) \times 0·6$

$$= 96 \, \text{N m}$$

Brake power dissipated in friction at the drum,

$$= \frac{2\pi n T}{1000} \text{ kW}$$

$$= \frac{2\pi \times 10 \times 96}{1000}$$

$$= 6·03 \, \text{kW}$$

(**Example**)

An indicator card taken from a slow-speed, six-cylinder, double-acting, two-stroke oil engine gave a diagram area of 18 cm² and a length of 7·5 cm. The calibration factor was 3 bar/cm, the cylinder bore 380 mm and stroke 1100 mm. Estimate the indicated power at 196 rev/min.

Solution

$$\text{Mean height of diagram} = \frac{\text{area of diagram}}{\text{length of diagram}} = \frac{18}{7·5} = 2·4 \, \text{cm}$$

IMEP = mean height of diagram × calibration factor

$$= 2 \cdot 4 \times 3 \times 10^5 \, \text{N/m}^2$$

$$= 7 \cdot 2 \times 10^5 \, \text{N/m}^2$$

Number of working strokes per second, $n_f = 2 \times 196/60 = 6 \cdot 53$

Total indicated power $= 6 \times \left(\dfrac{plAn_f}{1000} \right)$ kW

$$= 6 \times \left(\frac{7 \cdot 2 \times 10^5 \times 1 \cdot 1 \times \frac{\pi}{4} \times 0 \cdot 38^2 \times 6 \cdot 53}{1000} \right) \text{kW}$$

$$= 3519 \, \text{kW} \quad \text{or} \quad \textbf{3·52 MW}$$

Note the distinction in the above examples between n rev/s and n_f firing strokes/s.

Problems

21.21 The indicator diagram taken from a test on a low-speed oil engine gives the following ten mid-ordinates, 2·3, 2·46, 2·54, 2·57, 2·3, 1·32, 0·79, 0·38, 0·31, 0·18 cm. The calibration factor is 30 kN/m²/mm. Calculate the indicated mean effective pressure.   (455 kN/m²)

21.22 An indicator card taken from a test on a four-stroke, four-cylinder diesel engine gave the following eight mid-ordinates, 5·03, 4·4, 3·61, 2·56, 1·3, 0·56, 0·38, 0·28 cm.

The calibration factor was 50 kN/m²/mm. The engine bore and stroke were both 100 mm and the speed was 40 rev/s. Calculate the indicated power of the engine.   (71 kW)

21.23 A compression-ignition engine when tested gave an indicator card having area 31 cm² and length 7·5 cm. The calibration factor was 1·5 bar/cm. Calculate the MEP. If the mechanical efficiency of the engine was 80 per cent, what was the brake MEP?   (620 kN/m²; 496 kN/m²)

21.24 A 16-cylinder (four-stroke) dual-fuel engine works at one point solely on natural gas, neglecting the effect of pilot oil fuel. The calorific value of the gas is 38·5 MJ/m³ and the gas consumption 637 m³/h. The engine speed is 1000 rev/min, cylinder bore, 270 mm, stroke, 305 mm. An indicator card showed a mean height of diagram of 2·8 cm, when the calibration factor was 4·5 bar/cm. The net brake torque is 25 kN m.

Find (i) IMEP, (ii) IP, (iii) BP, (iv) mechanical efficiency, (v) brake thermal efficiency and (vi) fuel consumption in m³/brake kW h.

(12·6 bar; 2·93 MW; 2·62 MW; 89 per cent; 38·5 per cent; 0·24)

21.25 An engine under test is fitted with a brake drum of 1 m diameter. At a speed of 5 rev/s the dead load on the brake rope is 480 N and the spring balance reading 55 N. Calculate the BP of the engine. If the mechanical efficiency at this load is 80 per cent, what is the power lost in friction?

(6·67 kW; 1·67 kW)

21.26 The following data were obtained during a test on a double-acting, two-stroke, single-cylinder CI engine: MEP, $500 \, kN/m^2$; speed, 3 rev/s; brake torque, $1 \, kN \, m$; cylinder diameter, 200 mm; stroke, 250 mm. Calculate for this load (a) IP, (b) BP and (c) mechanical efficiency.

(23·6 kW; 18·9 kW; 80 per cent)

21.27 A six-cylinder, four-stroke engine runs at 3000 rev/min and uses petrol of density $0·82 \, Mg/m^3$ at the rate of 0·3 litres/min. The cylinder bore is 85 mm and stroke 100 mm. The average BMEP is $700 \, kN/m^2$. Find the BP and the fuel consumption in kg/brake kW h. (59·6 kW; 0·25 kg/brake kW h)

21.28 A single-cylinder oil engine, firing every other revolution, has a bore 150 mm and stroke 225 mm. When tested at 250 rev/min the indicated MEP was found to be $630 \, kN/m^2$ and the oil consumption 1·33 kg/h. The torque at the engine coupling was 165 N m. The calorific value of the oil was 40 MJ/kg. Find (a) the mechanical efficiency and (b) the thermal efficiency on the IP basis. (83 per cent; 35 per cent)

21.29 A diesel engine used 14 kg of oil per hour when running at 2000 rev/min. The dynamometer registered a torque of 260 N m. If the calorific value of the oil is 42 MJ/kg, find the thermal efficiency on a BP basis.

(33·3 per cent)

21.30 A three-cylinder, two-stroke oil engine gave the following results under test: speed, 50 rev/s; brake torque, 135 N m; fuel used, 13·5 kg/h; IMEP, $430 \, kN/m^2$. The calorific value of the oil was 42 000 kJ/kg. Calculate the brake and indicated thermal efficiencies. The bore and stroke are each 100 mm.

(26·9 per cent; 32·2 per cent)

21.31 During a constant load test on a gas engine the gas consumption was 43 m^3/h. The calorific value of the gas was $20 \, MJ/m^3$. The indicated thermal efficiency was 25 per cent and the mechanical efficiency 85 per cent at this load. Calculate the brake power developed.

The rate of water flowing through the cylinder jackets was 58 kg/min and the rise in temperature of the water was 21·2 K. Calculate the percentage of the heat supplied which was carried away by the cooling water.

(50·8 kW; 36 per cent)

21.32 The following observations were recorded during a trial of a four-stroke, single-cylinder oil engine: duration of trial, 30 min; oil consumed, 4·7 kg; calorific value of oil, 41 500 kJ/kg; area of indicator diagram, $8·5 \, cm^2$; length of diagram, 7·5 cm; calibration factor, 0·5 bar/mm; brake torque, 900 N m; speed, 220 rev/min; cylinder diameter, 300 mm; stroke, 450 mm.

Calculate the IP, BP and mechanical efficiency. Determine also the specific fuel consumption on a brake kilowatt-hour basis and the indicated thermal efficiency.

(33 kW; 20·7 kW; 62·8 per cent; 0·453 kg/brake kW h; 30·5 per cent)

21.33 A six-cylinder, four-stroke engine running at 2800 rev/min used petrol of density 0·8 Mg/m³ and calorific value 43·5 MJ/kg at the rate of 20 litres/h. The cylinder bore was 80 mm, stroke 96 mm.

For one cylinder the IMEP was found to be 840 kN/m² and assuming this to be the average for the engine estimate the IP and thermal efficiency.

If the average BMEP was 680 kN/m² find the mechanical efficiency, brake power and thermal efficiency, and the fuel consumption per kW of brake power output per hour.

(56·7 kW; 29·4 per cent; 81 per cent; 45·9 kW; 23·8 per cent; 0·35 kg/kW h)

Chapter 22

Fluids

....

Pressure and density have been mentioned already in various chapters but are dealt with here again in respect of fluids. This introductory chapter to the vast subject of 'fluids' covers only the fundamental topics of transmission of fluid pressure, forces on immersed bodies, buoyancy, the hydraulic press and pump, and a simple example of dynamic pressure.

22.1 Fluids

As we have already seen, a solid body offers considerable resistance to a change in shape. A *fluid* on the other hand is distinguished by its ability to flow and occupy any shape. A fluid may be a *gas* or a *liquid*. A gas always fills the container in which it is placed. A liquid may be contained in an open vessel, occupies a fixed volume and may present a *free surface* to the atmosphere.

22.2 Fluid pressure

Pressure is the intensity of a force measured per unit area of the surface on which the force acts. The pressure exerted by a fluid has the following important properties:

1. The pressure exerted by a fluid 'at a point', that is, on a very small area, is *the same in all directions*.
2. The direction of the resultant pressure exerted by a fluid at rest on a solid surface *is always perpendicular (normal) to the surface*.

Figure 22.1 illustrates the properties of fluid pressure. The pressure is the same at all points on the surface of a very small body immersed in the liquid shown and is everywhere normal to the surface of the containing vessel.

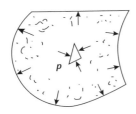

Fig. 22.1

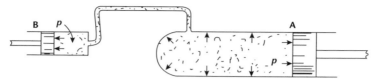

Fig. 22.2

22.3 Transmission of fluid pressure

The *principle of transmissibility of fluid pressure* states that the pressure at any point of a fluid at rest is transmitted without loss to all other points of the fluid. In Fig. 22.2 a force on the piston A in a *closed* cylinder *full* of water, for example, produces a pressure p exerted on the piston by the water. The same pressure is exerted on the container walls and also on the face of the second piston B. The pressure is said to be transmitted equally throughout the liquid.

Note that so far we have neglected any effects due to the weight of the fluid itself. A pressure may also be exerted by a fluid by virtue of its own weight. This pressure due to weight is often important for liquids but not for a relatively light gas. It is necessary therefore to revise our knowledge of density and relative density.

22.4 Density; relative density

The *density* of a liquid is its *mass per unit volume* and the SI unit is **kg/m³** (*see* also Ch. 14). The density of water is particularly important and must be remembered, thus

density of water = 1000 kg/m³ = 1 Mg/m³ = 1 t/m³ = 1 kg/litre

The *relative density* of a liquid is its density relative to that of water (at 4 °C). Average values of relative density for some liquids are as follows:

pure water	1·00
sea-water	1·026
mercury	13·6
petrol	0·75
lubricating oil	0·9
turpentine	0·87

For example, if the relative density of petrol is 0·8, its density is 0·8 × density of water, i.e. 0·8 × 1 = 0·8 Mg/m³ or 800 kg/m³.

Weight of a liquid

For a liquid of density ρ its weight per unit volume is ρg. Thus, for water its weight per cubic metre is

$$\rho g = 1000 \times 9\cdot8 = 9800 \, \text{N/m}^3 \quad \text{or} \quad \textbf{9·8 kN/m}^3$$

Specific weight of a liquid

The weight per unit volume (ρg) is often called the *specific weight* of the liquid and this term is used in some areas of engineering work. The specific weight of water is therefore $9.8 \, \text{kN/m}^3$.

22.5 Pressure at any depth of liquid

Consider the vertical column of liquid shown in Fig. 22.3, of vertical height h and cross-sectional area A. The weight W of the column is equal to mg where its mass m is given by the product *density* $\times$ *volume*.

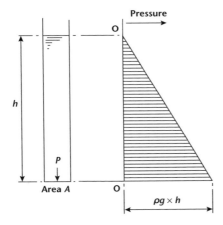

Fig. 22.3

Thus, if the volume of liquid in the column is V, then

$$V = A \times h$$

and

$$W = mg = \rho V g = \rho g A h$$

The pressure on the base of the column due to the weight of liquid is

$$p = \frac{W}{A}$$

$$= \frac{\rho g A h}{A}$$

i.e. $\quad \boldsymbol{p = \rho g h}$

Thus the pressure in a liquid at a particular point due to its own weight is given by the product

density $\times g \times$ depth of point below free surface

Since p is proportional to h, the pressure increases uniformly with depth. Figure 22.3 shows the variation with depth of the liquid pressure in the column.

Example

A ship's fuel oil storage tanks contain oil to a depth of 8 m. Calculate the load on a circular plate 450 mm in diameter in the base of the tank due to the weight of fuel oil. Relative density of fuel = 0·8.

Solution

Density of fuel oil, $\rho = 0·8 \times$ density of water

$$= 0·8 \times 1000 \, \text{kg/m}^3$$

$$= 800 \, \text{kg/m}^3$$

Pressure on plate $= \rho gh = 800 \times 9·8 \times 8$

$$= 62\,720 \, \text{N/m}^2 \quad \text{or} \quad \textbf{62·7 kN/m}^2$$

(If the density of the fuel is taken as $0·8 \, \text{Mg/m}^3$ then $\rho g = 0·8 \times 9·8 = 7·84 \, \text{kN/m}^3$ and the pressure is equal to $7·84 \times 8 = 62·72 \, \text{kN/m}^2$.)

Load on plate $=$ pressure $\times$ area

$$= 62·72 \times \frac{\pi}{4} \times 0·45^2$$

$$= \textbf{10 kN}$$

22.6 Total pressure at a point in a liquid

The weight of the air in the earth's atmosphere gives rise to a pressure p_a on the surface of the earth of about $101 \, \text{kN/m}^2$. The *total* pressure at the base of a column of liquid which has a free surface to the atmosphere, Fig. 22.4, is therefore given by

total pressure = atmospheric pressure + pressure due to weight of liquid

i.e. $$p = p_a + \rho gh$$

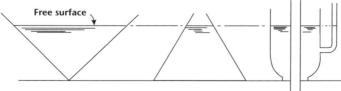

Fig. 22.4

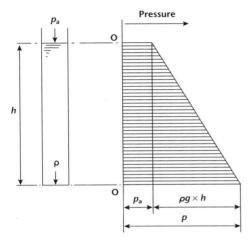

Fig. 22.5

The variation in total pressure with distance below the free surface of the liquid is shown in Fig. 22.5.

The total pressure at a point in a liquid having a free surface to the atmosphere varies only with the depth of the point below the free surface; e.g. the pressure at the base of each of the vessels shown in Fig. 22.4 is the same in each case and does not depend on the size or shape of the vessel.

22.7 Atmospheric pressure; gauge pressure

The measurement of atmospheric pressure has been dealt with in Chapter 16. As a reminder we note that

$$1 \text{ atmosphere} = 101{\cdot}325 \text{ kN/m}^2$$

$$= 76 \text{ cm of mercury}$$

$$= 760 \text{ mm of mercury}$$

$$= 10{\cdot}4 \text{ m of water}$$

(the above equals signs mean 'is equivalent to') and

$$1 \text{ millibar} = 100 \text{ N/m}^2$$

$$1 \text{ bar} = 10^5 \text{ N/m}^2$$

$$1 \text{ hectobar} = 10^7 \text{ N/m}^2 = 100 \text{ bar}$$

The pressure of a liquid may be measured by a pressure gauge of the Bourdon type or by a manometer (see p. 347). This method of measurement gives the *gauge* pressure p_g measured above the atmospheric pressure. The *absolute* pressure p is the sum of the gauge and atmospheric pressures. The atmospheric pressure is obtained from a barometer reading. Thus

absolute pressure = gauge pressure + atmospheric pressure

The term pressure used here always refers to absolute pressure unless otherwise stated.

Example

Figure 22.6 shows a manometer tube connected to a pipe-line containing water under pressure. The level of liquid in the tube is 0·6 m above the centre line of the pipe. Calculate the gauge and absolute pressures of the water in the pipe. Barometer reading = 750 mm Hg.

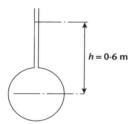

Fig. 22.6

Solution

Gauge pressure $= \rho g h = (1 \times 9\cdot8) \times 0\cdot6$

$$= 5\cdot88 \text{ kN/m}^2 \text{ (gauge)}$$

Now 760 mm Hg is equivalent to a pressure of 101·3 kN/m².

Hence

atmospheric pressure $= \dfrac{750}{760} \times 101\cdot3 = 100 \text{ kN/m}^2$

and

absolute pressure $= 100 + 5\cdot88 = \mathbf{106 \text{ kN/m}^2}$

Problems

22.1 A skin diver swims in sea-water (relative density of sea-water = 1·026), 20 m below the surface. What is the pressure on the diver due to the weight of water at this level?   (201 kN/m² (gauge))

22.2 A road tanker carries 30 000 litres of petrol of relative density 0·74. What is the total weight of fuel carried?   (218 kN)

22.3 A lubricating oil has a relative density of 0·91. Calculate its density (a) in tonne/dm³, (b) kg/litre and (c) Mg/dm³.

(0·00091 tonne/dm³; 0·91 kg/litre; 0·00091 Mg/dm³)

22.4 A rocket fuel of density 800 kg/m³ is transported to the surface of the moon. If the acceleration due to gravity on the moon is one-sixth that on the earth's surface, what is the specific weight of the fuel on the moon?

(1·31 kN/m³)

22.5 A tank contains fuel oil of relative density 0·72 to a depth of 5 m. Calculate the load on a valve plate of area 500 mm² in the base of the tank due to the weight of oil.

(17·6 N)

22.6 A hatch cover for a submarine has a surface area of 0·4 m². What is the total force on it due to the weight of water at a depth of 100 m? (402 kN)

22.7 The level of liquid in a manometer tube connected to a pipeline containing sea-water is 760 mm above the centre line of the pipe. Calculate the gauge and absolute pressure of the water in the pipe. Barometer reading, 770 mm Hg.

(7·64 kN/m²; 110 kN/m²)

22.8 Total thrust on an immersed plane surface

The *total thrust* on a plane surface immersed in a fluid is simply the force on the surface due to fluid pressure. If the surface is subject to fluid pressure on both sides the *resultant thrust* is the difference between the two total thrusts.

Consider the plane vertical rectangular surface shown partially submerged in Fig. 22.7. The free surface of the liquid is at OO. The pressure p on the wetted surface *due to the weight of the liquid* increases uniformly from zero at the free surface to a maximum value $\rho g h$ at the lowest edge of the surface. Since the pressure–depth graph is a straight line, Fig. 22.7(b), the average pressure p_{av} on one side of the wetted surface is one-half the maximum pressure. Thus

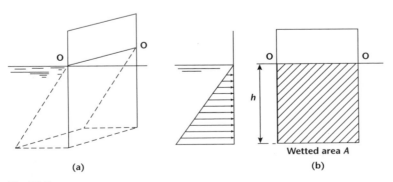

Wetted area A

(a)   (b)

Fig. 22.7

$$p_{av} = \tfrac{1}{2}\rho g h$$

$$= \rho g \times \tfrac{1}{2} h$$

$$= \rho g \times H$$

where H is the vertical depth of the centroid of the area of the *wetted* surface below the free surface. The total thrust P on a plane vertical surface of wetted area A is therefore

$$P = p_{av} \times A$$

$$= \rho g H A$$

$\rho g H$ is the pressure of the liquid at the centroid of the wetted area. The total thrust is therefore equal to the area multiplied by the pressure at the *centroid*.

This result can be shown to be true for any vertical plane surface. The total thrust does *not* act at the centroid of the area of the wetted surface but at a point on the surface known as the *centre of pressure*. We are concerned at this stage only with the calculation of the total thrust and not with problems involving the centre of pressure and moments due to the thrust. However, it should be noted that the centre of pressure is located at the centroid of the pressure–depth diagram for an immersed surface. In Fig. 22.7(b) the pressure–depth diagram is a triangle and the centroid of this diagram is at a height $\tfrac{1}{3}h$ above the base of the diagram. Thus the thrust in this case acts normal to the plane surface at a height $\tfrac{1}{3}h$ above its base.

Example

The vertical side of an oil tank is 3·6 m wide. The depth of oil in the tank is 2·4 m. Calculate the total horizontal thrust on the side due to the weight of oil in the tank. Density of the oil, 0·75 t/m³.

Solution
Depth of centroid of wetted surface below the free surface of the oil, $H = 1\cdot2\,\text{m}$.

Area of wetted surface of side of tank $= A = 3\cdot6 \times 2\cdot4 = 8\cdot64\,\text{m}^2$

Total horizontal thrust on side of tank $= \rho g H A$

$$= 0\cdot75 \times 9\cdot8 \times 1\cdot2 \times 8\cdot64$$

$$= 76\,\text{kN}$$

Example

A rectangular vertical sluice gate, 9 m wide of weight 6 t, is subjected to water pressure on one side to a depth of 6 m. The coefficient of friction μ between the gate and its guides is 0·1. Determine the force required to lift the gate and the power required if the speed of lifting is 50 mm/s.

Solution

Depth of centroid of wetted surface, $H = 3\,\text{m}$

Horizontal force on gate $= \rho g H A$

$$= (1 \times 9\cdot8) \times 3 \times (9 \times 6)$$

$$= 1588\,\text{kN}$$

Friction force at guides

$\quad = \mu \times$ normal force between gate and guides

$\quad = 0\cdot1 \times 1588$

$\quad = 158\cdot8\,\text{kN}$

Total force to lift gate = weight + friction force

$$= (6 \times 9\cdot8) + 158\cdot8$$

$$= \mathbf{218\,kN}$$

Power required = force (kN) × speed (m/s)

$$= 217\cdot6 \times 0\cdot05\,\text{kW}$$

$$= \mathbf{10\cdot9\,kW}$$

Problems

22.8 A vertical dam wall is 40 m wide and the top of the wall is 24 m above the floor of the dam. The water level is 4 m below the top of the wall. Calculate the total thrust on the wall. What is the thrust when the water just spills over the top of the dam?   (78·4 MN; 113 MN)

22.9 A closed vertical lock gate 12 m wide has sea-water on one side to a height of 8 m above the bottom. Calculate the total horizontal force due to the weight of water on one side of the gate. If the sea-water surface on the other side of the gate is 3·6 m above the bottom, what is the resultant horizontal force on the gate? Relative density of sea-water = 1·026.   (3·86 MN; 3·08 MN)

22.10 A water tank has vertical sides and is of rectangular section 4 m by 3 m. If the depth of water in the tank is 2 m, calculate the total thrust due to the weight of water (a) on the bottom, (b) on a side, and (c) on one end.   (235 kN; 78·4 kN; 58·8 kN)

22.11 A tank, 10 m square, contains fuel oil to a depth of 5 m. Calculate the total horizontal force on a vertical side due to the weight of oil. Density of the oil = 0·7 Mg/m³.   (858 kN)

22.12 A storage tank 1·5 m wide has two compartments separated by a vertical partition across the full width. One side contains water to a depth of 2·4 m and the other side contains oil to a depth of 1·2 m. Find the resultant horizontal force on the partition. Relative density of the oil = 0·8.

(33·8 kN)

22.13 A horizontal pipe of 3·6 m bore is just full of oil of density 900 kg/m³. The pipe is closed at the ends. What is the force on one end plate?   (162 kN)

22.14 A vertical lock gate used to close a canal of rectangular section 9 m wide is designed to withstand a resultant horizontal thrust of 1·2 MN. If the depth of water on the lower side is 3·6 m, what depth of water can be permitted on the other side?   (6·34 m)

22.15 A rectangular sluice gate 4·5 m wide weighs 10 kN and resists water pressure on one side. When the depth of water is 3·6 m the force required to lift the gate in vertical guides is 43·5 kN. Estimate the coefficient of friction between the gate and its guides.   (0·12)

22.16 A rectangular lock gate operates in vertical guides. The gate is 6 m wide and 6 m high. The water level is 3 m below the top of the gate on one side and 1·5 m on the other side. Find the resultant thrust on the gate. If the gate weighs 30 kN and the coefficient of friction between the gate and its guides is 0·15, find the force to lift the gate vertically. If 8 kW is employed, at what speed can the gate be lifted?   (330 kN; 79·5 kN; 0·1 m/s)

(22.9) Hydraulic machines

A *hydraulic machine* is a machine that uses water to transmit force or to do useful work. The term hydraulic is also applied where some other fluid is used, e.g. *oil-hydraulic*.

The hydraulic press

The *hydraulic* or *Bramah press* is illustrated in Fig. 22.8. Two cylinders X and Y are connected by a pipe and the whole apparatus is filled with a liquid. The piston in cylinder Y carries a load of weight W. This load is raised by applying a force P to the piston in cylinder X.

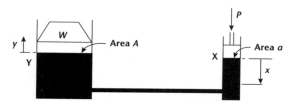

Fig. 22.8

If the liquid is assumed to be incompressible then the volume of fluid forced out of cylinder X is equal to the volume which flows into cylinder Y. Let a and A be the areas of the pistons X and Y respectively. Let piston X move downwards through a distance x, thus displacing a volume of liquid equal to $a \times x$. If piston Y rises through a distance y then the volume entering the cylinder is $A \times y$. Hence

$$a \times x = A \times y$$

i.e. $$\frac{y}{x} = \frac{a}{A}$$

Thus the distance moved by each piston is inversely proportional to its area.

The effort P required to raise the load W may be calculated as follows. Let the pressure on piston X be p; then the force on this piston is

$$P = p \times a$$

or

$$p = \frac{P}{a}$$

By the principle of transmissibility of pressure, the pressure on piston Y is also equal to p. Now, since the load on this piston is W, we have, therefore,

$$p = \frac{W}{A}$$

Hence

$$p = \frac{P}{a} = \frac{W}{A}$$

i.e. $$\frac{P}{W} = \frac{a}{A}$$

Thus the load on each piston is proportional to its area.

For an *ideal* machine with frictionless pistons the work done on the piston at X is equal to the work done by the piston at Y. This is shown as follows:

work done at X $= P \times x$

$$= \frac{a}{A} W \times x, \text{ since } P = \frac{a}{A} W$$

$$= \frac{a}{A} W \times \frac{A}{a} y, \text{ since } x = \frac{A}{a} y$$

$$= W \times y$$

$$= \text{work done at Y}$$

(**Example**)

A hydraulic press is to lift a mass of 10 t. The area of the piston in the load cylinder is $0.4 \, \text{m}^2$, and the area of the piston in the second cylinder is $0.05 \, \text{m}^2$. Calculate (a) the effort required to lift the load and (b) the distance moved by the effort when the load is raised 75 mm.

Solution

$$W = 10 \times 9.8 = 98 \, \text{kN}$$

(a) Effort $= W \times \dfrac{a}{A}$

$$= 98 \times \frac{0.05}{0.4} = \textbf{12.25 kN}$$

(b) Distance moved by effort $= x = y \times \dfrac{A}{a}$

$$= 75 \times \frac{0.4}{0.05} = \textbf{600 mm}$$

The reciprocating pump

The reciprocating or plunger pump consists essentially of a cylinder and a piston, or plunger, driven by an engine or motor, Fig. 22.9. The operation is as follows.

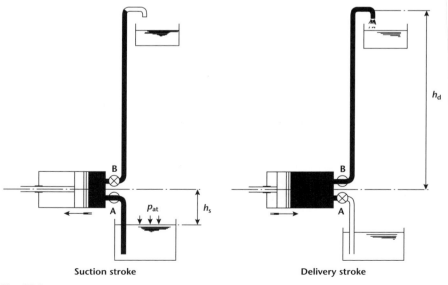

Suction stroke   Delivery stroke

Fig. 22.9

As the piston moves to the left, valve B is closed and valve A open. The atmospheric pressure (p_{at}) on the liquid in the supply tank forces the liquid into the increasing cylinder space. When the cylinder is full, valve A is closed and valve B is opened. As the piston now moves to the right, the force exerted by the piston drives the liquid through a vertical height h_d (the *delivery head*).

Since the cylinder pressure during the discharge stroke is $p = \rho g \times h_d$, it follows that the force P required on a piston of area A is

$$P = p \times A = \rho g h_d A$$

Similarly, a pump may discharge against a given pressure instead of raising the level of the liquid. For example, a boiler feed pump delivering water from an open tank to a boiler against a boiler pressure of p_g (kN/m^2 gauge) is performing the same task as if it were delivering the water to a height h_d (m) above the pump. Since $\rho g = 1 \times 9.8$ or $9.8\,\text{kN/m}^3$, therefore

$$h_d = \frac{p_g}{\rho g} = \frac{p_g}{9.8} \text{ metres}$$

Suction head

The depth h_s of the free surface of the liquid in the empty tank below the centre line of the pump cylinder is known as the *suction head* (usually expressed in metres of water).

In theory, the greatest suction head through which a pump will draw liquid corresponds to a height of liquid that will just balance the atmospheric pressure p_a exerted on the free surface. This is the case when

$$\rho g \times h_s = p_a$$

or

$$h_s = \frac{p_a}{\rho g}$$

For water, $\rho g = 1 \times 9.8$ or $9.8\,\text{kN/m}^3$ and $p_a = 101.3\,\text{kN/m}^2$. The maximum theoretical suction head is therefore

$$h_s = \frac{101.3}{9.8}$$

$$= 10.4\,\text{m}$$

The full theoretical lift is not attained in practice due to

- air leakage past the shaft seal and joints;
- losses due to the resistance to fluid flow of the valves, fittings, and pipework;
- the pressure exerted by water vapour (*see* Ch. 18).

The suction head or *lift* available in practice varies widely but may be about 5 to 7 m of water. It is increased by good seal design, fitting and maintenance, a large inlet pipe and valve, and a low flow rate. The available suction head is reduced by a low barometric pressure and a high water temperature.

22.10 Pumping power

The work done in lifting a mass m of liquid through a total height $h = h_d + h_s$ is $mg \times h$. If the time taken to do this is t, then the average rate of doing work, or average power, is

$$\frac{mgh}{t}$$

If the *rate* at which liquid is being pumped through a vertical height h (m) is $\dot{m}$ (kg/s), then the theoretical power required is

$$\frac{\dot{m}gh}{1000} \text{ kW}$$

This is the theoretical, or minimum, power required. In practice, additional power is necessary to provide for bearing or shaft friction and other losses in the pump itself, and also to provide for fluid losses in the pipeline. The fluid losses in the pipe are usually allowed for by adding to the theoretical head a suitable additional head to represent the loss. The losses in the pump itself are allowed for by using the idea of machine efficiency (*see* Ch. 6 and example below).

Example

A slow-acting pump raises 36 000 litres of water per hour through a height of 6 m. Calculate the average pumping power required. Neglect losses.

Solution
1 litre of water has a mass of 1 kg; therefore

$$\text{mass of water raised, } \dot{m} = \frac{36\,000}{3600} = 10 \text{ kg/s}$$

$$\text{Average power} = \frac{\dot{m}gh}{1000}$$

$$= \frac{10 \times 9 \cdot 8 \times 6}{1000}$$

$$= 0 \cdot 59 \text{ kW}$$

Example

A slow-acting reciprocating pump draws water from a level 0·6 m below the centre line of the cylinder and delivers it to a height of 1·8 m above the cylinder. The piston area is 5000 mm². Calculate the piston force required (a) during the suction stroke and (b) during the delivery stroke.

Solution

(a) Suction stroke:

force on piston = pressure × area

$$= \rho g h \times A$$

$$= (1 \times 9{\cdot}8) \times 0{\cdot}6 \times \frac{5000}{10^6}$$

$$= 0{\cdot}029 \, \text{kN}$$

(b) Delivery stroke:

force on piston $= (1 \times 9{\cdot}8) \times 1{\cdot}8 \times \dfrac{5000}{10^6}$

$$= 0{\cdot}09 \, \text{kN}$$

Example

A boiler feed pump is to supply 4000 litres of water per hour to a boiler against a gauge pressure of $1{\cdot}4 \, \text{MN/m}^2$. Neglecting losses, find the power required. If the efficiency of the pump is 65 per cent, find the power required. Neglect fluid flow losses.

Solution

1 litre of water has a mass of 1 kg; therefore mass of water pumped,

$$\dot{m} = \frac{4000}{3600} = 1.11 \, \text{kg/s}$$

Equivalent height h to which water is raised $= \dfrac{p_g}{\rho g} = \dfrac{1{\cdot}4 \times 10^6}{1000 \times 9{\cdot}8}$

$$= 143 \, \text{m}$$

Theoretical power $= \dfrac{\dot{m} g h}{1000} = \dfrac{1{\cdot}11 \times 9{\cdot}8 \times 143}{1000}$

$$= 1{\cdot}56 \, \text{kW}$$

Allowing for pump efficiency,

$$\text{power} = \frac{1{\cdot}56}{0{\cdot}65} = 2{\cdot}4 \, \text{kW}$$

Problems

22.17 A pump discharges 135 000 litres of oil per hour against a head of 20 m. Calculate the power required. Relative density of the oil = 0·77.   (5·7 kW)

22.18 A boiler feed pump supplies 2000 kg of water per hour to a boiler working at a gauge pressure of $1{\cdot}4 \, \text{MN/m}^2$. Calculate the pumping power required.
       (0·78 kW)

22.19 The pumps delivering cooling water to a power station supply $15\,000\,\text{m}^3$ per hour. If the total pumping power employed is $606\,\text{kW}$, against what head can the pump discharge? Neglect losses.   ($14.9\,\text{m}$)

22.20 A reciprocating pump is to be used to pump oil out of a tank. What is the greatest theoretical depth below the pump at which the inlet to the suction may be placed? Relative density of the oil $= 0.72$.   ($14.4\,\text{m}$)

22.21 A reciprocating pump discharges water to a height of $50\,\text{m}$ above the pump. What is the pressure is the cylinder?   ($490\,\text{kN/m}^2$)

22.22 A horizontal reciprocating pump draws water from a point $7\,\text{m}$ below the centre-line of the cylinder and delivers it to a height of $30\,\text{m}$ above the pump. If the power available at the piston is $4\,\text{kW}$, what is the delivery in m^3/h?   ($39.7\,\text{m}^3/\text{h}$)

22.23 A pump is used to raise $360\,\text{m}^3$ of water per hour to a height of $60\,\text{m}$. If the efficiency of the pump is 70 per cent, find the power required.

($84\,\text{kW}$)

Fig. 22.10

22.24 A hydraulic jack of the type shown in Fig. 22.10 is used to drive the piston of an extrusion press. The pressures on the two sides of the piston are 42 and $0.7\,\text{MN/m}^2$. The diameter of the piston is $150\,\text{mm}$ and the diameter of the piston rod $50\,\text{mm}$. The lower pressure acts on the piston-rod side of the piston. Calculate the net force exerted by the jack. If the piston moves through a stroke of $0.6\,\text{m}$ in $20\,\text{s}$, find the average power developed.

($732\,\text{kN}$; $22\,\text{kW}$)

22.11 Buoyancy

It is well known that a piece of metal placed in water will sink, a piece of cork placed below the surface will bob to the surface, and a steel ship having a large volume of empty space inside the hull will float. We shall now investigate the conditions which must be satisfied in order that a body immersed in a fluid will float, sink or rise.

For simplicity we consider a solid body of rectangular section which floats on the surface of a liquid, Fig. 22.11. The body is in equilibrium; hence the total downward force due to the weight of the body is equal to the resultant upthrust exerted by the liquid on the body. This resultant upthrust of the liquid is termed the *buoyancy*.

Let the body be of weight W and the density of the liquid ρ. Let the volume of body immersed below the free surface be V_d and the depth of the lowest surface h.

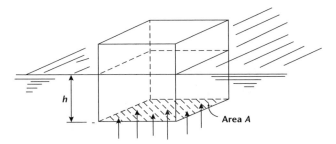

Fig. 22.11

The fluid pressure at depth h is $p = \rho gh$ and the upward force due to this pressure is

$$P = p \times A$$

$$= \rho gh \times A$$

where A is the area of cross-section of the body on which the upthrust acts.* Now $h \times A$ is the volume V_d below the surface which is displaced by the solid. Hence the upthrust due to this fluid is

$$P = \rho g \times V_d$$

The condition for floating is

upthrust = weight of body

i.e. $\rho g V_d = W$

But $\rho g V_d$ is the *weight of fluid displaced by the body*. Hence the condition that a body shall float is that the weight of the displaced fluid shall equal the weight of the body.

The above argument is true also when the body is completely immersed in the fluid. The volume of fluid displaced is then equal to the total volume V of the body. The condition for a body to float completely immersed is

upthrust = weight of solid

$$\rho g V = \rho_s g V$$

where

$$\rho_s = \text{density of the solid}$$

Thus

$$\rho = \rho_s$$

i.e.   *density of liquid = density of the solid*

---

* In buoyancy problems we are usually concerned with a pressure difference due to the weight of liquid column between the level considered and the free surface. The pressure calculated in this way is therefore a gauge pressure. It is unnecessary to convert to absolute pressure since the pressure of the atmosphere also acts downwards on the body.

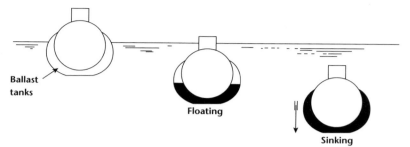

Fig. 22.12

Hence if the density of the solid body is less than that of the liquid it will float or rise when immersed completely. If the density of the solid is equal to that of the liquid, the solid will float submerged. If the density of the solid is greater than that of the liquid, it will sink.

For example, a submarine will be designed to float on the surface of the water. By flooding its ballast tanks, Fig. 22.12, its total weight is increased without changing its volume. Hence the submarine will sink. The weight of water in the tanks may be adjusted to allow the submarine to float completely immersed. If the tanks are emptied the submarine will rise to the surface.

Similar arguments apply to liquids that do not mix one with the other. For example, an oil that has a relative density less than that of water, and does not mix with it, will always rise to the surface of the water.

The above results are summarized in the *principle of Archimedes* which states:

The resultant upthrust on a body (solid or liquid) immersed in a fluid is equal to the weight of fluid displaced by the body. This upthrust acts through the centroid of the displaced fluid.

Example

A submarine floats in sea-water with three-quarters of its volume submerged. Calculate the mass of water required in the ballast tanks in order that it may float completely submerged. The total volume of the submarine is $1100\,\text{m}^3$. Density of sea-water = $1 \cdot 026\,\text{Mg/m}^3$.

Solution

Weight of submarine with tanks empty

= weight of water displaced by three-quarters of its volume

$= \rho g V_\text{d}$

$= 1 \cdot 026 \times 9 \cdot 8 \times (\tfrac{3}{4} \times 1100)$

$= 8295\,\text{kN}$

When completely submerged,

upward thrust of water $= \rho g V$

$$= 1.026 \times 9.8 \times 1100$$

$$= 11\,000\,\text{kN}$$

If W = weight of water in ballast tanks when submerged then

$$8295 + W = 11\,000$$

so that

$$W = 2700\,\text{kN}$$

Therefore

$$\text{mass of water} = \frac{2700 \times 1000}{9.8} = 276\,000\,\text{kg}$$

$$= 276\,\text{tonnes}$$

(22.12) ## Apparent weight

As we have seen, a solid body when completely immersed in a fluid will sink if its weight is greater than that of the volume of fluid it displaces. The body may be suspended by a spring balance or cord to prevent it from sinking; the reading on the balance or the tension in the cord is then the *apparent weight* of the body in the fluid.

The tension in the spring balance or cord, together with the upthrust of the fluid, must balance the weight. The apparent weight of the body is therefore the difference between the true weight of the body and the upthrust.*

Thus, if a body of weight W and volume V is prevented from sinking by the upward pull T in a cord, Fig. 22.13, then

apparent weight $= T =$ weight of body − upthrust

$$= W - \rho g V$$

The *apparent loss of weight* is $\rho g V$, the weight of water equal to the volume of the body.

Measurement of the apparent weight gives a simple method of determining the relative density of a solid; thus:

---

* The density of air, and therefore the apparent weight of a body in air, is proportional to the atmospheric pressure. However, the air density is small ($1.2\,\text{kg/m}^3$, roughly) compared with that of most liquids (480–$1000\,\text{kg/m}^3$, say). The buoyancy of the *air* is therefore neglected in this chapter, i.e. we assume the apparent weight of a body in air to be almost equal to the *true weight* in a vacuum.

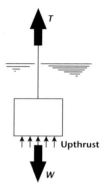

Fig. 22.13

$$\text{relative density of a solid} = \frac{\text{mass of solid}}{\text{mass of an equal volume of water}}$$

$$= \frac{\text{weight of solid}}{\text{weight of an equal volume of water}}$$

$$= \frac{\text{weight of body in air}}{\text{apparent loss of weight on immersion in water}}$$

Alternatively the method may be used to find the volume of a solid of known weight, or if a solid of known volume and weight is suspended in a fluid, the relative density of the fluid may be found.

Example

A metal component weighs 18 N in air. Its apparent weight when suspended completely immersed in oil of relative density 0·8 is 13·5 N. What is the volume of the component?

Solution

Apparent weight = weight in air − upthrust

$$13\cdot5 = 18 - \rho g V$$

$$= 18 - (0\cdot8 \times 1000 \times 9\cdot8\,V)$$

Therefore,

volume of metal $V = 0\cdot000\,57\,\text{m}^3 = \textbf{0·57 litres}$

Example

A balloon of volume 2240 m³ and containing hydrogen gas has a total weight of 9 kN. Calculate the tension T in the mooring cable (Fig. 22.14). Density of air = 1·23 kg/m³.

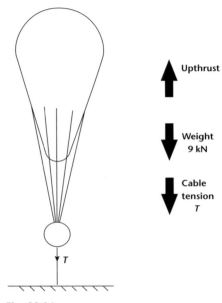

Fig. 22.14

Solution

Tension in cable T = upthrust due to air pressure − weight of balloon

$$= \rho g \times \text{volume of balloon} - 9000\,\text{N}$$

$$= (1{\cdot}23 \times 9{\cdot}8 \times 2240) - 9000$$

$$= 18\,000\,\text{N}$$

$$= 18\,\text{kN}$$

(Example)

A metal sphere, 50 mm in diameter, is suspended in oil of density $700\,\text{kg/m}^3$. Its apparent weight is 4·9 N. Find the density of the metal.

Solution

Volume of sphere $V = \frac{4}{3}\pi r^3$

$$= \frac{4 \times \pi \times 0{\cdot}025^3}{3}$$

$$= 6{\cdot}55 \times 10^{-5}\,\text{m}^3$$

Let ρ_m be the density of the metal.

Weight of sphere in air $= \rho_m g V$

$$= \rho_m \times 9 \cdot 8 \times 6 \cdot 55 \times 10^{-5}$$

$$= \rho_m \times 6 \cdot 42 \times 10^{-4} \, \text{N}$$

Upthrust due to oil $= \rho g V$

$$= 700 \times 9 \cdot 8 \times 6 \cdot 55 \times 10^{-5}$$

$$= 0.449 \, \text{N}$$

Apparent weight = weight of body in air − upthrust in oil, hence

$$4 \cdot 9 = \rho_m \times 6 \cdot 42 \times 10^{-4} - 0 \cdot 449$$

Therefore

$$\rho_m = 8330 \, \text{kg/m}^3$$

Problems

22.25 A ship displaces 10 000 tonnes of sea-water. Calculate the volume of the ship below the water line. Density of sea-water $= 1 \cdot 026 \, \text{Mg/m}^3$. $(9750 \, \text{m}^3)$

22.26 A rectangular tank, of floor area $0 \cdot 4 \, \text{m}^2$, floats upright in water, partially immersed, when it contains 9 litres of water. The bottom of the tank is then 150 mm below the surface of the water. Calculate the weight of the tank.
(500 N)

22.27 A submarine floats in sea-water with 80 per cent of its volume submerged. Its total volume is $1400 \, \text{m}^3$. Calculate (a) the dead weight of the vessel and (b) the mass of sea-water required in the ballast tanks in order that it may float completely submerged. Density of sea-water $= 1 \cdot 026 \, \text{Mg/m}^3$.
(11·3 MN; 282 tonnes)

22.28 An iron sphere 20 mm in diameter is suspended completely immersed in oil of relative density 0·7. What is its apparent weight? Relative density of iron = 7·86.   (0·3 N)

22.29 A solid iron cylinder, diameter 150 mm, height 1 m, relative density 7·86, is suspended in water. Two-thirds of its volume is submerged. Calculate the tension required in the cable to support the cylinder.   (1250 N)

22.30 The volume of a complicated casting is to be estimated by weighing it in air and then completely immersed in water. The weights are found to be 27 and 15 kN respectively. What is the volume of the casting? $(1 \cdot 23 \, \text{m}^3)$

22.31 A 'dracone' is a sausage-shaped plastic container for the transport of oil. If a dracone is full of oil of relative density 0·84, what percentage of its total volume lies below the surface when floating in sea-water? Density of sea-water $= 1 \cdot 026 \, \text{Mg/m}^3$. Neglect the weight of the plastic skin.   (82 per cent)

22.13 Steady flow: equation of continuity

In the solution of engineering problems dealing with the flow of liquids (and gases) it is useful to assume an ideal fluid with no viscosity and that the rate at which liquid flows through a pipe or into and out of a container is constant, i.e. the mass of liquid passing a given point per second is constant and the flow is said to be *steady*. In steady flow the flow pattern and velocity of the particles *do not change with time.*

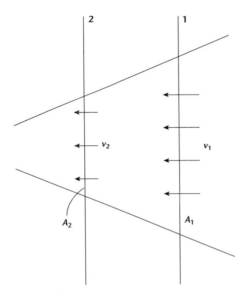

Fig. 22.15

Consider a liquid flowing through a pipe of varying section, Fig. 22.15. If the flow is steady the mass rate of flow $\dot{m}$ across any two sections is the same. At section 1 let the area be A_1 and the mean speed of the liquid past the section be v_1. Then the volume of liquid passing the section per second is given by

$$\dot{Q}_1 = A_1v_1$$

and the mass flow rate is

$$\dot{m} = \rho\dot{Q}_1 = \rho A_1v_1$$

where ρ is the density of the liquid, assumed constant. Similarly for section 2

$$\dot{m} = \rho\dot{Q}_2 = \rho A_2v_2$$

hence

$$A_1v_1 = A_2v_2 \quad \text{or} \quad Av = \text{constant}$$

This is called the *equation of continuity* and applies to any liquid that has steady flow. The density may be taken as constant since liquids are usually considered to be incompressible, i.e. they do not change under pressure. The same equations apply to gas flow except that gases are compressible at low pressures and density changes would have to be taken into account. At high pressures gases also may be considered incompressible (constant density). For example, the air at intake to jet-planes, even at speeds well below the speed of sound, is almost incompressible and more so the gases in the exhaust jetstreams of planes and rockets.

(22.14) Power of a jet

If a jet of fluid has a velocity v then the kinetic energy – the energy of motion – of a particle of mass m is $\frac{1}{2}mv^2$. If $\dot{m}$ is the mass rate of flow per second then the energy per second or *power* of the jet is $\frac{1}{2}\dot{m}v^2$. Thus

$$\text{power} = \tfrac{1}{2}\dot{m}v^2$$

If ρ is the density of the fluid and A the area of cross-section of the jet, then

volume of fluid per second = area of section × velocity

i.e. $$\dot{Q} = Av$$

and

mass of fluid per second, $\dot{m} = \rho \times$ volume per second

$$= \rho\dot{Q}$$

$$= \rho Av$$

Hence

$$\textbf{power} = \tfrac{1}{2}\dot{m}v^2 = \tfrac{1}{2}\rho Av^3$$

(22.15) Velocity head

Consider a tank containing liquid to a height h above a small opening as shown in Fig. 22.16. A particle of mass m at the free surface will have potential energy mgh measured relative to the outlet. If the velocity of the jet issuing from the opening is v then the kinetic energy of the particle will be $\frac{1}{2}mv^2$. From the law of conservation of energy, we have

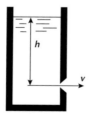

Fig. 22.16

loss of PE = gain of KE

i.e. $mgh = \frac{1}{2}mv^2$

so that

$h = v^2/2g$

or

$v = \sqrt{(2gh)}$

Thus the kinetic energy of a jet of liquid or any flowing liquid can be expressed as a *head* of liquid. The quantity $v^2/2g$ is called the *velocity head* and is the kinetic energy per unit weight. The unit of v is the metre per second and $g = 9.8\,\text{m/s}^2$, so that the head h is in metres.

It is not necessary for any particle to have actually fallen from the free surface to the outlet before issuing out. If a particle is already at the outlet in the tank, it possesses pressure energy and this in turn may be converted into kinetic energy. If p is the *gauge* pressure, i.e. the pressure above that of the atmosphere into which the jet issues, then

$p = \rho gh$

hence

$$v = \sqrt{(2gh)} = \sqrt{\left(2g \times \frac{p}{\rho g}\right)} = \sqrt{\frac{2p}{\rho}}$$

Example

A piston pumps 5 litres of glycerine along a pipe at a velocity of 3 m/s. Find the kinetic energy of the glycerine. Density of glycerine = $1.26\,\text{Mg/m}^3$.

Solution

5 litres = $0.005\,\text{m}^3$

Mass of $0.005\,\text{m}^3$ of glycerine = $1.26 \times 10^3 \times 0.005 = 6.3\,\text{kg}$

Kinetic energy = $\frac{1}{2}mv^2$

$\qquad = \frac{1}{2} \times 6.3 \times 3^2$

$\qquad = 28.4\,\text{J}$

Example

A water-wheel is driven by a jet from a nozzle of 25 mm diameter in a water main under a gauge pressure of $700\,\text{kN/m}^2$. Neglecting losses find the power supplied to the wheel.

Solution

$$p = 700 \times 10^3 \, \text{N/m}^2$$

$$h = \frac{p}{\rho g} = \frac{700 \times 10^3}{1000 \times 9 \cdot 8} = 71 \cdot 4 \, \text{m}$$

$$v = \sqrt{(2gh)} = \sqrt{(2 \times 9 \cdot 8 \times 71 \cdot 4)} = 37 \cdot 4 \, \text{m/s}$$

$$\text{Power supplied} = \tfrac{1}{2} \rho A v^3$$

$$= \frac{1}{2} \times 1000 \times \frac{\pi}{4} \times 0 \cdot 025^2 \times 37 \cdot 4^3$$

$$= 12\,840 \, \text{W}$$

$$= 12 \cdot 8 \, \text{kW}$$

Problems

22.32 Water is pumped along a pipe with a velocity of 10 m/s. Express this as a velocity head in metres of water and give the corresponding pressure in kN/m². (5·1 m; 50 kN/m²)

22.33 A jet of water 20 mm in diameter has a velocity of 20 m/s. Find the power of the jet in kilowatts. (1·26 kW)

22.34 The discharge from a small outlet in the side of a tank containing liquid is 16 kg/s when the head of liquid above the outlet is 1·1 m. Find the velocity and power of the jet. (4·64 m/s; 173 W)

22.35 A jet of water issues from a small orifice in the side of a tank. The head of water above the orifice is 500 mm. Find the velocity of the jet. (3·13 m/s)

22.36 A jet of water 25 mm in diameter issues from a nozzle under a head of 15 m. Find the velocity and energy per second of the jet. (17·1 m/s; 1·24 kW)

22.37 A nozzle delivers a jet of water 15 mm diameter. The power of the jet is 10 kW. Find the velocity of the jet. (48·5 m/s)

22.38 A nozzle discharges a jet of water at the rate of 10 litres/s with a velocity of 20 m/s. Find the power of the jet. (2 kW)

22.39 A jet of water of section 30 mm by 20 mm issues from a nozzle fed by a main. If the gauge pressure of the supply is 850 kN/m², find (i) the mass of water supplied per second and (ii) the power of the jet. (24·8 kg/s; 21 kW)

22.16 Dynamic pressure

A dynamic pressure occurs whenever the velocity of a fluid is suddenly changed in magnitude *or* direction, as for example when a jet of fluid strikes a plate or when the flow of liquid in a pipe is affected by the sudden closure of a valve or change in pipe section. The calculation of a dynamic pressure depends on the application of Newton's second law, which states that the rate of change of momentum (or the momentum per second) is equal to the applied force and takes place in the direction of the force. It should be noted that it is change in momentum that is used to calculate the force and not the change in energy (*see* Ch. 9). We shall restrict our considerations of dynamic pressure to those due to a jet impinging on a fixed flat plate.

(22.17) Impact of jet on fixed plate

Jet normal to flat plate

Consider a jet of fluid, flowing with velocity v and a mass rate of flow $\dot{m}$ per second, and impinging on a fixed flat plate normal to its surface as shown in Fig. 22.17. The fluid is assumed to spread out across the plate without any of it splashing back, and the diameter of the jet is assumed to be small compared with the plate dimensions. After impact, the jet has no velocity in the direction of its original motion and therefore its momentum normal to the plate has been destroyed by the force applied by the plate. Hence the *average* force of the plate on the fluid is equal to the change of momentum per second, i.e. the final momentum per second minus the initial momentum per second. We wish to know the force of the jet *on the plate*; hence

jet force normal to plate F = *initial* momentum per second − *final* momentum per second

$$= \dot{m}v - 0$$

$$= \dot{m}v$$

$$= \rho A v^2$$

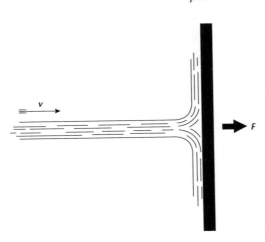

Fig. 22.17

since $\dot{m} = \rho Av$, where ρ = density of fluid and A = area of section of jet. Thus

$$F = \dot{m}v = \rho Av^2$$

This force may be written

$$F = \rho Av^2 = \rho(2g)A\frac{v^2}{2g} = 2\rho Agh$$

where $h = v^2/2g$ is the velocity head corresponding to velocity v. The static force produced by a head h of liquid on an area A is ρAgh (*see* page 494). Hence the dynamic force produced by the jet is twice the static force due to a head h of liquid.

Jet inclined to flat plate

In this case the jet strikes the plate at an angle θ to the normal, Fig. 22.18. The velocity of the jet can be resolved into two components,

component normal to plate = $v \cos \theta$

component parallel to the plate = $v \sin \theta$

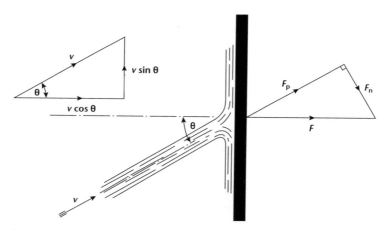

Fig. 22.18

On impact the velocity normal to the plate is destroyed causing a dynamic force and the other component of velocity is left unchanged. Thus

jet force normal to plate, F = loss of momentum per second normal to plate

$$= \dot{m}v \cos \theta - 0$$

$$= \dot{m}v \cos \theta$$

$$= \rho Av^2 \cos \theta$$

i.e. $F = \dot{m}v \cos \theta = \rho Av^2 \cos \theta$

This total thrust, F, may be resolved into two components parallel and perpendicular to the jet; thus, referring to Fig. 22.18,

thrust parallel to jet, $F_p = F \cos \theta = \rho A v^2 \cos^2 \theta$

thrust normal to jet, $F_n = F \sin \theta = \rho A v^2 \cos \theta \sin \theta$

For futher work on jets impinging on curved blades, see the authors' book *Applied Mechanics*.

Example

A jet of water leaves a nozzle with a velocity of 20 m/s and impinges on a large fixed flat plate. The discharge is 6 litres/s. Find the force on the plate.

Solution

Discharge $\dot{Q}$ = 6 litres/s = 0·006 m³/s

$\dot{m}$ = 0·006 × 1000 = 6 kg/s

Force on plate = $\dot{m}v$ = 6 × 20 = **120 N**

Example

A jet of oil 5 mm diameter impinges on a fixed plate at an angle of 60° to its surface. The jet velocity is 45 m/s. Find the normal force on the plate and the force in the direction of the jet. Density of the oil = 900 kg/m³.

Solution
Referring to Fig. 22.18,

$\theta = 30°$; $v = 45$ m/s; $\rho = 900$ kg/m³

Normal force, $F = \rho A v^2 \cos \theta$

$$= 900 \times \left(\frac{\pi}{4} \times 0{\cdot}005^2 \right) \times 45^2 \times \cos 30°$$

$$= \mathbf{31\,N}$$

Force parallel to jet = $F \cos \theta$

$$= 31 \times 0{\cdot}866 = \mathbf{27\,N}$$

Example

A circular flat disc of diameter 120 mm and mass 2 kg is hinged as shown in Fig. 22.19 so that it is free to swing in a vertical plane. A jet of oil of relative density 0·85 strikes the disc normal to its surface at a point 80 mm vertically below the hinge. If the disc is at 20° to the vertical in the equilibrium position, find (i) the force of the jet and (ii) the velocity of the jet. The diameter of the jet is 10 mm.

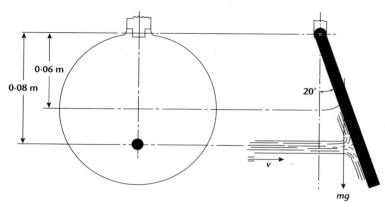

Fig. 22.19

Solution

The disc is held in equilibrium by two equal and opposite moments about the hinge exerted by the weight of the disc (mg) and the force of the jet. Equating these moments,

$$\frac{\text{jet force normal to disc} \times 0.08}{\cos 20°} = mg \times 0.06 \sin 20°$$

$$\text{(normal) jet force} = \frac{2 \times 9.8 \times 0.06 \sin 20° \cos 20°}{0.08} = 4.72\,\text{N}$$

This is the *total* force of the jet on the disc with components parallel and perpendicular to the direction of the jet. The jet force normal to the disc is given by $\rho A v^2 \cos 20°$; hence

$$\rho A v^2 \cos 20° = 4.72$$

i.e. $0.85 \times 1000 \times \left(\frac{\pi}{4} \times 0.01^2\right) \times v^2 \times 0.9397 = 4.72$

so that

$$v^2 = 75.3$$

$$\therefore \quad v = \textbf{8.7 m/s}$$

Example

A jet of oil, density $880\,\text{kg/m}^3$ and $36\,\text{mm}$ in diameter, moving at $10\,\text{m/s}$, impinges on a fixed flat plate inclined at $60°$ to the axis of the jet as shown, Fig. 22.20. Find the mass of water per second flowing upwards along the plate. Neglect all losses.

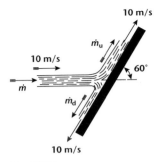

10 m/s

$\dot{m}_u$

10 m/s

60°

$\dot{m}$

$\dot{m}_d$

10 m/s

Fig. 22.20

Solution

$$\dot{Q} = Av = \frac{\pi}{4} \times 0.036^2 \times 10 = 0.01 \, \text{m}^3/\text{s}$$

Total flow rate, $\dot{m} = \rho\dot{Q} = 880 \times 0.01 = 8.8 \, \text{kg/s}$

The jet velocity remains constant in magnitude at 10 m/s for both streams of fluid up and down the plate since there is no friction. There is no change of momentum parallel to the plate and we can therefore equate the momentum per second before and after impact in this direction. If $\dot{m}_u$ and $\dot{m}_d$ are the mass flow rates up and down the plate respectively then

$$\dot{m}_u + \dot{m}_d = \dot{m} = 8.8 \, \text{kg/s}$$

and

$$\dot{m}v \cos 60° = \dot{m}_u v - \dot{m}_d v$$

i.e. $8.8 \times 10 \times 0.5 = (\dot{m}_u - \dot{m}_d) \times 10$

$$\dot{m}_u - \dot{m}_d = 4.4$$

hence

$$\dot{m}_u = 6.6 \, \text{kg/s}$$

Problems

22.40 A length of pipe running full of oil (relative density 0·9) varies in diameter from 50 mm to 30 mm. The rate of flow is 7 litres/s. Find the mass flow rate and the velocity of the oil at the two sections.

(6·3 kg/s; 3·56 m/s; 9·9 m/s)

22.41 In an experimental apparatus a jet of benzene is arranged to impinge normally on a fixed flat plate and the force of the jet is measured. If the force is 25 N when the diameter of the jet is 5 mm, find the velocity of the jet. Relative density of benzene = 0·88.

(38 m/s)

22.42 A jet of oil 50 mm diameter impinges on a large flat plate at an angle of 60° to the surface of the plate with a velocity of 30 m/s. Find the normal force on the plate and the force parallel to the jet. Relative density of the oil = 0·92.

(1·4 kN; 1·2 kN)

22.43 A flat plate of mass 10 kg and 100 mm square is hinged along one edge so that it can swing freely in a vertical plane. A jet of water impinges normally on the plate at a point 60 mm below the hinge in the middle of the plate. If the area of section of the jet is 200 mm² and its velocity is 10 m/s, find the angle made by the plate with the vertical when in the equilibrium position.

(14·2°)

22.44 A thin jet of oil (relative density 0·8) moving at 25 m/s strikes a fixed plate normal to its surface. The rate of flow of the jet is 5·5 litres/s. Find (i) the mass flow rate and the initial kinetic energy of the jet per second and (ii) the force on the plate. (4·4 kg/s; 1·38 kJ/s; 110 N)

22.45 A jet of water 60 mm in diameter moving with velocity 10 m/s strikes a smooth flat plate inclined to the axis of the jet at an angle of 50°. The plate is stationary. How does the mass of liquid flowing distribute itself on the plate? Neglect all losses.   (23.2 kg/s and 5·05 kg/s in opposite directions)

Index